OCEANOGRAPHERS AND THE COLD WAR

CEANOGRAPHERS AND THE COLD WAR

DISCIPLES OF MARINE SCIENCE

JACOB DARWIN HAMBLIN

UNIVERSITY OF WASHINGTON PRESS
SEATTLE

This book is published in memory of
MARSHA L. LANDOLT (1948–2004),
Dean of the Graduate School and Vice Provost,
University of Washington, with the support of
the University of Washington Press Endowment.

© 2005 by the University of Washington Press

Designed by Pamela Canell

All rights reserved. No part of this publication may be reproduced or transmitted in any form or by any means, electronic or mechanical, including photocopy, recording, or any information storage or retrieval system, without permission in writing from the publisher.

University of Washington Press
www.washington.edu/uwpress

Library of Congress Cataloging-in-Publication Data
Hamblin, Jacob Darwin.
Oceanographers and the cold war : disciples of marine science / Jacob Darwin Hamblin. — 1st ed.
  p. cm.
Includes bibliographical references and index.
ISBN 0-295-98482-1 (hardback : alk. paper)
 1. Oceanographers—United States—Biography.
 2. Oceanography—History—20th century. I. Title.
GC30.A1H36 2005   551.46'0973'09045—dc22
                                            2004021922

The paper used in this publication is acid-free and 90 percent recycled from at least 50 percent post-consumer waste. It meets the minimum requirements of American National Standard for Information Sciences—Permanence of Paper for Printed Library Materials, ANSI z39.48–1984.♾

FOR MY WIFE, SARA

CONTENTS

Preface *ix*
List of Abbreviations *xiii*
Introduction *xvii*

1. Beginnings of Postwar Marine Science and Cooperation *3*
2. Oceanography's Greatest Patron *32*
3. The International Geophysical Year, 1957–1958 *59*
4. The New Face of International Oceanography *99*
5. Competition and Cooperation in the 1960s *140*
6. Oceanography, East and West *177*
7. Marine Science and Marine Affairs *217*
8. Conclusion *259*

Notes *267*
Bibliography *307*
Index *333*

PREFACE

In browsing these pages, the reader will notice a very loose usage of the term "oceanography." The book's subtitle reflects an even more vague term: marine science. The coverage here is not limited to any particular branch of marine science, though often some fields dominated at the expense of others. Because the book is about politics, patronage, and communities in many different branches of science pertaining to the sea, I did not wish to splinter the discussion by needlessly separating the scientists as they might have done themselves. Thus the book runs the risk of painting a picture with rather broad brushstrokes; however, I have made an effort to be consistent with one of the themes of the book, which is the effort of international organizations and leading scientists to define the field broadly. Although some may take oceanography to mean simply the study of the chemical composition and physical dynamics of the sea, this book does not conform to that narrow definition. This raises another issue: in discussing the subject, should we use "oceanology," "oceanography," "hydrography," or some other term? The reader will discover that oceanographers in the Soviet Union were typically called oceanologists, for reasons that are discussed in the text. I have kept this usage on occasion, but generally I use the term "oceanographer" to describe them all.

There are a few other points of usage. I tend to use "Soviet" rather than "Russian," but the reader should be aware that this is inconsistent with what most scientists used in the documents I examined for this study. Readers outside the United States may object to my using "Americans" to describe only the citizens of the United States of America, not all the people on the two American continents; I do so for convenience, as there is no easy alternative for myself or the reader. I also use the terms "East" and "West." These are terms of convenience with geopolitical connotations and do not have any real geographic meaning. In my discussion, the East refers to the Soviet

Union and its political allies, and the West refers to the United States and its political allies (which puts Japan, rather counterintuitively, into the Western category). I discuss the characteristics of oceanography in East and West in some detail. I do not use "North" and "South" very often, but prefer to speak of industrialized countries and those of the developing world. These terms follow the usage of the people described in this book. Another term loosely employed is "military," which most accurately would mean land forces, while "naval" would describe sea forces. Sticklers will be disappointed to find that I use the term more generally, as most Americans do, to describe all kinds of armed forces; for example, I treat funding by the U.S. Navy as a kind of military patronage.

Unfortunately, this story is extraordinarily acronym-rich. When possible, I have made an effort to ease the reader's suffering by using real words instead of acronyms. Thus I use "Scripps," instead of SIO (and instead of a worse alternative, spelling out each time Scripps Institution of Oceanography), and "Woods Hole," instead of WHOI. Also, UNESCO has been changed to Unesco, purely as a matter of style; I am not breaking new ground here, as this form has appeared occasionally even in official publications. Occasionally I use full names when I might have left the acronym, as in the case of the National Academy of Sciences or the National Science Foundation. In all cases I have done things that sacrifice consistency for the greater good of ease of reading. Acronyms tend to collide with each other on the page, standing out and diminishing the flow. They also confuse, as in the case of the IOC and the ICO, which were very different bodies but were involved in very similar things and occasionally are mentioned on the same pages of this book. The reader is in good company if confused; I found documents that had been filed incorrectly in major archives because of the closeness of these two acronyms. With ship names, typically I have eliminated words such as "HMS," "USS," "R/V," or other designations beyond the name itself. With individuals, I tend to avoid professional or honorific designations such as "Dr.," "Academician," "Sir," "Lord," and so on, except in cases where a title enhances the reader's ability to understand who the person is (i.e., government and military titles and ranks). This is all a bit informal, but I suspect the reader can get used to it.

This book has been made possible through the help of colleagues, patrons, family, and friends, not necessarily in that order. It began as a dissertation at the University of California, Santa Barbara. I thank my dissertation committee: Lawrence Badash, Michael A. Osborne, and Fredrik Logevall, for their guidance in that process. I thank them and all the faculty and graduate students who advised me and critiqued my work during the early stages, especially participants in the History of Science Colloquium and the Cold War History Group (now the Center for Cold War Studies). I must also thank the Institute on Global Conflict and Cooperation for a major dissertation fellowship, which allowed me to conduct research in England, France, and (less glamorously) Massachusetts. I am indebted to the Center for History of Physics at the American Institute of Physics, for its Grant-in-Aid for History of Physics and Allied Sciences. I also thank my colleagues at the Centre Alexandre Koyré, who kindly hosted me in Paris as a postdoctoral fellow while I continued my research at Unesco and rewrote the entire manuscript. Special thanks go to Jacqueline Ettinger at the University of Washington Press for taking an interest in the manuscript and seeing it through.

I also would like to thank the many people who have commented upon or critiqued my work as it appeared in published or draft form, or otherwise encouraged the writing of this book. They include Lawrence Badash, Michael A. Osborne, Fredrik Logevall, Benjamin C. Zulueta, Zuoyue Wang, Peter Neushul, Ronald Rainger, Ronald E. Doel, Roy Macleod, Naomi Oreskes, David van Keuren, Helen Rozwadowski, Kurk Dorsey, Dominique Pestre, Amy Dahan Dalmedico, Margaret Rossiter, Walter Lenz, Harry N. Scheiber, Dean C. Allard, Roger Stuewer, David C. Engerman, and a number of very helpful anonymous referees. I wish to extend my gratitude to several people connected to the Intergovernmental Oceanographic Commission in Paris: Gary Wright for his continuous enthusiasm for my project; Alexei Suzyumov for discussing with me aspects of Soviet oceanography; Ray Griffiths for his reminiscences and his family's hospitality in lovely Saint Cloud; and Warren Wooster for kindly letting me interview him on his birthday. I owe a special debt to Warren Wooster and Ray Griffiths, who provided detailed comments on the entire manuscript.

I am grateful to the staffs at all the archives visited. In particular, a few individuals made my work much easier and more pleasant: Jens Boel and Mahmoud Ghander at the archives of Unesco; Janice Goldblum and Daniel Barbiero at the National Academies Archives; Deborah Day at the Scripps Institution of Oceanography; and Melissa Lamont at the archives of the Woods Hole Oceanographic Institution. In addition, I especially thank historian Margaret Deacon for her assistance with her father's papers at the Southampton Oceanography Centre and for her family's hospitality during a very rainy English November.

Naturally, I would like to thank my family, especially Les, Sharon, and

Sara Hamblin, and Paul and Cathy Goldberg. The love and support of friends from our Santa Barbara and Paris days have been much appreciated. I also thank the late John Coleman, whose encouragement was always heartening, and whose tenacious refusal to call anything but five-card draw (no wilds) inspired both ire and admiration. Rest in peace.

Most of all, I thank my wife, Sara Goldberg-Hamblin.

ABBREVIATIONS

AAAS	American Association for the Advancement of Science
AMSOC	American Miscellaneous Society
CICAR	Cooperative Investigations of the Caribbean and Adjacent Regions
CIM	Cooperative Investigations in the Mediterranean
CINECA	Cooperative Investigations of the Northern Part of the Eastern Central Atlantic
CNO	Chief of Naval Operations (United States)
CNRS	Centre National de la Recherche Scientifique (France)
COMSER	Commission on Marine Science, Engineering, and Resources (United States)
COSPAR	Committee of Space Research
CSAGI	Scientific Committee for the International Geophysical Year (Comité Speciale de l'Année Géophysique Internationale)
CSK	Cooperative Study of the Kuroshio and Adjacent Regions
EAMFRO	East African Marine Fisheries Research Organisation (United Kingdom)
ECOSOC	Economic and Social Council (United Nations)
FAO	Food and Agriculture Organization
FCST	Federal Council for Science and Technology (United States)
FOA	Foreign Operations Administration (United States)
GEBCO	General Bathymetric Chart of the Oceans
GEOSECS	Geochemical Ocean Sections Study
IACOMS	International Advisory Committee on Marine Science
IAPO	International Association of Physical Oceanography
ICA	International Cooperation Administration (United States)

ICES	International Council for the Exploration of the Sea
ICITA	International Cooperative Investigations of the Tropical Atlantic
ICNAF	International Commission for Northwest Atlantic Fisheries
ICO	Interagency Committee on Oceanography (United States)
ICSU	International Council of Scientific Unions
IDAB	International Development Advisory Board
IDOE	International Decade of Ocean Exploration
IGC	International Geophysical Cooperation (1959)
IGOSS	Integrated Global Ocean Station System
IGY	International Geophysical Year (1957–58)
IHB	International Hydrographic Bureau
IIOE	International Indian Ocean Expedition (1959–65)
IOBC	Indian Ocean Biological Centre (Cochin, India)
IOC	Intergovernmental Oceanographic Commission
IUGG	International Union of Geodesy and Geophysics
LEPOR	Long-Term and Expanded Programme of Oceanic Exploration and Research
LOFAR	Low Frequency Analysis and Recording
MATS	Military Air Transportation Service
NAS	National Academy of Sciences (United States)
NASCO	National Academy of Sciences Committee on Oceanography (United States)
NATO	North Atlantic Treaty Organization
NEL	Naval Electronics Laboratory (United States)
NIO	National Institute of Oceanography (United Kingdom)
NOAA	National Oceanic and Atmospheric Administration (United States)
NORPAX	North Pacific Experiment
NRC	National Research Council (United States)
NSF	National Science Foundation (United States)
ONR	Office of Naval Research (United States)
OST	Office of Science and Technology (United States)
PIPICO	Panel on International Programs, ICO (see above)
POG	Pacific Oceanographic Group
PSA	Pacific Science Association
PSAC	President's Science Advisory Committee (United States)
SCAR	Scientific Committee on Antarctic Research

Abbreviations

SCOR	Scientific Committee on Oceanic Research
SEATO	Southeast Asian Treaty Organization
SOC	International Coordination Group for the Southern Oceans
SOFAR	Sound Fixing and Ranging
SOSUS	Sound Surveillance System
TENOC	Ten Years in Oceanography (report by the United States Navy)
TPO	Technical Panel on Oceanography for the IGY (United States) (see above)
UMC	Upper Mantle Committee
UMP	Upper Mantle Project
Unesco	United Nations Educational, Scientific, and Cultural Organization
USNC-IGY	United States National Committee for the International Geophysical Year

INTRODUCTION

In late 1963, not long after replacing his assassinated predecessor, President Lyndon Baines Johnson addressed the United Nations with an unorthodox plan for world peace. Rather than focusing on nuclear disarmament, containment of communism, or turning away from superpower posturing, he made an unexpected suggestion. He pointed to the long tradition of moral codes at sea, where people worked together for common objectives regardless of political boundaries. Scientists in particular, he said, were engaged in cooperative ventures that promised to break down animosities and ease global tensions. "Because of this tradition," Johnson asserted, "it appears that positive actions to bring about a peaceful world would be effective if based on scientific activities related to the world's ocean areas."[1] Like presidents before him, Johnson looked to science as a way to ease the tensions of the Cold War and to solve mankind's pressing problems. Yet he singled out oceanography, not nuclear physics or space technology, subjects that thus far had monopolized the public's imagination.

The new president's remarks baffled government-sponsored think tanks, because they knew that most oceanographic scientific work since 1945 was funded through the nation's defense expenditures. They wondered: what boundaries were transcended, what tensions eased, what problems addressed during almost two decades of research on undersea warfare? American oceanography, one of these think tanks insisted, "has never been conceived as an opportunity to lessen international tensions and attain President Johnson's objective to end the cold war."[2] Quite the contrary, oceanic science dealt with problems such as submarine acoustics, fleet operations, and sea-launched nuclear missiles. Of all the guiding principles at their disposal, Johnson and his speechwriters had chosen a scientific activity that was unsurpassed in its interconnections with the American military-industrial complex.

Despite the apparent contradiction, Johnson's words on the tradition of internationalism in oceanography were not entirely misplaced. In addition to military projects, scientists also had undertaken large-scale international ventures such as the International Geophysical Year, the International Indian Ocean Expedition, and numerous data exchange programs with political allies and with the Soviet Union. This was one of the great paradoxes of oceanography during the first two decades after World War II. Support for research was based on its usefulness for making war on other nations. At the same time, oceanography retained an identity that tied it closely to international cooperation.[3] That contradiction invites an exploration of the international context of oceanography during the Cold War. The science was young, having matured hastily from intense funding during the 1950s and 1960s, as the United States looked increasingly to science and technology as a cornerstone of power in the world. Oceanography's accelerated adolescence through military funding is one reason that historian of science Eric L. Mills has written that its history provides "a virtually unexploited opportunity to link the advance of knowledge with an understanding of how and why science is done by people, with human motives, with human aims."[4] Recent studies by scholars in the United States agree; they point out the personal and institutional links formed in the early postwar period and demonstrate the consequences for the military, American universities, and for scientists themselves.[5] The growth of oceanography under the care of military establishments in the quarter century after World War II is only now receiving due attention from historians.

How does international cooperation fit into a military framework? Oceanography was a Cold War science, tied to geopolitics as much as any other scientific field. Its most crucial component was international cooperation, which was not merely the domain of a few pious souls who wished naïvely to see everyone work together. The major figures in international oceanography were also the leaders of national institutions; the people who attended international congresses often were the same people who attended top secret military and foreign policy briefings. This is the first study to examine the parallel trajectories both of oceanography's "Cold War" side and of oceanographers' international focus, taking into account the role of the Navy, United States foreign policy, and the activities of scientists all over the world, including developing countries. Despite the seemingly isolated strength of American science, the most ambitious efforts in oceanography during the 1950s and 1960s were international; consequently, American oceanography cannot be understood without taking into account its role in conflict and in cooperation with the other nations of the world.

OCEANOGRAPHY AND INTERNATIONAL COOPERATION

The lack of national borders at sea, the indiscriminately hostile environmental conditions, and the global scope of observations have long lent oceanography the reputation of being an inherently international endeavor. Just as often history reveals the ocean as a conduit of power, a "terrain" as fiercely contested as any other. Mastery of sea-lanes, coastal areas, and long-range trade routes have shaped, or even defined, the power structures of entire civilizations. The study of the sea has long contributed to national, often military or propagandistic, enterprises. Around 1768, for example, Benjamin Franklin and Timothy Folger developed the first chart of the Gulf Stream because merchants spent weeks longer traveling toward the colonies than they did sailing back to England. French ships used such charts to expedite shipments of arms and supplies from Europe during the American Revolution.[6] The famous *Challenger* expedition of 1872–76, in which British scientists circumnavigated the globe and collected biological specimens and hydrographic data, initially met with universal praise from scientists in other countries. But when some of the results were first published in an American journal, British scientists were furious, feeling that they had a natural claim to work on the collection first. To compete for prestige, Norway, Germany, France, Austria, and Russia all funded oceanographic cruises in the wake of the *Challenger*'s.[7] The use of such cruises to demonstrate power and prestige extended into the twentieth century, as in the case of the *Meteor* expedition. Forbidden by the Treaty of Versailles to send naval vessels to foreign ports, the German Admiralty in 1919 decided to outfit a scientific vessel to show a German presence in foreign countries. The scientific leader of the 1925–27 expedition, Alfred Merz, felt that Germany's destiny could be achieved by scientific greatness.[8]

Despite nationalistic tendencies, there were also numerous examples of cooperation during the era prior to World War II. These often were practical in nature. Countries with common economic interests in the North Atlantic recognized the need for cooperation in the early twentieth century, establishing the International Council for the Exploration of the Sea (ICES) in 1902. Its purpose was to encourage and coordinate oceanographic activities, particularly those related to fisheries. Also, nations wishing to standardize surveying methods and establish universal symbols in nautical charts formed the International Hydrographic Bureau (IHB) in 1921.[9] In the aftermath of the *Titanic*'s tragic sinking in 1912, the United States Coast Guard established the International Ice Patrol to keep track of the icebergs that appeared in the North Atlantic each spring. Although the Ice Patrol halted its activities during World War I, it carried on during the interwar period, putting into practice the latest methods of dynamic oceanography to track icebergs along the currents of the North Atlantic.[10]

During and after World War II, scientists in the United States and elsewhere forged strong bonds with government patrons. One effect of military patronage was the primacy of a few fields closely related to naval questions. Although scientists had explored physical oceanography and the relatively new fields of marine acoustics and marine geophysics during the interwar period, none of this work attracted significant funding from the U.S. Navy. The situation changed during the course of World War II, and by the late 1940s oceanography became one of the beneficiaries of the explosion of funding opportunities for science under the auspices of the Office of Naval Research, founded in 1946. The Navy played a critical part not only in supporting research with money but also in logistical support for major expeditions. Oceanographers and the Navy came to rely on each other, particularly because of the Navy's own competition with other armed services. Facing strategic obsolescence, it cast its lot with scientists, who assured the Navy that it could renew its relevance by focusing on the submarine threat and by developing an alternative nuclear deterrent at sea. Navy leaders learned the importance of a continuous flow of environmental data, to ensure the efficient use of its military technology. In addition, the Navy accepted international cooperation as a part of its mission to expand its sources of data.

The first major effort to put cooperation into practice on a large scale, with participation transcending Cold War boundaries, would be the International Geophysical Year (IGY) of 1957–58. It did not begin as an oceanographic enterprise, but oceanographers played a part, and their projects were the most ambitious they had ever attempted. To justify projects such as the IGY, the National Science Foundation emphasized nonsecrecy and data sharing with all nations. It reasoned, perhaps foolishly, that although such openness would benefit all nations, the United States was in the best position to translate shared data into innovative technology. As part of its own IGY program, the Soviet Union issued a timely challenge to that assumption by launching the first artificial satellite, *Sputnik*. Soviet scientists also stepped up their work in other domains, especially oceanography, making *Sputnik* a symbol of scientific and technological competition across disciplines.

The Soviet challenge during the IGY split the American oceanographic community into two camps. Some thought cooperation ought to continue but shifted their focus away from "easing tensions" and latched onto another goal, namely, promoting marine science in poorer countries. The late 1950s and early 1960s saw the birth of a new project even more ambitious than the IGY: the International Indian Ocean Expedition (IIOE). New bodies, such as the Scientific Committee on Oceanic Research (SCOR) and the Intergovernmental Oceanographic Commission (IOC), adopted a pragmatic vision for science, hoping to use science to address the world's problems, particularly its food shortages. The IOC brought more nations into oceanographic work, soliciting the participation of developing countries.

Others were more reluctant to promote international cooperation after the launch of *Sputnik*. Troubled by the geopolitical challenge posed by the Soviet Union, and reluctant to accept the new development-oriented outlook of international oceanography, these scientists retreated into national projects and tried to turn the international community against projects designed by the Soviets. Oceanographers routinely used the threat of Soviet leadership in science to attain congressional support for national programs in oceanography. This pursuit of scientific leadership often was self-defeating, as when Americans abandoned initiative in the Upper Mantle Project to pursue national projects. The Americans eventually chastised themselves when they realized that their insistence on a "first"—in this case their attempt to drill into the mantle during the failed Project Mohole—forced them to abandon their leadership position in international projects.

Many Western scientists felt increasingly disillusioned with cooperation by the late 1960s. They were squeezed between two forces: the agenda of the Soviet Union and the needs of the developing world. Their frustrations culminated in the North Atlantic Treaty Organization (NATO) Science Committee, a body that was partly international, but excluded the Soviet Union, and did not have to sell its research on the grounds of economic development. The Subcommittee on Oceanography was among its most successful activities. The Soviets, meanwhile, wanted a tougher IOC that could compel scientists to do certain projects. Against the official Soviet position were the scientists who wanted a free hand, claiming that intellectual autonomy and problem-solving were more important than endlessly recording more and more data. Attitudes toward Soviet science, usually informed by Cold War prejudices, turned increasingly negative in the face of its uninspired research programs, its wish to compel extensive surveys, and, perhaps more important, the fact that Soviet scientists were out of step, conceptually, with many of the new ideas about the oceans that appeared toward the end of the 1960s.

To the dismay of many American oceanographers, international programs by the late 1960s catered to the world's economic needs. This was due partly to the efforts of Unesco and its IOC. But in addition, attitudes toward oceanography were changing in the United States. Through the active support of President Johnson, American oceanography had achieved what many had been wanting since the launch of *Sputnik:* a Marine Sciences Council, to focus all American efforts into a single government advisory body that answered to the president of the United States. As they had when courting the Navy, scientists had gained a powerful ally, and they hoped to use international cooperation as a way to ensure that the recommendations of an international scientific body should decide the agendas of large-scale projects. But the council, which saw its zenith of influence under President Johnson, adopted "marine affairs" as its primary subject, abandoning science for its own sake. The 1970s were dubbed the International Decade of Ocean Exploration, and the Marine Sciences Council won the argument for marine science by focusing on economic development. But it remained to be seen if this would be at the expense of science itself.

By the end of the 1960s, cooperation had become an inextricable component of oceanography, for better or for worse. Scientists had convinced their patrons, first the Navy and then many other branches of the government, that cooperation could address their needs while keeping scientists happy by not subjecting cooperative work to security classification. But with expanding support for international cooperation, the price was high: Americans had to fight for control of their projects against world politics, they were held accountable to claims that science benefited the economy, and perhaps worst of all, they had to confront the falsehood of their own belief in the universality of science, as Cold War tensions divided oceanographers not only politically but also scientifically.

The inclusion of the term "Cold War" in the title of this book is intended to signal the importance of geopolitical considerations in the development of marine science after World War II. It is intended to enhance the argument of the book, not to define the years "covered" by it; the book itself ends in the early 1970s. International marine science was shaped by a confluence of scientific, military, and diplomatic efforts in the heyday of international cooperation in the 1950s and 1960s. The subsequent era, beginning

Introduction xxiii

in the 1970s, differed in a number of ways. To name a few: expeditions declined in importance in favor of unmanned stations; plate tectonics became the dominant paradigm in Western marine geophysics, while Soviet scientists were prevented from publishing on the subject; both the President's Science Advisory Committee and the Marine Sciences Council were dismantled, eliminating the key liaison offices between scientists and the government and replacing them with weaker bodies. The zenith of oceanographers' influence in government had come and gone. In addition, my analysis ends in the early 1970s because one of the overriding themes of the subsequent period is far better known and might obscure the analysis of the earlier period. To be specific: the most significant change in international marine science in the 1970s was the importance placed upon environmental issues, sparked by devastating oil spills in the late 1960s and controversies over marine pollution of various kinds. Environmental controversy played a much smaller role during the earlier period; any book about international marine science in the 1970s and beyond will inevitably (and justifiably) showcase the rise of environmental consciousness at the expense of other themes. Although such a book would be fascinating, the present book tells a very different, and no less fascinating, story.

DISCIPLES OF MARINE SCIENCE

The premise of this book is that oceanographers in North America and northern Europe made international cooperation the common denominator for a host of activities that otherwise might have appeared incongruous or even conflicting. They sought support where they could find it, altering their purpose to appeal to whoever was listening, creating "disciples of marine science" wherever possible. Their strategy for doing this was to expand the definition of oceanography, or to embrace preexisting broad understandings of it, to include an endless number of scientific disciplines, to gather traditional support constituencies under one roof, and to extend the community of marine science to every country of the world. American scientists understood that oceanography was a collaborative enterprise not only between nations but also between disciplines and that its interdisciplinary character could provide a broad base of support both at home and abroad. Consider the term itself, "oceanography," which implies an emphasis on the measurement of the sea, not the scientific study of it, as "oceanology" might. For years, scientists in many other nations (such as the Soviet Union) called their subject "oceanology," leaving "graphy" work to a different set of specialists. Americans, while often admitting that "oceanology" was more proper, kept the term "oceanography," not only from the inertia of common usage, but also because its broad definition helped to attract funds from a wide range of sources. Particularly when money was so forthcoming from the U.S. Navy for oceanography in the 1950s, American scientists had little incentive to insist on explicit boundaries between disciplines.

Even prior to the period covered in this book, Americans had begun to adopt a broad definition of oceanography. By World War II there were two major institutions for oceanography in the United States, one on the Pacific Coast and one on the Atlantic. In 1903, a group of marine biologists formed a research institution near San Diego, on a tight budget provided by a few philanthropic individuals, and in 1924 it became the Scripps Institution of Oceanography. The Woods Hole Oceanographic Institution in Woods Hole, Massachusetts, owed its beginning in 1930 to several grants from the Rockefeller Foundation.[11] Neither of these institutions confined itself to oceanography as a narrow discipline. Their leading researchers barely considered themselves oceanographers at all; their doctoral degrees were in biology, chemistry, geology, and physics. They studied subjects as diverse as sea life, oceanic chemical processes, seafloor topography, meteorology, and the transmission of sound under the sea. Henry B. Bigelow, Woods Hole's first director and one of the founders of modern oceanography in the United States, claimed that oceanography could only be defined "as the study of the world below the surface of the sea." But then he added, broadening the definition further, that it also included the relationship of the sea with the atmosphere. Expeditions prior to World War II often focused on biological oceanography, gathering data on sea life, or on marine chemistry and physics, observing the ocean's temperature and salinity with a view to understanding the sea's dynamics.[12] But after World War II, many different fields turned to the sea to solve their pressing problems. This was particularly so for marine geology and geophysics, the latter owing its growth largely to the marine investigations begun in the 1930s and funded by the Navy during and after World War II.[13] In 1942 three prominent scientists of the sea, Harald Sverdrup, Martin W. Johnson, and Richard Fleming, published an influential book, *The Oceans,* which included more than just current patterns and ocean dynamics. This work, which described oceanography as a broad field embracing an array of subjects, represented a new standard of inquiry in the United States.[14] America's oceanographic institutions defined their field broadly and viewed their subject itself as spanning the entire globe. The need for collaboration, across both disci-

Four directors of Woods Hole Oceanographic Institution, 1960. From left: Paul Fye, Henry Bigelow, Columbus Iselin, and Edward Smith. Courtesy Woods Hole Oceanographic Institution Archives

plinary and national lines, was an integral part of this vision of oceanography.

Cooperation, American scientists learned, was also politically attractive. The question of "easing tensions," a phrase widely used during the IGY (1957–58) to advertise the benefits of international cooperation in science, provided the initial motivation for this study.[15] By building personal relationships with colleagues in the Eastern bloc, speaking a common intellectual language that rose above politics, some Americans claimed to be easing the tensions of the Cold War. At the same time, they were pursuing a scientific tradition that emerged strongly after World War II, namely, the social responsibility of science.[16] But closer examination yields a different picture, one of American scientists using "easing tensions" to advertise the project to the public while at the same time promoting various goals to their sponsors, some scientific, some not, some peaceful, some not, some drawing scientists together, some driving them apart along Cold War lines. The character and goals of their projects depended on who was listening.

Over the next decade or so, to about 1970, oceanographers constructed permanent international scientific bodies amid developments that seemed to indicate a trend, not toward peaceful cooperation, but toward complementing American military and foreign policy activities. Developments that seemed to indicate the importance of these military or foreign policy links included: (a) the massive support for "basic" scientific research by the U.S. Navy during the late 1940s and 1950s; (b) the role of scientists in the crisis of strategic roles within the armed services; (c) the growth of oceanography in the United States and the dominance of it by acoustics, ocean dynamics, and geophysics; (d) the pervasive recognition of an important relationship between science policy and foreign policy; (e) the development of a federal policy connecting international cooperation to American scientific and technological superiority; and (f) the Soviet Union's express challenge to American scientific and technological leadership not only in space but also at sea. Yet during the same period, scientists laid the foundations of international and intergovernmental machinery for coordinating oceanographic research, enlisting the participation of developing countries, gaining the endorsement of the American government at the highest level, securing the financial backing of the Navy, and rallying the support of scientists around the world for American-backed plans.

What were scientists' motivations for cooperation? One was the redemptive value of being part of an international community. Historians of science (and scientists) are well attuned to pleas for support of "basic" science, or its moral equivalents, "pure" science, "fundamental" science, and "unfettered" research. In addition to providing the "capital" for future technology, as leading science policymaker Vannevar Bush once wrote, basic research was something that scientists did to maintain their integrity as scholars and their reputations as scientists rather than engineers. The idea of an international community helped to preserve these notions for scientists working under military patronage. Oceanographers in the 1950s and 1960s conducted work that became known to scientists worldwide, had foreign colleagues with whom they interacted, and complained at restrictions that constrained the free flow of knowledge from one country to another. This was especially clear in regard to classification policies of the U.S. Navy; scientists did not feel that their connection to the military violated their freedom of inquiry as long as the Navy allowed them to pursue their own ideas

Introduction xxvii

and *to permit research to be known outside the United States.* When it refused the latter, marine scientists complained bitterly, more than they ever did when the Navy tried to "direct" their research toward specific applications. In the 1960s, this situation remained virtually unchanged. As the implementation phase of the *Polaris* missile and its successors promised to provide long-term support for oceanography, Henry Stommel appealed to Woods Hole director Paul Fye not to transform the institution to a purely military enterprise under a veil of secrecy; the scientists, he wrote, had international reputations to maintain.[17]

International cooperation also gave scientists opportunities to solicit funding from a broad range of sponsors while promoting their own scientific goals. When such goals were validated by international scientific bodies, they were more defensible against interference by sponsors. As scientists sought patronage beyond the military, international scientific bodies became useful sources of authority to justify scientific work whose applications or utility were not readily apparent to sponsors.

Oceanographers adopted "development" as a rhetorical strategy after the IGY, because the old justification, "easing tensions," had lost its credibility as a selling point: what tensions were eased with the launch of *Sputnik* under the auspices of the IGY? Soviet and American activities in Antarctica and renewed efforts by American scientists to open congressional purse strings by fomenting anxiety over Soviet oceanographic activities, both at the close of the IGY, did little to reinforce the notion that the project had eased tensions. Development, however, had potential. If developing countries, and international organizations such as Unesco, could be convinced to participate, scientists could widen the scope of their observations and create even larger projects than the IGY, all under the vague promise of helping to understand the practical problems of the oceans that affected all humanity. This certainly helped to acquire funding, but addressing such problems also obscured the difference between basic and applied research, a difference that had served scientists so well in dealing with the Navy. Selling science in this way opened up a host of new problems: conflicts with more genuine fisheries organizations, conflicts between fields of marine science fighting for dominance (biologists seemed to think their work was relevant to fish, too), and the constant headache of governments expecting scientists to make good on their promises of practical results.

Motivations for oceanographic cooperation were many, and they were not limited to science. Some were based upon defeating communism, gaining strategic advantages, or defending a garrison state; others were based upon the spread of scientific inquiry to other nations, or upon using science to help solve humanity's most pressing problems. Often, one person could embrace all these goals, even when they seemed to contradict one another. Scientists often adapted to new selling points with genuine zeal; some oceanographers took pride in working for the Navy, just as others hoped to see their work contribute to ending world hunger. American oceanography and the beginnings of marine science in many countries were born into this paradox. There was no single underlying motivation. This book does not attempt to define a meta-motivation for cooperation, because it would never stand up to historical criticism. However, this book does endeavor to demonstrate how scientists used international cooperation to appeal to diverse interests and gain supporters and advocates. Or, as the title of this book suggests, they used international cooperation to cultivate "disciples" of marine science. The strategy they most often used was to adopt a broad, inclusive definition of oceanography, often employing the term "marine science," with its broad applicability and potentially wide appeal.

The subtitle of this book is taken from the reflections of a British scientist discussing the merits of a Unesco training course; he said that the purpose of cooperation was not necessarily to discover new theories or to create new Ph.D.'s, but rather it was to generate "interested young men and women who will be disciples of the marine sciences in their own countries."[18] Many countries of the developing world, heretofore relatively disinterested in oceanography, were counted among the disciples, with scientific communities and government sponsors that began to look to the ocean as a significant component of scientific health and economic well-being. Other disciples were the governments that took on major financial commitments to participate in international cooperative projects. These governments began to appreciate the sea as a source of food, of minerals, and even as a future area in which to claim national sovereignty. In the United States, President Johnson was not the first disciple in government, but he certainly was the first president to insist that, if scientists were making promises of economic benefits, they ought to deliver on them. Perhaps the most problematic disciples were the scientists themselves who, despite the all-inclusive appellation "marine science," rarely acted as a unit and even more rarely were comfortable with sharing responsibilities, money, and research priorities. At the same time, even the greatest skeptics of large-scale international cooperation, such as British physical oceanographer George Deacon, admitted that defining marine science broadly was probably the best means to achieve financial support and endorsement by various sectors of government. But

Introduction

Deacon, like so many others, bitterly fought the readjustments in relative power between physical oceanography, marine geophysics, and the ever-threatening biological sciences. Oceanographers adopted strategies ensuring support for their work and used international cooperation to do so in ways coinciding with their own interests, whether those be muting the problems of military patronage with free data exchange, finding global social problems to justify research trajectories, or ensuring their autonomy at home by pointing to the activities and recommendations of an international scientific community. This book tells the extraordinary story of how this was accomplished amid the dangerous backdrop of the Cold War.

CORRECTIONS

Pages 172–73: An error in the original edition of this book mislabeled James H. Wakelin, Jr., Assistant Secretary of the Navy, as an admiral. Before moving to civilian life, Wakelin's highest rank in the Navy was Lieutenant Commander. Technical constraints disallowed making the correction for this edition.

The photos on pages 157 and 203 are now part of the Special Collections and Archives, UC San Diego.

OCEANOGRAPHERS
AND THE COLD WAR

1 BEGINNINGS OF POSTWAR MARINE SCIENCE AND COOPERATION

While on a fellowship in Japan in 1953, marine geologist Robert S. Dietz observed, "The time has come when a 'showing of the flag' can be more effectively done in many parts of the world by a vessel engaged in scientific pursuits than by a man o' war."[1] He was writing to scientists at the Scripps Institution of Oceanography in San Diego, California, who were planning an expedition to cross the Pacific Ocean and visit ports in Japan. Dietz did not specify precisely how he thought marine science could influence relations between the United States and Japan, but he believed that science could accomplish something that traditional diplomacy and military power could not. Perhaps he sensed, as many Americans did, that the status gained by science since the war gave it a unique role in international affairs and that scientists could serve in roles as diverse as military advisors, espionage agents, atomic diplomats, and harbingers of world government. International scientific cooperation seemed poised to play an important role in extending military power, pursuing foreign policy, and building contacts between scientists.[2]

Given the long-standing needs of science to coordinate the collection and interpretation of environmental data, studies of the ocean traditionally invited international cooperation; at the same time, they were heavily dependent upon military funding, implicitly challenging the premise that the research was truly "international."[3] This chapter traces a couple of basic changes that established the trajectory of scientific cooperation in marine science after World War II. One change was an emphasis on physical oceanography and marine geophysics, both in ascendancy in the Cold War era but having their roots in the 1930s. The other was in the patronage for oceanography, which shifted from private and philanthropic enterprises to military organizations, particularly the U.S. Navy. Both of these trends fed off of each other, allowing oceanographers to conduct ambitious interdisciplinary expeditions in the early 1950s and helping to shift the focus of scientific activity from northern Europe to two institutions in the United States. However, American oceanographers did not simply retreat into military patronage under a veil of secrecy. They pressed for wider participation by other countries, to experiment with synoptic investigations and to share data, expanding the scope of a vaguely defined field infused with money in the early years of the Cold War.

NEW DIRECTIONS IN COOPERATION AND RESEARCH BEFORE WORLD WAR II

After World War II, cynics saw science taking a turn for the worse, seduced by military patronage and deprived of its democratic, international image. Cooperation tempered science's militaristic image and tied scientists to loftier goals, such as the advancement of science and the amelioration of global political pressures. This conception of science would later become a major focal point of the International Geophysical Year, which was trumpeted as a means to use science's international, nonpolitical character to "ease the tensions" of the Cold War through cooperation. In a well-known 1962 article, Michael Polanyi outlined some characteristics of the global mentality in what he termed the "republic of science." To be part of it, scientists not only had to exercise freedom of inquiry but also needed to participate in a community as large as the total number of scientists. Autonomous activity and coordination were the soul of Polanyi's republic, because "scientists, freely making their own choice of problems and pursuing them in the light of their own personal judgment are in fact cooperating as members of a closely knit organization."[4] Combining individuality and universal interdependence, Polanyi's "republic of science" was based on freedom of personal inquiry and close coordination with everyone else. Adherents to the view embodied by Polanyi's article embraced the notion of a scientific community transcending national borders. Scholars of science and colonialism have pointed out some of the flaws in this vision, particularly when scientists felt on the periphery because of their distance from cores of scientific activity in Europe and North America. Many felt that the "tyranny of distance" from the dominant centers of intellectual activity made their communities weak, subordinate to the leadership of Europe or the United States. Lack of access to institutions, to funding, and to data undercut many scientific communities' efforts to participate as equals. As historian George Basalla once noted, "colonial science" persisted even in the absence of formal colonial relationships.[5] None of these limitations, however, seemed to extinguish scientists' faith that such an ideal could and should exist.

Before World War II, oceanographers tried to approximate the "republic of science" ideal in the Pacific region through the first four Pacific Science Congresses. In the 1920s, these congresses brought together scientists from countries bordering on the Pacific Ocean to coordinate research and, more important, to appeal collectively to home governments for funding. They reasoned that governments might be more sympathetic if local scientists could demonstrate the international significance of such patronage. The first congress passed dozens of resolutions calling upon governments to support surveys both at sea and on land. Only by working together, they reasoned, could they tackle the most important scientific problems of the Pacific. Oceanography was a major component, and at the second congress, in Melbourne and Sydney, Australian scientists organized a committee on the physical and chemical oceanography of the Pacific, consisting of representatives of each country. Coordinating research, outlining future areas of critical importance, and preventing duplication of effort all would remain hallmarks of international oceanography for years to come.[6]

Some scientists also believed that these meetings served the interests of world peace. Members of the Pacific Science Association (PSA) wanted to cross the threshold of political divisions between Europeans, Americans, and Asians.[7] Scientific advance would not be their only goal. Perhaps by the act of cooperation, they could attempt to promote positive international relations. In the words of Yale geologist Herbert E. Gregory, the 1923 congress in Melbourne demonstrated that "friendship and science held equal place." In 1926, the president of the National Research Council of Japan, Prince Joji Sakurai, noted that the most remarkable thing about the congress in Tokyo was the "genuine warmth of feeling which pervades it." Several scientists echoed these sentiments and hoped that the meetings would foster not only science but also understanding.[8] In this respect, scientists consciously took up politics to promote a consciousness that was global in scope.

The congresses never matched the idyllic model of Polanyi's "republic." The Americans who dominated them were disappointed with the scientific efforts of their foreign colleagues. Thomas Wayland Vaughan, director of Scripps, chaired the first standing committee on oceanography and set the standard for the papers presented. Only the Japanese and the Canadians compared well, while most others failed to impress in thoughtfulness or rigor. American marine biologist Carl L. Hubbs scorned the Dutch working "day" that lasted from nine until one. He wrote to his wife after the 1929 meeting in Java that "the Dutch certainly fall down on details of administration, the natives are exasperatingly stupid and lazy, and neither have any real sense of time." He felt that only the Americans made a strong showing and that some papers, notably those of the Soviets, were so bad that they should be struck from the congress's printed volume.[9] But these Americans nonetheless viewed the congresses positively: they were a step in the right direction, toward creating high international standards and establishing interdependent relations between scientific communities in the Pacific.

Because most of the Asian participants were colonial Europeans, whose scientific communities had strong ties to those in their home countries, the sense of an international scientific community in the Pacific Ocean region may come as no surprise.[10] With indigenous peoples, Western scientists had rather different relationships, usually lacking the basic element of trust in the science itself. An important exception to this rule was Japan, whose tradition of biological research already had entered a period of flourishing.[11] Hubbs was impressed by the quality of the Japanese papers delivered at the 1929 Java congress, and it was clear that scientists from Japan prioritized the congresses, often sending more delegates than the Americans did. When Hubbs visited Japan just after the Java congress, he was charmed by the enthusiasm of the scientists he met and by their desire to establish reciprocal arrangements for sharing literature and specimens.[12] The Japanese seemed to make promising partners in scientific cooperation. Scientists had high hopes for the Pacific community, linked by an international scientific body that attempted, formally and informally, to create lasting cooperative bonds between scientists.

International scientific cooperation in the Pacific in the 1930s failed to meet the aspirations of the 1920s. Many intellectuals tried to forge closer relations between the interests of science and the needs of society, concluding that the "internal" (science) and the "external" (society) could not easily be separated.[13] But such insights, when pitted against the economic and political strains of the 1930s, failed to provide effective tools by which internationally minded scientists could shape the world around them. The Great Depression had a devastating effect on international cooperation. Scientific conferences were an expensive luxury for scientists with falling salaries and limited research grants. In the Pacific region, Japan's militarism in Asia put severe strains on the spirit of cooperation. Participation in the congresses declined dramatically during this period, and the only ones held during the 1930s were in Canada and the United States, in which scientists from North

Beginnings of Postwar Marine Science 7

America made up the vast majority of participants. The realization of their helplessness in the face of international strains actually prompted the PSA to add a new emphasis on the social sciences, to promote the application of brainpower to practical human relations.[14] Still, the PSA's desire to use scientific cooperation as a means to ease international tensions, and even to apply scientific methods to social problems, established among oceanic scientists not only an ideal of social responsibility but also a feeling that science might be able to contribute to international peace.

These cooperative sentiments, ambitious as they were, inspired only a small portion of oceanography before World War II. Oceanographic endeavors generally were not cooperative, and coordination of expeditions was rare. This is not to say that they were all purely nationalistic enterprises, but simply that they were funded and executed by individual countries, often with support by navies. In the 1920s and 1930s, the British sent the *Discovery* and *Discovery II* to the Antarctic, Denmark sent the *Dana* to the Indian and Pacific oceans, and Germany sent the *Meteor* on voyages in the Atlantic. The *Meteor* was supported by the German Admiralty, which reasoned that the vessel could serve as a symbol of peacetime German power and prestige in foreign ports.[15] This may appear to be a naïve expectation or at least a tall order for a small research vessel, but in light of Robert Dietz's later feeling that American scientific vessels could accomplish similar purposes as warships without the negative feelings that went with ostentatious displays of military power, Germany's interwar strategy does not seem out of the ordinary.

It should come as no surprise that governments were at least somewhat receptive to scientists' appeals to support work in oceanography and related sciences. After all, scientists already had proven their worth in assessing national resources. During and after the First World War, American (and European) general staffs relied upon geological advice for military planning, particularly in regard to trench and tunnel construction, and at Versailles geologists were called in to remap Europe according to its natural borders and mineral wealth.[16] In 1929, the United States Navy invited the Dutch geodesist Felix Andries Vening Meinesz to conduct gravity experiments aboard an American submarine. The Navy believed that Vening Meinesz's work on gravitational compensation on the ocean floor (studies in isostasy) might aid in the discovery and exploitation of fuel resources, which the Navy wanted in order to avoid dependence upon imports during wartime.[17] The resulting "S-21 expedition" revealed the complexities of gravity anomalies beneath the seafloor and gave Vening Meinesz's work a wide audience. Naomi Oreskes has called Vening Meinesz's expedition the birth of marine geophysics in America.[18] Because of its source of patronage, the expedition was also an early step in creating a bond between oceanography and the Navy.

Already by the 1930s, the field of oceanography in the United States was evolving toward physical oceanography, geology, and geophysics. The two major institutions, Woods Hole and Scripps, were still small, relying on the patronage of private donors and philanthropic organizations, and much of their work was in the area of marine biology. But both directors, Henry Bryant Bigelow and Thomas Wayland Vaughan, viewed oceanography as a broad subject, taking an interest in a wide spectrum of fields including biology, physics, chemistry, and marine geology. At Woods Hole, Bigelow and others, including future directors Columbus Iselin and Edward Smith, were inspired by the work of Norwegian physicist Vilhelm Bjerknes, whose studies of the interactions of air and sea were laying the foundations of modern meteorology. At Scripps, Vaughan wanted to conduct expeditions in the Pacific, to make major contributions to theories about oceanic processes like those being made by Bjerknes, and to develop physical oceanography on the scale that it had assumed in Norway and other European countries. To succeed him as director, Vaughan chose a countryman and student of Bjerknes, Harald Sverdrup, who worked on ocean circulation and air-sea interactions and attempted to analyze them in quantitative terms using physical principles. Under Bjerknes's influence, Sverdrup defined his field broadly and attempted to develop an understanding of large-scale processes, particularly the relationship between the ocean and the atmosphere. When he took the helm of Scripps in 1936, the institution took on a new strength as a center of physical oceanography.[19]

Many scientists also felt that the sea provided endless opportunities for marine geologists and geophysicists using new techniques for exploring the depths. One of these enthusiasts was Richard Field, a geologist at Princeton University. It was Field who, according to British physicist Edward Crisp Bullard, jump-started the study of marine geology in the 1930s. He did not do much of the work himself. "His technique," Bullard later recalled, "was to get hold of young men and persuade them that this was worth doing and, in fact, got hold of Harry Hess and Maurice Ewing and myself, and persuaded all three of us that the future of geology lay in the oceans."[20] Hess, another Princeton geologist, had worked with gravity measurements aboard submarines, like Vening Meinesz, and was continuing his research in the Caribbean Sea and in Venezuela. Ewing was then at Lehigh University, applying techniques he had learned while helping the oil industry conduct prospecting surveys. Instead of using TNT on land, however, Ewing developed a way to put it on the bottom of the ocean, explode it, and record the seismic waves to determine the speed of sound under the ocean bottom and thus gain some idea of its structure. Ewing's pioneering work opened up new areas of research not only in submarine geology and geophysics but also in marine acoustics, as undersea explosions allowed studies of the transmission of sound in seawater. In the late 1930s, Field invited Bullard to the United States and, despite Bullard's protestations about seasickness, took him to sea just days after his arrival, showing him all the work being done on underwater sound. After his experience in the United States, where he witnessed Ewing's work firsthand, Bullard returned to Cambridge University where he began a major thrust in seismic work at sea. British geophysicst Maurice Hill later noted that the influence of Ewing's work, along with that of Vening Meinesz and his pendulum apparatus in gravity surveys, spurred the modern era of marine geophysics.[21]

World War II, despite putting a stop to some research, had a catalyzing effect on American oceanography, driving it further toward physical oceanography and marine geophysics. During the war, Harald Sverdrup and his Scripps colleagues Martin Johnson and Richard Fleming published *The Oceans*, a book outlining the major features of oceanography in an interdisciplinary fashion but with a strong focus on physical principles. Even the chapters by Johnson, a biologist, emphasized the importance of the distribution and ecology of marine organisms in their relation to physical parameters such as light, temperature, and ocean currents.[22] Just before the United States' entry into the war, W. Maurice Ewing took a post at Woods Hole and began work with Columbus Iselin on the effects of ocean conditions on the transmission of sound, which the Navy supported because of its applications in submarine detection. Throughout the war, scientists helped the Navy understand the properties of marine acoustics and assisted in technical questions related to mine sweeping and the execution of amphibious landings. It was an awakening for both the scientists and the Navy, as each realized how much it could benefit from the other, and many of American oceanography's future leaders, such as Ewing and Roger Revelle, were deeply involved in it. The end result was exciting for scientists, who felt that the Navy had been converted to the idea of fundamental research and was prepared to provide a blank check for all future work in marine science at Woods Hole and Scripps. Revelle, working at the Navy Bureau of Ships during the war and the Office of Naval Research after its creation in 1946, was instrumental in convincing the Navy to support such research.[23]

After the war, deep-sea expeditions began to proliferate, as scientists felt they now had the means to explore the ocean depths in ways that they never could before. In Britain, Bullard considered studies of the ocean floor, through coring, topography studies, and of course geophysical techniques, to be the most likely line of work to advance knowledge in geology. In Ewing's seismic work, explosives and instruments had been placed on the ocean bottom. Bullard believed that both could float in the water, a few hundred feet below the surface, treating the water as another layer in which the sound waves must travel. This would make possible the study of the earth's crust even in the deeper parts of the ocean, eliminating the need to put explosives and recording instruments on the bottom. Although Cambridge made a brief study of the matter, all work on this subject had been halted by the war, and Britain's resources were severely diminished in the immediate postwar period.[24] Meanwhile, a Swedish oceanographer, Hans Petterssen, began to experiment with a hydrophone (a sound listening instrument that, unlike a geophone, was placed floating in the water rather than on the ocean bottom) and reported that he was able to detect rock layers some hundreds of feet below the ocean bottom. Bullard and his colleague, B. C. Browne, tried to secure two ships to do this kind of work, one to fire explosions and the other to listen, but for a couple of years failed to persuade anyone in the country to provide a ship for a major effort. In oceanic studies, he lamented, the British were "worse off than the Swedes and the Danes (to mention only small nations)."[25]

By the early 1950s, Europeans had made slow but sure efforts to conduct deep-sea research. Hans Petterssen embarked upon the Swedish Deep Sea Expedition in 1947, using an innovative geological tool, a piston corer developed by his compatriot Börje Kullenberg, to take cores of ocean bottom sediment up to twenty meters long. Denmark, for its part, launched the *Galathea* expedition in 1950, dredging for marine life in the Pacific Ocean.[26] Although in January 1947 Britain made its first expedition, led by Maurice Hill and P. L. Willmore, it was a minor one limited to the continental shelf. Its goal was to compare the use of hydrophones with that of geophones, and ultimately it demonstrated the feasibility of conducting seismic refraction experiments in deep water without having to place a geophone on the ocean floor. The British scientists, barely able to obtain access to a single ship, had to develop a method using a sono-radio buoy to record the results of the explosions, rather than rely on a second ship. Their first deep-sea expe-

Beginnings of Postwar Marine Science

dition in 1949 was limited, again the result of ship availability, to a circle with a thirty-mile radius near the Rockall Bank. The first long-range (and long-term) expedition began in 1950 aboard the *Challenger II*, and over the course of two years the ship circumnavigated the globe conducting many investigations, including seismic refraction studies.[27]

Despite these European efforts, the center of activity for oceanography shifted decisively from northern Europe to the United States, whose economy had rebounded dramatically because of the war and whose scientists were benefiting from it. More important, key oceanographers (Roger Revelle in particular) who had worked for the Navy during the war now exercised influence in the newly created (1946) Office of Naval Research, which supported research in the ocean sciences without an aggressive concern for its applications. In the United States, conditions for research were vastly better for planning expeditions, which were instigated in the Pacific Ocean by Scripps and in the Atlantic Ocean by Woods Hole. Deep-sea expeditions had the strong support of the Navy, and the Americans could afford the luxury of conducting refraction experiments with two ships, in addition to a great deal of other scientific work conducted on the cruises. The Navy sponsored a major expedition to the Marshall Islands to assess the effects of atomic bomb tests (the Bikini Scientific Resurvey, for Operation Crossroads), and scientists were able to conduct their own work on these voyages. When Scripps director Roger Revelle launched the MIDPAC expedition in 1950, his facilities were well funded, and the expedition conducted a broad range of scientific activity.[28]

Despite the early reliance on national and Navy-sponsored expeditions, there were close informal interactions and sharing among oceanographers just after the war. Bell Telephone Laboratories, an American research organization, loaned its first portable crystal chronometer to Vening Meinesz, who shared it with Bullard for use on his own voyages. When Bell turned over ownership of the chronometer to Woods Hole, Ewing wrote to Bullard that he should continue to use it if necessary. Bullard shared his ideas with Ewing about using a hydrophone in water rather than a geophone on the ocean bottom, hoping that the Americans could explore the idea even if the British scientists were slow in starting. He wrote of another instrument, an accelerometer developed by a Cambridge colleague, B. C. Browne, that "is probably still secret but I expect you can get details...."[29] Bullard himself left Cambridge briefly during the late 1940s to take up a position in Canada, but he spent his summers at Scripps developing a new method for measuring heat flow through the floor of the ocean. Petterssen visited Scripps in 1949 and gave lectures about the techniques used on the Swedish Deep Sea Expedition and discussed the tools to be used by the Americans on their upcoming voyage, including Bullard's heat flow instrument, developed with a colleague at Scripps, Arthur Maxwell.[30] When the Americans, under the leadership of Revelle, embarked upon their MIDPAC expedition in 1950, they employed the latest techniques built upon collaboration with colleagues in the United States and abroad: they dredged corals, conducted seismic work using Ewing's ideas and recent techniques of the British and Swedes, measured heat flow with Bullard's method, and utilized the Kullenberg apparatus to take cores as long as twenty-four feet in water 2,200–2,800 fathoms deep (however, they relied primarily on gravity corers, which produced only about three or four feet but could be taken more quickly than the Kullenberg cores, which also tended to be heavily mixed in the upper layers). The results of all of these studies yielded new and surprising results. "We used to bet a bottle of beer on what we would find in any particular core," Revelle wrote to Bullard, "the chances always seemed about even that we would get . . . ooze, red clay, volcanic ash or various modifications and combinations of these."[31]

By the end of the 1940s, American oceanographers were positioned to take a leadership role in oceanography, riding the wave of resources that the Navy seemed willing to pour into both Scripps and Woods Hole. This had a major effect upon not only the growth of oceanography but also its focus. The result was particularly striking at Scripps, where there were numerous opportunities to shift the institution's focus more clearly toward marine biology and even fisheries research. The first major government grant provided to Scripps, three hundred thousand dollars by the State of California, was to research the origins of a recent decline in the sardine catch.[32] But Scripps director Sverdrup and his protégé in Washington, Revelle, had other plans for the institution, which would define projects broadly enough to attract the Navy's attention as well, keeping physical oceanography and marine geophysics very much at the forefront of research. The appointment of Revelle to the directorship in 1950 was bitterly opposed (in vain) by some, largely because of his emphasis on a particular kind of patronage and a particular brand of oceanography, oriented toward the military.[33] But Sverdrup and Revelle objected to this characterization; their vision of military patronage would save oceanography from having to justify the economic benefits of research, and it would go hand in hand with international cooperation. They still corresponded, shared instruments, and discussed results with their colleagues in Europe. Sverdrup wrote that the war had done more than reveal

Beginnings of Postwar Marine Science 13

military uses of oceanography; it had also revealed what could be accomplished through cooperative efforts. In addition to the many informal contacts made, he pointed to the swell forecasting network set up in England during the war, in which observations of many countries were pooled to provide a broad picture of ocean conditions. With military funding, even more ambitious projects among nations seemed possible.[34]

SCIENTIFIC AND TECHNICAL ASSISTANCE PROGRAMS

Oceanographic cooperation, however, rested largely upon informal links between individual scientists. By the late 1950s, this would change considerably, and it is worth exploring the origins of more official connections between scientific research and international collaboration. The first explicitly cooperative enterprises after World War II were often directed at the developing world and were not large-scale cooperative ventures between industrialized countries. Such enterprises were not a new phenomenon. In the 1930s, philanthropic organizations such as the Rockefeller Foundation took pride in encouraging research that would have lasting economic and social consequences for developing countries. Occasionally such efforts met with resistance, particularly when scientists and technicians were insensitive to local traditions. Nevertheless, often these organizations proceeded on the assumption that the development of science and that of economic growth complemented each other and that it was something needing to be taught by industrialized countries.[35]

Scientists typically did not cooperate with their colleagues in the developing world in the same manner as they did with Europeans. Without a Western imprimatur, the science seemed to lack credibility. When his studies of the floor of the Caribbean Sea brought Princeton geologist Harry Hess to Venezuela in the 1930s, the survey projects that ensued were organized and implemented by Hess with the support of the local government. Hess felt that his work could be beneficial both to science and to the mineral development of Venezuela (indeed, Venezuela was to become one of the great petroleum-producing countries of the twentieth century). He also hoped that his work might stimulate local scientific activity, but Venezuelan scientists made no independent intellectual contribution to these surveys. A Caracas newspaper summed up the situation thus:

> Dr. Hess said there is plenty of exploratory work to keep new students busy; he is always interested in the exchange of international scholars, and he would

like to see more Venezuelan boys mapping their own country—while pursuing their studies at Princeton, of course.[36]

When war broke out in Europe in 1939, Hess lamented that the Venezuelan government was unable to maintain its support for surveying projects. He implored geologists there to keep abreast of latest developments, at the very least by subscribing to American journals, lest they fall behind after so much progress. He was deeply troubled by the extent to which the war halted scientific development in Latin America, slowing the eventual inclusion of its scientists in the international community.[37] His suggestions underlined the view that significant scientific activity in Latin America could be possible only when guided and led by others, especially Americans. Perhaps Hess fits historian Lewis Pyenson's characterization of many Western scientists as Kulturträger, bearers not only of culture but also of all the trappings of civilization, including the knowledge and methods of science, across national borders.[38]

Marine science became part of large-scale efforts to bring science to the developing world after World War II. The United Nations Educational, Scientific, and Cultural Organization (Unesco) coupled its humanitarian focus with scientific and economic development. Under Unesco sponsorship, international scientific bodies such as the International Council of Scientific Unions acquired a humanitarian mandate, abandoning an exclusive commitment to science for its own sake and addressing problems of importance to the human condition.[39] Unesco's Department of Natural Sciences tied the aims of science explicitly to the service of mankind, using it to help many countries recover from the devastation of World War II, through international scientific and technical cooperation. Here, the oceanic sciences would play an important role, largely because saltwater fisheries could provide a significant source of the world's food. The United States also adopted policies to utilize scientific aid for humanitarian objectives. President Harry Truman made support for the United Nations an important element of his foreign policy. True, it was a forum for seeing the world through crises and for coordinating international aid for the benefit of war-torn countries. But it also was an element of Truman's strategy of achieving the containment of communism, while waving the banner of internationalism. Truman and his successors struggled to contain the Soviets economically in programs for "development," and both the United States and the Soviet Union provided funding, foodstuffs, and scientific and technical assistance to countries that were weak or nonaligned politically, ostensibly for development's sake, but also in a struggle for political influence.⁴⁰

In addition to monetary aid, Truman's strategy of containment included government-sponsored connections between science and international relations. In the fourth point of his inaugural address in 1949, President Truman affirmed his administration's desire to incorporate science into this policy:

> We must embark on a bold new program for making the benefits of our scientific advances and industrial progress available for the improvement and growth of underdeveloped areas. . . . The United States is pre-eminent among nations in the development of industrial and scientific techniques. The material resources which we can afford to use for the assistance of other peoples are limited. But our imponderable resources in technical knowledge are constantly growing and are inexhaustible.

Truman invited other countries to pool their resources with the United States and the United Nations to work together in programs of technical assistance. Although the United States would develop its own "Point Four" programs based on this speech, Truman insisted that it be a "worldwide effort for the achievement of peace, plenty, and freedom."⁴¹ The success of the Marshall Plan in Europe had proven that the United States could help to build up and stabilize countries. But there would be no Marshall Plan for the developing world. In the 1930s, the United States learned that money alone could not prevent enemy advances into strategic areas such as Latin America.⁴² Truman's secretary of state, Dean Acheson, felt that only highly developed societies could make efficient use of capital. He believed that "capital loans in advance of technical and managerial competence are not only a waste but a disadvantage" to the borrowing country. First, the United States, under the auspices of cooperative programs in the style of the United Nations, must make the poor productive.⁴³

Unesco's efforts to wed science and development fell short of American expectations. American officials accused it of being sluggish in promoting scientific programs. One observer of Unesco operations in South Asia complained that its Science Cooperation Office in New Delhi served simply as a clearinghouse for scientific information, having no active role in promoting scientific activity. Scientists in Thailand complained that they "were being surveyed to death," caught up in extended studies of their capabilities and potential. To critics, Unesco was slow to promote tangible results.⁴⁴ In May

1951, the State Department urged American diplomats to pressure Unesco to concentrate more substantially in assisting the underdeveloped nations of the free world, in order to accelerate their educational and scientific development. The Americans felt that the organization's effectiveness had been dissipated by its wish to emphasize its nonpolitical character.[45] Rather than focus on areas vulnerable to communist influence, Unesco tried to provide fair representation to a wide range of nations. Consequently, it spread itself thinly and had far-ranging difficulties in administering its hundreds of projects.

Unesco and Point Four programs attempted to export knowledge in a number of fields, including oceanography, and to create long-lasting scientific foundations in countries in the developing world, particularly Latin America and Asia. In Latin America, the practice of setting up a local knowledge base was systematized, and the Americans set up a *servicio* for each development scheme. The State Department sent an expert to manage each *servicio*; the expert would attempt to design research programs, set high standards, and eventually leave it to the new local experts trained by him.[46] Several of these experts were fisheries specialists who organized, planned, and conducted research programs designed to expand and modernize commercial fisheries. After this instruction, the new local experts were expected to add their own sophisticated research in biological or physical oceanography to the fisheries investigations.[47] The overall conception of Point Four was that it was a temporary measure to stimulate local activity. The expert's duty was, in one author's words, "to work himself out of his task," by creating knowledge and skill equivalent to his own, so that eventually his services were not needed.[48] However, this was more easily said than done.

Disillusionment with programs for development was not long in coming. One fisheries specialist, Charles B. Wade, found that it was impossible to carry out his duties in Peru. First, funding was not stable, making the long-term establishment of a Peruvian oceanographic community unlikely. Because the director of Peru's fish and game service resigned before Wade arrived, there was no one to provide funds for a marine biological laboratory or a research vessel. Second, the facilities were not conducive to ambitious research projects. The biological section of the fish and game service consisted of three persons (one of whom was on leave in the United States doing graduate work) and was wholly contained in a single room in the technological laboratory. Wade complained that "the girls"—the two remaining biologists were women—simply worked on whatever research ideas came to their heads, without effort to coordinate with other scientists. The difficulty in changing the situation was compounded, no doubt, by the fact that Wade spoke no Spanish.[49]

Tying marine science to development appeared unrealistic. It was premised upon the notion that scientific communities of developing countries would, with a little help, begin to conduct American-style scientific work independent of American help and supervision. But in Wade's case, he had to start at a far more fundamental level than he expected, training the Peruvian biologists in applying quantitative analysis, using methods no more sophisticated than mean, median, mode, and probable error. He had to convince them that fisheries research required gathering large amounts of raw data to be reduced mathematically. He had to persuade them that the need for accurate data and adequate samples required serious physical labor and that results needed to be reasonably accurate, if others were to use them. By the end of the commercial fishing season, he felt that he had made some progress.[50] Nevertheless, the next year another advisor arriving on the scene was equally appalled: "I found that the staff did not have sufficient grasp of the principles, were inclined to treat inspection as a routine procedure, simply a matter of 'smelling fish.'"[51] The advisors met outlooks they had not anticipated, making their task of developing a viable scientific community impossible.

Using science as an instrument of foreign policy yielded frustrating failures, as the quality of work seemed less important than the simple act of "cooperation." Failures to accomplish scientific activity were repeated in other Latin American countries, especially where it was difficult to convince local officials of what science actually entailed. Uruguay had a unique organization, the Servicio Oceanográfico y de Pesca, which was less a scientific body than a government monopoly of the fishing industry. One of the American advisor's first recommendations was to change its name, as there was no oceanographic research of any kind being pursued.[52] In Colombia, the same advisor found that local practices (in this case, fishing by dynamite) were depleting the fish population. Yet there was no way to study the problem because no one, including the fisheries inspectors, had the appropriate training. In addition, there was no government agency, such as a fish and game service, which might logically support a program of fisheries development. The advisor complained:

> Any Point IV technician assigned to Colombia would be entirely on his own without the possibility of training anyone to continue a project. Except in rare instances, this lack of [a] national technician to carry on and develop a program practically nullifies the work of the Point IV technician.[53]

The dim prospects of promoting long-term scientific activity called into question the legitimacy of scientific cooperation in service of foreign policy.

Both the United States and Britain tried to promote new programs in Asia, many of them related to marine science. The United States had its hands full trying to reconstruct the Japanese government in its own image, and most of its scientific efforts in the region were directed toward building up the Science Council of Japan and testing atomic weapons in the Pacific. South Asia was not initially a high priority in American foreign policy, and it was only after the French were ousted from Vietnam in 1954 that the United States began to treat South Asia as a major battleground in the Cold War it waged with the Soviet Union.[54] But it still took an active interest in the area, seeking to align as many countries with the West as possible, particularly as many of them made the transition from colonial possessions to newly independent nations. Asian nations, while welcoming assistance, were wary of the imperial component of such "cooperation." Immediately following World War II, British scientists had looked to the Indian Ocean, with its untapped resources, as one of its most important regions of study for the future development of the British Empire. They convened a conference to survey the critical needs of the British Commonwealth, and not surprisingly, empire-minded scientists on the Oceanography and Fisheries Committee trumpeted the necessity of using science to develop its imperial territories. One of its members declared:

> If ever a sea or ocean deserved the title of "Mare Nostrum," which Mussolini in his pride applied to the Mediterranean Sea, it is the Indian Ocean, surrounded as it is by Members of our great Commonwealth of Nations or countries that are under British Protection.[55]

The British could claim India, Ceylon, Burma, the Malay Archipelago, Australia, and many possessions in Africa, all of which surrounded the Indian Ocean. Cooperation, as envisaged by the British, was an imperial relationship. Over the next decade, such imperial impulses were blurred by the growing Cold War confrontation, as a great deal of aid and external political influence originated from Washington and Moscow, not merely from traditional colonial powers.

Asian nations resisted the inclusion of former imperial powers in cooperative arrangements because, despite being laden with the relatively new ideological conflict of the Cold War, these programs had the potential to reestablish colonial relationships. The British designed a cooperative program, the Colombo Plan, to connect countries in South Asia with donor countries in various assistance programs, and the politics of this organization was particularly revealing of the anti-imperial dimensions of scientific and technical assistance. In 1955, after a major military defeat of French forces in Vietnam the previous year, French diplomats began to request that France become a member of the Colombo Plan. They argued that their country was the only member of the newly created Southeast Asian Treaty Organization (SEATO) that was not also a member of what many perceived as the military organization's economic counterpart. But to Asian members, there were already too many non-Asians in the Colombo Plan, and they staunchly opposed bringing in the recently ousted colonial power. When a British official informed a French diplomat that nothing could be done to bring France into it, the Frenchman treated him to "an exhibition of French diplomacy at its most petulant, remaining glued to his chair, maintaining long silences, fixing me with sinister stares and otherwise giving a passable imitation of Napoleon I at a bad moment, all of which made me very uncomfortable."[56] The political stakes in technical cooperation were high. During the early 1950s, the British were in fact inundated with requests to join, from those who recognized the political ramifications of scientific and technical assistance. Despite the fact that it was originally a Commonwealth plan that was extended to focus on South and Southeast Asia, countries such as Korea, Afghanistan, and even Jordan tried to take part. Much of this was due to pressures by the United States State Department, which was eager to include as many nations as possible in plans, such as the Colombo Plan and Point Four, that appeared poised to align countries economically with the West. For this reason it was most supportive of the inclusion of Japan, which hoped to attain regional status as a donor country, eventually joining the Colombo Plan with the understanding that it would teach other nations the finer points of rice farming and fishing.[57]

This kind of cooperation, pushed by the Foreign Office and the State Department, had little hope of accomplishing any scientific aims. There is no evidence that either the United States or Britain devoted serious attention to ensuring that these programs were as focused on doing worthwhile science as they were in cementing relations between states. Some critics, on the one hand, characterized this kind of foreign aid as an invalid means to spend American taxpayers' money; it was not helping other countries, they argued, and it might even be doing harm. On the other hand, the programs were criticized from within government for doing too much "good" when they ought to be more focused on combating communism. Within the White House's International Development Advisory Board (IDAB), some of the staff felt that the impetus for Point Four, namely, the struggle against communism, had been forgotten by those implementing it and that the foreign policy arguments were "vague and considered secondary to a substantially 'do-good' objective." In late 1950, a committee chaired by Nelson A. Rockefeller tried to rectify this by formulating a strategy that framed all aid for foreign development in terms of the East-West struggle. Members of IDAB recognized in early 1951 that "there lies here an opportunity of using the [Rockefeller] Report (assuming that it lives up to its promise) as a kind of 'Magna Carta' around which to rally congressional and public opinion into a new, non-partisan unity regarding a strong foreign policy." The committee wished to affirm the need for foreign aid to bolster both military and economic strength in the free world. They hoped that scientific and technical assistance would "increase the productivity of the members of the free world bloc (especially the 'underdeveloped' areas on the outside periphery of the iron curtain) so vigorously in the Western atmosphere of political freedom, that Soviet areas on the inside periphery of the 'curtain' will eventually break the hold of Soviet imperialism."[58]

This kind of focus appears to lend credence to the assertions of historians that the scientific aspects of cooperation in the Point Four programs were simply tacked onto American foreign policy as an afterthought. The State Department, they claim, harbored no enthusiasm for the programs beyond their potential anticommunist value. Scholars Walter Isaacson and Evan Thomas have observed that Secretary of State Dean Acheson regarded Point Four as political rhetoric, and not as a real presidential mandate to be carried out.[59] Acheson himself later wrote that expectations had been too high for the Point Four programs. The budget placed limitations on Point Four's effectiveness, and "although the program continued to do a creditable job, it remained the Cinderella of the foreign aid family."[60] Indeed, perhaps the only value of Truman's Point Four was to wed American foreign policy firmly to the objectives stated in the United Nations Charter, a purpose the president may have hoped to achieve without an effective implementation strategy.[61] This goes a long way in explaining why science, and even international cooperation, succeeded far more when tied to the military. The Navy expected to get good science for its money; the State Department cared little for the science itself.

Scientific cooperation, despite the often-misdirected nature of the Point Four programs, was supposed to have a genuine purpose in foreign policy, namely, to develop the scientific and technical foundations of developing countries.[62] Internal critics within the United States government felt that the programs underemphasized the reciprocal nature of cooperation. They stood in stark contrast to a 1950 State Department report by a committee chaired by Lloyd Berkner, which recommended the establishment of science liaison offices in key embassies throughout the world, to utilize the scientific output of nations other than the United States. The report affirmed that "scientific developments are increasingly recognized to be essential to economic welfare; and from economic welfare stems the political security and stability of any nation."[63] The logic itself was sound: a country needed to acquire knowledge to aid its development; yet to be a real participant in the international scientific community, a country also needed to export knowledge. The liaison system and Point Four were not connected, because one was for Europe and the other was for the developing world. But these programs, and the rhetoric of helping the developing world or of using science to fulfill foreign policy aims, would resurface again and again for oceanography. In the early 1950s, such impulses were embryonic, and a far stronger influence on oceanography, and certainly a less disingenuous one, came from the U.S. Navy.

PACIFIC COOPERATION IN THE EARLY FIFTIES: TRANSPAC AND NORPAC

While governments floundered with programs for scientific and technical assistance, cooperation began to flourish under the aegis of military patronage. The early Pacific expeditions by Scripps demonstrated a bond between oceanographers and the Navy rather than scientists' enrollment in the new policy of showing support to the developing world. Often there was little connection between science done under military patronage and science done for foreign policy purposes. For example, while Point Four technicians struggled in vain to design a permanent structure for oceanic research in Peru, American oceanographers working on the Navy's dollar were meeting informally with Peruvian scientists. When Scripps scientists conducted an expedition to the eastern South Pacific in 1952 with Navy sponsorship, they made an arrangement with Peru to share data. There was one man in the Peruvian Navy who had been trained at Scripps, so the Americans made use of his services. Often in developing countries military institutions claimed the best scientists, particularly in oceanographic research, which required considerable training and (often) internationally standardized levels of precision. In the case of the Shellback expedition, cooperative arrangements were carried out with the Peruvian Navy in an unofficial capacity.[64]

Military sponsorship trumped foreign policy aspects during the early 1950s. One example was the TRANSPAC expedition of 1953. American scientists were not blind to the connections being forged between science and world affairs during the late 1940s and early 1950s, and some oceanographers attempted to capitalize on it. The National Research Council (NRC) of the National Academy of Sciences instigated a policy of showing support for scientific institutions in foreign countries, particularly in the developing world. At Scripps, a number of scientists hoped to do this by reestablishing some of the links that had been broken during World War II in the scientific community of the Pacific region. The two American scientists who most actively sought to do this were Scripps marine biologists Carl L. Hubbs and Claude ZoBell. The first Pacific Science Congress to be held in a developing country after the war was to be in Manila, the Philippines, in 1953. Many scientists in Asia hoped that the congress would attract scientists from abroad and that the publicity might resuscitate their faltering institutions. Hubbs and ZoBell urged the NRC to help it promote an expedition to cross the Pacific and to make port in Manila for the congress.[65]

Scientists in Asia welcomed the possibility. The Manila Oceanographic Institute invited Americans to visit and inspect the facilities and develop contacts with the staff. Its scientists hoped that a visit by a crowd of American scientists might help raise the profile of the institute, which was suffering from financial neglect by the government. Because Unesco was to finance a symposium on physical and biological oceanography, marine science promised to be a major component of the congress. If judged successful, it could provide a major stimulus to scientific activity in the region. Here was a way that oceanographers could contribute to foreign policy, to show support for scientific activity in a developing country and perhaps discuss techniques and ideas. The NRC urged Scripps to "transport to the meeting as large a group of U.S. oceanographers and biologists as could be justified by the program of work which would be undertaken on your proposed expedition." Taking a step further, it encouraged Scripps to show its support for other oceanographic institutes in the region. Harold J. Coolidge of the NRC felt that the finest institute in Southeast Asia, operated in Nhatrang, Vietnam, by French scientists since 1922, was in danger of being abandoned by both French and Vietnamese officials. Coolidge urged Scripps to send scientists also to Nhatrang, as another show of support for Pacific science. The visit, he claimed, would be "a great inspiration to those who are struggling to prevent the Institute from being turned into barracks for a nearby school, which would lead to the termination of its effective use for scientific research of any kind."⁶⁶ In the Philippines and in Vietnam, the NRC hoped to use scientist-to-scientist contacts to boost existing local science, affirming its relevance to the international scientific community as a whole, a much-needed mark of prestige for scientists who relied on government patronage.

For similar reasons, the NRC encouraged the expedition to go to Japan. With its rich history of scientific activity and many institutions devoted to marine sciences and fisheries research, it would not be difficult for American oceanographers to make fruitful contacts with scientists in Japan. And because the United States was seeking opportunities to consolidate its relationship with Japan, a scientific venture would be timely. Coolidge felt that the Americans should assist Japanese colleagues by letting them familiarize themselves with new equipment and techniques. Such a show of support, he wrote, "would be making an important contribution to international good will and would help us to obtain further cooperation from Japanese scientists in joint researches in the field of oceanography."⁶⁷ It was in view of this visit that marine geologist Robert S. Dietz wrote from Japan that, because of Japan's dependence on marine food, and its "marine 'fixation' without parallel anywhere in the world," cooperation with Japan could reap important cultural and scientific benefits.⁶⁸ Although the cultural benefits likely would be intangible, scientific cooperation could lead to the long-term exchange of ideas that could provide benefits both to Japan and to the United States.

The efforts of Hubbs and ZoBell to spearhead a major biological expedition seemed well positioned to further the cause of biological science and its economic connections to fisheries research. The recent grant by the State of California for research on the decline in the sardine catch appeared to necessitate, despite the intentions of its director, Roger Revelle, a major commitment to studies bearing directly on fisheries.⁶⁹ A biological expedition showing support to institutions in Asia, where the United States already was emphasizing the connections between science and economic development, might have tied biological science at Scripps closely to international cooperation, economic development, and American interests abroad. Perhaps Hubbs was trying to achieve something for marine biology that he felt had been lost at Scripps since Revelle became director in 1950. Revelle had made no secret of his conviction that physical oceanography and marine geophysics, with clear avenues of patronage by the U.S. Navy, should become the dominant objects of study. It was this conviction, not to mention his repeated efforts to cast biological questions in terms of physical processes and statistical analyses, that had caused Hubbs and several others strongly to oppose his appointment as director. After being appointed, Revelle alienated Hubbs and others by moving swiftly to diminish the role of biology at Scripps and strengthen the relationship between oceanographers and the Navy. Revelle's appointment was a major setback for those who opposed connecting American oceanography too much to the Navy.[70]

Revelle's unabashed commitment to the Navy should come as no surprise. Revelle had spent years in the Navy himself trying to cultivate a respect for basic research there, and he felt that those efforts had met with great success. Naval support already had launched the Bikini Scientific Resurvey (1947), the MIDPAC expedition (1950), the Northern Holiday expedition (1951), the Capricorn expedition (1952; also associated with a bomb test), and other projects, all of which not only had provided data to the Navy but also had produced significant results for scientists at Scripps. Revelle felt no great love for Scripps's biologists, who not only had tried to keep him from the directorship but also had threatened to redirect the focus of Scripps toward applied fisheries research, which he felt was not the true mission of a research institution. Granted, Navy support smacked of applied science as much as fisheries research, but Revelle did not view it that way. Instead, he recognized the Navy's interest in all things having to do with physical oceanography and marine geophysics, and he knew that through the Navy, basic research in these fields had the potential for long-term support.[71]

Now, three years after Revelle's appointment, the marine biologists were struck another blow. The expedition did not transpire as they had envisioned. In early 1953, the Office of Naval Research (ONR), which provided funding for the expedition, intervened in its planning. One of its liaison officers wrote that the northernmost Pacific "has been sadly neglected," despite the need to collect data of vital interest to the Navy in the area. He advised scaling down the proposed expedition, concentrating more vigorously on a smaller area to reap scientific results of utility to the Navy. Totally undermining the original conception of Hubbs, ZoBell, and NRC's Harold Coolidge, the idea to send a shipload of scientists to the Pacific Science Congress was scrapped. In a feeble gesture of compensation, the Navy offered to fly a select few scientists by military transport from Japan to Manila for the duration of the meeting. No arrangements were made to visit the institute in Nhatrang. The ONR felt that, if international contacts were desirable, then the scientists should concentrate on Japan, where developing contacts had the potential to create immediate opportunities to share data and scientific information.[72] The TRANSPAC expedition was reoriented to fit the Navy's requirements.

Beginnings of Postwar Marine Science

As an ONR scientist wrote, the eventual track to be followed by the ship conformed not to a plan set forth by the scientists but, rather, to

> ... a coverage of the North and Northwest Pacific Ocean areas which is least known hydrographically and oceanographically to the U. S. Navy. It is the opinion of this Office that subject cruise is of extreme importance to the U. S. Navy and should be supported in every detail. It is of particular value to those activities concerned with undersea warfare.[73]

Although TRANSPAC evolved into a military-oriented mission, some biological studies were indeed carried out, but not on the scale planned. To the biologists, it was a repeat performance of the MIDPAC expedition, whose purpose initially had been a biological one, but was dominated by physical and geophysical work. Hubbs, no doubt frustrated by this turn of events, decided not to take part in the expedition.[74]

Despite the Navy's conspicuous role in the expedition, TRANSPAC was something of a sensation in Japan. Prior to the arrival of the vessel, *Spencer F. Baird*, marine geologist Robert S. Dietz traveled throughout the country promoting what really was not a joint expedition, but just a brief visit to Japanese ports on an otherwise military-oriented expedition. Still, he was confident that he had drummed up some enthusiasm:

> I have shown the Mid-Pacific Expedition film in many parts of Japan. At least a couple thousand people have seen it. This is not too much to our credit for there are "taxan" (Jap. for *beaucoup*) people over here and almost anything draws a crowd. But anyway I'm sure that the Expedition will find that Scripps is a "household word" at least among the scientific fraternity when they arrive.[75]

Interest in the American visit went far beyond the "scientific fraternity" and reached the highest levels of Japanese society. Among those to see the film of the 1950 expedition was Emperor Hirohito, who even invited Dietz to visit his marine zoological laboratory and to have a personal audience. Dietz was delighted to find that Hirohito was well acquainted with oceanography, having published a number of works of his own in marine biology.[76] Given the emperor's biological interests, it seems a bit ironic that neither Hubbs nor any other prominent biologist would be aboard the American ship. All the same, the indications of cultural respect for science bespoke of more intangible advantages for cooperation not limited to Japanese government support.

The level of scientific and governmental interest evident at the Japanese reception of Scripps's ship, the *Spencer F. Baird*, contrasted sharply with the experience of American scientists in Latin America on programs sponsored under Point Four. Kanji Suda, chief of the Japanese Hydrographic Office, prepared a reception committee in Tokyo consisting of scientists from the Science Council of Japan, the Oceanographic Society, the Fishery Society, and the Geological Society. Also included were the mayor of Tokyo and representatives of the Asahi Press. Similar receptions were planned for the *Baird*'s other ports of call in Hakodate and Kobe. The message sent to Scripps by the expedition leader, Warren Wooster, on September 29, 1953, showed that Suda had made good on his promise to drum up interest throughout the local community:

> During official stay every minute was planned with tours of research institutes, discussions with scientists, banquets, and receptions. Thousands of people visited the ship. If the Tokyo reception is proportionally greater none of us will survive the ordeal.[77]

The reception in Tokyo was indeed greater. Emperor Hirohito himself invited the Americans to visit his laboratory on the palace grounds. This provided opportunities not only for the Americans but also for the Japanese scientists who came along. They themselves had never met the emperor and now found themselves raised in status, enjoying a personal audience. As gifts, Wooster and the other scientists presented Hirohito with samples dredged from the nearby Bayonnaise rocks and with specimens connected to the emperor's own scientific interests, namely, sea slugs (nudibranchs) from the California coast. The emperor was delighted to discuss scientific matters with his guests from across the Pacific Ocean. Such a level of enthusiasm for American science, in addition to Japan's reliance on the sea for its food, indicated that Japan might be fertile territory to develop a cooperative scientific relationship.[78]

Because of the existing interest in marine science in Japan, the Americans saw that their presence might stimulate a cooperative relationship. The scientists of the TRANSPAC expedition sent a clear message to the Japanese that they would not be able to use the United States as a scientific crutch for their own development, but rather that the Japanese would do their own work and perhaps share results. The director of the Nagasaki Marine Observatory asked Wooster if his ship might visit Nagasaki and conduct some work in the East China Sea. Wooster declined, claiming that knowledge of the area "will come chiefly from the intensive systematic investigations which you and your colleagues are carrying out so successfully."[79] The

Japanese respected this attitude, and as a token of willingness to be part of a reciprocal relationship, presented their American colleagues with some scientific instruments for use on their expedition.[80] Suda hoped that the port receptions would bolster enthusiasm for investigating the problems of the Pacific and "may be effective for the development of oceanography in Japan." To back up these sentiments, the Japanese Hydrographic Office planned to send its own ship to make observations along a parallel course to the *Baird*. Suda wrote to Revelle back at Scripps: "I hope that this will become a epoch [*sic*] of the cooperative studies of N. P. O. [North Pacific Ocean] by both country [*sic*] . . . We have never [had] such [a] good chance to see the first order of observation of the world."[81] The Japanese revealed a willingness not only to continue research of their own but also to reciprocate any assistance by contributing their own work and data.

The success of TRANSPAC revealed a potential for cooperation that went far beyond the technical and scientific assistance programs that had seemed so one-sided. With Japan, foreign scientists seemed willing and able to conduct their own work, thus creating a mutually productive situation. At the same time, here was an expedition that was both military-sponsored and international. It was the first evidence that American oceanographers might be able to pursue both and that Navy sponsorship would not necessitate sacrificing international cooperation. Realizing this, Revelle moved to consolidate the relationship with the Japanese. In December 1953, he wrote to geologists at Tokyo University, suggesting that Japanese and American scientists should publish their findings together. Revelle believed that one article, on the Bayonnaise rocks, should be submitted to a Japanese journal. Another, on the Jimmu Seamount, should be submitted to a Western journal such as *Science, Deep-Sea Research,* or the *Bulletin of the Geological Society of America*. In addition, they might send a note to *Nature* announcing the results of the joint effort. This article would contain petrographic and topographic descriptions that would include a discussion of structural trends, analyzed jointly by American and Japanese scientists.[82] Such publications would show the world that Japan too was pursuing marine science research of the highest order and that the United States and Japan had created a mutually beneficial cooperative relationship. Writing to Suda, Revelle affirmed that Japan could be "justly proud of the careful and comprehensive research being carried out by her marine scientists" and that continued progress would depend on the free exchange of data and people exemplified by the TRANSPAC expedition.[83]

This experience with Japan inspired American oceanographers, particularly Revelle, to attempt an extensive survey that would entail sharing data. Revelle felt that Japan's cooperation could make possible a series of observations in the North Pacific that would be undertaken at the same time in various sections, providing a scientific "snapshot" of the whole region. In April 1954, he announced that, because the following year would see at least nine oceanographic vessels of various countries operating in the North Pacific, scientists should attempt Revelle's "long-dreamed-for, but heretofore unrealizable, result—a truly synoptic hydrographic study of the entire North Pacific."[84] The NORPAC project, as it was called, began from informal discussions between Scripps oceanographer Joseph Reid and the head of Canada's Pacific Oceanographic Group (POG), John P. Tully. They realized that a cooperative survey between Scripps and POG would produce more significant results than the individual programs planned by each group.[85] Once again it was in the strategic North Pacific where the Navy had insisted that the TRANSPAC expedition go. Although NORPAC too would enjoy Navy sponsorship, the express purpose of the project would be to obtain data in such physical and biological oceanographic detail as to understand better the problem of fisheries in the Pacific region.[86] At the Fifth Pacific Tuna Conference in November 1954, scientists affirmed the need for more knowledge of oceanic circulation and of oceanic processes in general. On NORPAC, the vessels of each country would obtain meteorological data and take measurements of plankton net hauls, temperature, salinity, oxygen, and phosphate, all at standardized depths.[87] Because American scientists planned to coordinate their activities with Canadians, the inclusion of the Japanese, who had recently shown an unprecedented enthusiasm for cooperation, seemed the logical next step.

With Japanese cooperation, NORPAC became the largest project of its kind ever undertaken, and given the region to be studied, the Navy was more than willing to sponsor it. But by this time, Revelle was learning that promoting research for its fisheries applications, which he previously had resisted doing, was a preferable alternative to trumpeting military applications to scientists in foreign countries. In international expeditions, the Navy's sponsorship was an asset only for its financial support. Press releases from Scripps avoided indications that the Navy had any interest in the work beyond its contribution to basic research. It emphasized that the real significance of the NORPAC area, which was about 50 percent larger than the North American continent, was the fact that it provided almost half of the world's commercial food fish haul. Before NORPAC, information about the region had come from solitary oceanographic cruises that had taken months, producing data on oceanic conditions that might have changed substantially between the first and last measurement stations. What NORPAC offered was a rapid, intensive survey in a much shorter interval of time.[88]

To a far greater degree than they had on TRANSPAC, the Japanese rose to the scientific challenge. They sent about fifteen ships from various Japanese agencies, plus about ten smaller ships operating close to the Japanese home islands. The only assistance the Japanese requested was that they might borrow equipment if their own instruments failed.[89] For his part, Revelle already had begun to integrate Japan into his overall vision of international cooperation. That summer, Revelle traveled to Japan both to firm up the details of NORPAC and to discuss Japan's role in an even more ambitious project, the International Geophysical Year (IGY). He urged the Japanese to send representatives to the meeting for the IGY in Brussels in September 1955, to help the world's oceanographers plan the event.[90] Revelle himself was getting more deeply involved in international cooperation and was excited about the opportunities it provided for oceanography. Japan was his first success; in a very short time, it had gone from being a country with a struggling marine science community to a full-fledged participant in the most ambitious projects in the history of oceanography.

Revelle saw in Japan an opportunity to create a foreign scientific community with goals identical to those of American oceanographers. For the meeting in Honolulu in February 1956, Revelle found funds for a number of Japanese scientists to attend. He wanted to include more of the younger Japanese scientists, who would not normally have been brought along by the senior ones. These younger Japanese scientists had played a conspicuously greater role in NORPAC than their older colleagues, and Revelle hoped for their continued participation in joint projects. Moreover, according to Scripps scientist Joseph Reid:

> Dr. Revelle feels that we must try to break some of the younger people free from the traditional path of classical oceanography to which the Japanese have adhered. They should be allowed to meet and talk to people in their own field from this country and in a sense be educated away from the traditional methods of [Professor Koji] Hidaka, not only for their own good but in the long run for the good of physical oceanography and our own best interests.[91]

The need to "break" younger scientists from traditional Japanese oceanography was an understandable concern for Americans attempting to establish standardized procedures along American lines. It was also understandable in light of Scripps's sharp focus on physical oceanography and marine geophysics, an emphasis it hoped to extend to Japan, where the biological elements of fisheries oceanography had strong roots. During the TRANSPAC expedition, Scripps had taken pains to ensure that one of the scientists on the *Baird* was Noriyuki Nasu, a Japanese-born graduate student who was studying at Scripps. Not only would the expedition provide fieldwork for Nasu's degree, but also he would be an ideal liaison between American scientists and their Japanese colleagues once the *Baird* arrived in Japan. But Scripps had no intention of keeping him. Trained in American methods and familiar with American practices, Nasu was an example of the kind of young scientist whom Americans wished to send back to Japan to ensure Japanese involvement in the international scientific community and continued Japanese appreciation of American attitudes toward research.

After the TRANSPAC and NORPAC expeditions, American and Japanese scientists met to discuss the potential for future collaboration. First, they needed to meet in order to identify which areas and subjects were worth pursuing, to ensure close interaction and avoid duplication. Second, they had to discuss publication. This came eventually in the form of an atlas of the North Pacific Ocean that displayed the distribution of ocean properties, invaluable for future research "as well as being in itself the first quasi-synoptic atlas of any large area of the ocean."[92] This was quite a task, as participants had to agree on standard nomenclature. Third, scientists needed to discuss specific cooperative efforts to undertake. In the wake of NORPAC, a similar expedition was planned for the equatorial Pacific Ocean region, from Central America to the Philippines. The EQUAPAC expedition, as it was called, would involve participation by the United States, Japan, and France.[93] Much of this impulse was absorbed into the International Geophysical Year of 1957–58, which aimed to extend synoptic data collection over the entire earth.

By the early 1950s, marine science in the United States took on a reorientation of focus toward physical oceanography and marine geophysics, due mainly to the influx of funding from the U.S. Navy. At the same time, it was beginning to explore one of the great trends of marine science for the next two decades: international cooperation beyond North America and northern Europe. The experience with Japan was not only financed but also shaped by the needs of the Navy, while other scientists would have preferred a more explicitly biological expedition with ties to foreign policy, showing support to scientific communities of the developing world. What made TRANSPAC and NORPAC crucial was the fact that the rhetoric of supporting fisheries and the developing world remained, even in cases where clearly the greatest benefits would be in the realm of physical oceanography and military planning. Playing with this rhetoric would become a recurring feature of international cooperation. The United States and other Western powers did have foreign policy goals of using science to build up developing countries, and oceanographers were conscious of them. However, despite adopting some of the rhetoric, the cooperative programs in oceanography of the 1950s were much more a part of the military's interest in oceanographic research. What cannot be emphasized enough is that the Navy's priorities reigned supreme in these ventures, and leading scientists typically were pleased to keep it that way. The Navy was bringing the center of activity in oceanography to the United States, it was providing new ways for geologists and physicists to work on fundamental problems, and it seemed willing to support cooperative activities that made synoptic studies possible. The evolution of the Navy's support for science, the issues raised by it, and the role of international cooperation are all subjects of chapter 2.

2 OCEANOGRAPHY'S GREATEST PATRON

Scientists' participation in cooperative ventures depended upon the acquiescence of their government patrons, particularly those groups within the United States Navy that actively promoted research. This chapter reveals the context in which oceanographers promoted their strategy of international cooperation even among those who, by the nature of the Navy itself, might seem ill suited to international endeavors.[1] It traces the relationship between the Navy and the oceanographic community, particularly with regard to research that was international in scope, and reveals how the needs of each complemented and contradicted the other. This relationship is crucial in the history of the American military in an era dominated by the need to adjust to the requirements of constantly changing technology. Not only did the Navy provide logistical support for such symbols of peaceful scientific cooperation as the IGY, but also it was the leading patron of scientific research after the Second World War.[2] Scholars have devoted considerable attention to developing sophisticated conceptions of the reasons for sustained support for science by the military and other government patrons. Advocacy of international science and the Navy's own motivations for pursuing international research have received comparatively little attention.[3]

Although there were opportunities to connect science to foreign policy, these prospects paled in comparison to the scale of work made possible by the Navy. This was one of the reasons Revelle fostered that relationship at the expense of others. Scientists at Woods Hole also cultivated the Navy's appreciation for the connections between oceanography and the military uses of the sea. After TRANSPAC and NORPAC, Revelle recognized that the Navy was not averse to cooperative projects that might serve scientific as well as military goals. The Navy's logistical support for these expeditions, and for the larger ones that followed, was not part of its blind support for basic research, but neither was international cooperation a façade covering purely military work. These projects were the result of a partnership between oceanographers and the Navy that developed most strongly during the Navy's strategic reorientation during the early 1950s.

That the large-scale international projects emerging in the 1950s and 1960s were born of the bond between science and the Navy, rather than the cooperative spirit of existing international organizations, is one of the main points of the present study. The Navy was the first significant disciple of marine science in the postwar era. Its relationship with scientists in the 1950s was one of mutual advantage. The oceanographers wished to conduct research, and the Navy wished to understand its workplace, the sea, which it considered the crucial battleground of the next war. Significantly, it admired scientists' strategy of focusing on long-term national security goals such as countering the submarine threat, developing an alternative (submarine-based) nuclear deterrent, and investing in new technologies. That strategy, however, was interwoven with an international outlook based upon openness and data sharing as the means to pursue science efficiently and with an assumption that the United States could transform such science into new technology the fastest. The Navy learned that international cooperation in oceanography was also a useful means to provide constant operational information for its existing defense technology, but it also strove to keep much of its own data secret. This would prove to be an uncomfortable point of conflict with scientists, making international cooperation even more attractive if it could promise openness of scientific results. The scientists' strategy provided a confluence of means both for scientists wanting international cooperation and for the Navy wanting operational data. This mutual understanding did not always work smoothly; the Navy's unwillingness to release data, and thus make information available to interested marine scientists throughout the world, was a greater source of conflict than the question of whether the military controlled research priorities.

THE PARTNERSHIP BETWEEN SCIENCE AND THE NAVY

The Navy made a dramatic change in its perception of the value of oceanographic work for military purposes during the Second World War. Prior to the war, scientists felt that the Navy's understanding of their usefulness came only in spurts and funding came only for specific projects.[4] During the war, at least initially, the Navy's knowledge and appreciation of science was limited, but it turned to civilians to help it tackle some of the difficult problems it could not solve on its own. Under director Columbus Iselin, scientists at Woods Hole worked on marine acoustics, using the physics of sound transmission to transform techniques of harbor mining and submarine detection. These scientists established a record of usefulness for national security by addressing the Navy's most acute difficulties in conducting undersea warfare. The wartime report by Woods Hole's W. Maurice Ewing, "Sound Transmission in Seawater," remained the standard text on the subject within the Navy for many years.[5] By the end of the war, the Navy had committed to funding scientists at both Woods Hole and Scripps on a continuing basis for an indefinite period. As oceanographer Roger Revelle put it, the Navy gave these institutions tenure. Since the specific military applications were classified, usually even beyond the scientists' security clearance, many simply pursued their own ideas and officials used the research as they saw fit.[6] For oceanographers, increased patronage was a blessing, and it promised to give a few fields (especially geophysics, geology, and physical oceanography) a significant boost.[7]

In the immediate postwar years, questions of patronage for science were dominated by concerns about military control.[8] Skeptics feared that the military might not accept the scientists' view that science should be free to pursue its own end, regardless of technological applications. The threat of being governed by military priorities loomed over numerous disciplines in the United States. As military support for science expanded, no civilian funding agency experienced any comparable growth. The scientific community found itself being accused, like Faust, of striking a deal with the devil. It seemed divided between those researchers who chose to accept funding from the military and those who refused to do so, thus condemning themselves to working with older and cheaper techniques. In oceanography, expeditions were very expensive; those who refrained from making them could hardly expect to be leaders in the field. Few scientists in any discipline could ignore the issue, and in 1947 a public debate over military funding of science erupted between several noted scientists and intellectuals in the pages of the *American Scholar* and the *Bulletin of the Atomic Scientists*. The opening salvo was launched by physicist Louis Ridenour, who scoffed at the concept of "intrusion" by the military into the scientific realm. He felt that scientists should be glad to have the opportunity to do research, and in no way were they forced to take money from the military. Ridenour wrote that the military may be guilty of seduction, but it was not guilty of rape.[9]

Others were more alarmed by what they perceived as the inherent dangers of a militarized scientific community. Several well-known scientists and intellectuals spoke out against it, particularly novelist Aldous Huxley, mathematician Norbert Wiener, and physicist Albert Einstein, the last even likening postwar American society to Germany on the eve of the First World War.[10] These men hoped that the scientific community could take control over itself and not allow itself to be led astray by the money the military offered. Others in the debate felt that the concerns for their profession could be addressed without taking the military out of the equation. Vannevar Bush, who had managed the American research efforts during the war, thought that most of the dangers could be avoided if scientists could persuade the military of the imperative of supporting "basic" research (he meant work unconnected to any specific technological application) as well as engineering research for military needs. If that particular imperative could be met, then the danger posed by a military hand on the purse strings would be negligible.[11]

The view of military funding as a benefit with manageable risks was shared by many involved in oceanography, especially because so many leaders of this field had either served in the Navy during World War II or worked as civilians under the military. Many scientists, including Revelle, who became director of Scripps in 1950, felt that they owed a great deal to the armed services, particularly to the Navy. Many oceanographers became involved in postwar military work, including nuclear tests in the Pacific Ocean, out of a sense of obligation. Revelle recalled, "I'd been in the navy for eight years, and all of us had been involved with World War II. We felt that we ought to do what we could to help the United States government."[12] Next to this sense of obligation, many felt that the dangers of military patronage were small. However, as Bush noted, the danger was there; scientists had to protect their professional interests. One way would be to convince the military to support basic research. Oceanographers, through the efforts of Revelle and others during and after the war, accomplished this early on. Later, they would look to international cooperation as yet another strategy to offset some of the negative effects of military patronage, while still serving the needs of the Navy.

The fountain of money from which most of the funding now came forth, the Office of Naval Research (ONR), also submitted a comment to the *Bulletin of the Atomic Scientists*. Far from trying to control science, it claimed, ONR had taken the lead in supporting basic research after the war. Representatives from ONR said that it had been an expensive, difficult, and even pioneering task to integrate that practice into the Navy, but they felt it had a special responsibility to do so since no civilian agency at that time was equipped to do it.[13] Roger Revelle later recalled that ONR took pride in supporting basic research when it might have chosen projects that seemed more practical. The office had learned that "the way to support research was to support the research that researchers wanted to do instead of dreaming up projects for them to do."[14] Indeed, this line of reasoning entered into Ridenour's initial essay. Because freedom of inquiry was the hallmark of its attitude, he claimed that the Navy had a remarkably sophisticated approach to research.[15]

On the whole, however, the Navy of the late 1940s was not so scientifically sophisticated. Those who championed the cause of basic research had been in the minority for years. Few considered research itself to be a critical component of the Navy's mission. Rear Admiral Harold Bowen later wrote that he credited his stint within the Navy's research establishment to the fact that he had made so many enemies as a flag officer that the Navy effectively banished him to the Naval Research Laboratory, making him its director in 1939.[16] Later Bowen was instrumental in the establishment of ONR, but he already was a maverick within the Navy by that time. He even tried to explain to others that basic research had value separate from engineering, but it came out sounding ridiculous. He recalled his testimony to a congressional appropriations committee, during which he said that "if you knew what you were doing it wasn't research."[17] Years later, Secretary of Defense Charles E. Wilson would turn the statement around to say with disdain, "Basic research is when you don't know what you're doing."[18] Basic research was among the first things that the Navy wished to cut when trying to trim costs as the war in Korea began to heat up in late 1950.[19] Many Navy leaders chafed at attempts by science advocates such as Vannevar Bush to put scientists' priorities before military ones. To them, Bush's attempts to form an elite leadership of experts to ensure the pursuit of basic research were just parts of his technocratic vision of American leadership. They struck many military leaders as a bid for power and as a bad omen that the military might lose control of the very projects they funded. Each of the armed services, including the Navy, often resented and resisted the new power of scientists in government.[20]

The Navy leadership's attitude toward science, particularly oceanography, was about to change in the scientists' favor. The atomic age was promising an era dominated by nuclear physics and strategic bombers, not of naval task forces; the atomic bomb brought about changes in technology and strategy that had made the hard-won lessons of conventional warfare in World War II seem irrelevant.[21] The military services themselves entered a period of intense rivalry in which the Army and Navy resented the Air Force's new role as the centerpiece of national security strategy, while the Air Force branded the others as redundant or even obsolete. The Navy especially was handicapped by these attitudes. In 1949 President Harry Truman appointed Louis Johnson as secretary of defense in order to end the rivalry through strong centralized leadership. To do so, Johnson eliminated programs that he considered wasteful duplication of effort, which meant a reappraisal of American global strategy and the Navy's role in it. In Johnson's view, most of the Navy's functions had been usurped by the Air Force. One of his first actions was to cancel, without notifying either Secretary of the Navy John Sullivan or Chief of Naval Operations Louis Denfield, the already-begun construction of the aircraft carrier *United States*, the "supercarrier," one of the Navy's prized projects. Sullivan resigned in disgust, and adding insult to injury Johnson announced a $353 million cut in Navy Department funds. Denfield was removed from his post for his role in the subsequent "Revolt of the Admirals," during which Navy leaders tried in vain to show Congress that the next war could not be won by the Air Force alone.[22] With new leaders more suited to the secretary of defense's outlook, the struggle to convince Washington that the Navy occupied a crucial position in modern American military strategy was faltering badly.

This background is essential for understanding the Navy's fruitful relationship with science in the 1950s. It had less to do with its "sophisticated" vision and more to do with ensuring its prominent role in national defense strategy. One of the ways that it did so was to emphasize conventional threats. Intelligence reports concluded that the Soviet Union already had the capability to launch submarine-based rockets against major American ports, built from designs captured from the Germans at the end of the war.[23] The submarines themselves were equipped with snorkel tubes, allowing them to remain submerged and undetectable by radar for long periods. Here was a research need that was decidedly different from the science of atomic bombs; the Navy needed classical physics to facilitate the manipulation of a water environment that was largely unknown. To address the problem, the new chief of Naval Operations, Admiral Forrest Sherman, put scientists from the Massachusetts Institute of Technology (MIT) under contract to study how to counter the threat. Project Hartwell (named for a restaurant at which the scientists often dined) drew together scientists from elite institutions throughout the United States. These scientists recommended long-term planning to protect American oceanic shipping and to conduct antisubmarine warfare. The Navy received their report with enthusiasm, praising the mutually beneficial relationship between science and the Navy and calling the report the "bible of undersea warfare."[24]

What made the Hartwell summer study unique was that its participants expanded the scope of their study to include the whole mission of the Navy. They were not content to focus on a narrow problem; instead, they put a more system-oriented approach to national security into practice. They recommended a number of changes, from the level of technical training for officers to the future development of tactical atomic weapons for use by American submarines. With war erupting in Korea in 1950, the Hartwell scientists felt a heightened sense of resolve to address exactly what it would take to fight a war halfway around the world in the 1950s and beyond. Rejecting the view that atomic weapons could only be land- or air-based weapons, they urged the Navy to plan an integration of them into their battle plans.[25] What emerged from the Hartwell project was a blueprint for scientific and technological work and new goals for the Navy in the coming years. Acting on the scientists' recommendations allowed the Navy leadership both to focus on a conventional threat (submarines) and to place the Navy into America's strategic vision in the nuclear age.

A key Navy leader who took the Hartwell recommendations to heart, and who would later be an instrumental supporter of international scientific operations, was Admiral Arleigh Burke. After participating in the "Revolt of the Admirals," he had been transferred to the Defense Research and Development Board for much the same reasons that Admiral Bowen earlier had been "banished" to the Naval Research Laboratory. He seemed unlikely to cause much trouble in a position of secondary importance, namely, scientific research.[26] But he was convinced that the Navy's future would be defined by its ability to accommodate and take advantage of new technology, and thus there needed to be parallel trajectories for science and for the Navy. He and a few others decided that the Navy could ensure its relevance by developing new ways to detect enemy submarines and even by providing an alternative deterrent to bombers and land-based ballistic missiles. By developing submarine-launched missiles, Burke and his supporters reasoned that the United States' nuclear deterrent might become both hidden and mobile, decreasing the chance that the Soviets would attack and expect to destroy the American ability to retaliate. As director of Strategic Plans and later as chief of Naval Operations (CNO) for three terms in the 1950s, Burke became one of the forces behind the Navy's drive to develop a national strategy based largely on submarine and antisubmarine warfare.[27]

The scientists' recommendations shaped the Navy's operational requirements for the entire decade. As early as 1951 the assistant CNO for undersea warfare, Frank Akers, began to implement the recommendations of the Hartwell scientists. He was particularly interested in a research program dubbed Project Jezebel that would lead to the long-range detection of submarines by low frequency analysis and recording (LOFAR) to be developed over several years. This technology would improve upon the sound fixing and ranging (SOFAR) that developed from W. Maurice Ewing's discovery of a deep sound channel during the war. Not only would LOFAR detect the location of distant sounds, but it also would enable listeners to classify them and thus perhaps identify the source of the noise.[28] By networking the listening stations together into a sound surveillance system (SOSUS), the Navy could provide a protective barrier against submarine infiltration over a huge area. Science and technology became, more than ever, fundamental to naval strategy. Civilian scientists were active in helping the Navy coordinate information for the stations, and even in setting them up, under Project Caesar, at their pilot locations in the Atlantic Ocean. Oceanographic surveys by civilians were needed both to find out whether or not arrays could be laid at particular locations and to ascertain the projected quality of the equipment's performance.[29] Shortly after the Hartwell scientists delivered their recommendations to the Navy, the chief of Naval Research proposed a series of Undersea Warfare Symposia to keep civilian scientists abreast of naval developments and technological requirements.[30] The Hartwell scientists' insistence that a submarine should not be perceived merely as an antishipping weapon was borne out when in 1952 Naval Intelligence reported that the Soviets were experimenting with launching guided missiles from submarines. The missiles, presumed to be similar to German-designed V-1 or V-2 rockets, in the future could be fitted with atomic warheads. The director of Naval Intelligence was convinced that these submarine-based missiles would be the most probable means of employing atomic weapons against the continental United States in wartime and that "it is believed that if the Soviets so desired V-1s could be launched against our coastal cities at any time."[31] Such developments brought the need to detect undersea threats sharply into focus.

Submarine detection was the impetus for a great deal of marine research. Some of this was applied research to improve the efficiency of the LOFAR equipment. For example, low frequencies gave a better chance of detection, so the Navy strove for the lowest frequencies possible in its equipment. Because low frequencies required larger and more expensive transducers, one area of research was in finding ways to improve quality at higher frequencies. Also, some of the more difficult problems with such listening networks were not in the actual detection but in the classification of noise. "[Long-range detection] is not worth a damn unless you have classification," insisted Captain J. S. McCain of the Navy's Antisubmarine Plans and Policies Group, touting the benefits of LOFAR. Unless the detection mechanism could isolate and quantify momentum waves and natural resonance, it would be difficult to differentiate attack submarines from natural phenomena such as whales or even large schools of fish. Even when clearly a submarine had been detected, other sounds could be discerned—shaft and blade speeds, engine explosion rates, gears and blowers, and cavitation speeds—to provide the "signature" of a submarine conducting particular operations.[32] Here were avenues of civilian research for both biological and physical oceanography, pure and applied. Through ONR, the Navy contracted such research out to civilians in industrial laboratories, such as Bell Laboratories and Hudson Laboratories, and in oceanographic centers such as Woods Hole and Scripps.

All of these needs heralded new research opportunities for marine scientists. The Navy wished to prioritize the North Atlantic initially, because it conducted its most active peacetime forward deployment missions there. Because that region provided access to Soviet deepwater ports, Navy leaders assumed that the most significant threat to American sea-lines of communication in wartime would come from the Soviet submarines in this region.[33] Consequently, Woods Hole enjoyed ongoing support from the Navy to continue oceanic research in the Atlantic. In addition, scientists at the fledgling Lamont Geological Observatory (founded in 1949) had been active in setting up a pilot SOSUS in Bermuda, and the Navy recognized their efforts by giving them their own research ship. Use of this ship, the *Vema*, enabled some of the most intensive oceanographic surveys in the Atlantic, organized by Lamont's director, W. Maurice Ewing (formerly at Woods Hole).[34] The Navy needed the expeditions as badly as the scientists did, and there were numerous observations and experiments to be conducted, some primarily for the Navy and some mainly for the scientists, on the same cruises. One oceanic phenomenon that kept ships at sea in the Atlantic Ocean was the thermocline, discovered before World War II, which obstructed the effectiveness of sonar. Because of great variations in water temperature in certain parts of the ocean, the boundaries between temperature layers would act as walls to the active sonar pings from surface ships. A thermocline could mask the presence of a submarine. This phenomenon provided the ultimate research agenda for both the scientists and the military. Using an instrument called the bathythermograph (or BT), scientists could map the locations of thermoclines for the Navy's operational use without impeding whatever other research they prioritized for themselves.[35]

Scientists based on America's Pacific Coast also took advantage of the Navy's pressing needs. As noted earlier, shortly after taking the helm of Scripps in 1950, Roger Revelle planned the institution's first major oceanographic expedition, MIDPAC. The expedition was an eye-opener both for science and for undersea warfare. Scientists found that the heat flow through the seafloor was at least as high as on land, which, Revelle wrote in a letter to *Nature*, could be explained by a convection (rather than conduction) hypothesis of heat flow within the earth's interior.[36] They also dredged very young corals—Upper Cretaceous, only about eighty million years old—from the peaks of undersea mountains. This meant that somehow these mountains had transformed from islands to deep mountains in eighty million years, whereas scientists had assumed that the mountains had sunk over a period of two or three billion years. In addition, seismic research revealed that the sediments of the ocean floor were only a couple of hundred meters thick, whereas scientists had expected thousands of meters. Revelle recalled years later that "everything turned out to be different than anybody had thought it would be before."[37] These were early signs of the youth and mobility of the ocean floor, concepts that soon would transform how scientists viewed the earth.

The Navy in turn drew upon the work of scientists to improve the effectiveness of naval technology. Revelle insisted that the MIDPAC studies did not benefit the Navy directly. "They were fundamental discoveries about the ocean," he said, "but they didn't tell you very much about how a submarine could behave."[38] This was not altogether true. Although the scientists may have taken away from MIDPAC puzzling insights about oceanic processes, the Navy took away something far more basic. H. William Menard, then a scientist at the Naval Electronics Laboratory (NEL), recalled when he compiled the data from MIDPAC that "it became obvious that almost all of the sea floor was an endless expanse of hills."[39] Although somewhat interested in the peculiar scientific results of the MIDPAC expedition, the Navy was far more concerned with the severe challenge that this hilly seafloor might pose to the effectiveness of submarine detection. Its interest in expeditions to map the contours of the ocean basin was ignited anew during the efforts to enhance antisubmarine capabilities. Over the next several years, scientists on both coasts organized expeditions that provided the Navy with a constant flow of new operational information, while aiding the scientific community immensely.

The Navy recognized that its technology required constant attention and that scientists could provide the means to ensure such attention. Research on expeditions was perceived largely as operational, even if it had nothing to do with the improvement of the LOFAR devices themselves. Simply improving knowledge of the ocean floor was a way to improve the effectiveness of the technology. For the oceanographers, it was the most fundamental of research; the development of so many new findings and ideas about the ocean floor only reinforced that perception. Yet for the Navy it was closer to operational data collection than to the definition of basic research popularized by Vannevar Bush in his 1945 report *Science, the Endless Frontier*. It was more similar to troubleshooting existing technology than providing what Bush called scientific "capital" for the development of new technology.[40] Most of the data from these expeditions were subject to security classification during the 1950s. Without classified clearance it was impossible to look at bathymetric charts of the ocean floor, to say nothing of bathythermograph and geophysical data. Although the Navy gave scientists a fairly free hand in conducting research, the results of the work often remained under lock and key, a major point of contention between scientists and the Navy. But they each appreciated basic research as a component of both scientific advance and national security, and that convergence of mission served them both well during the 1950s.

WHY INTERNATIONAL COOPERATION?

Not only did the Navy recognize the importance of science, but also it came to appreciate the need for international cooperation. The most obvious reason for international cooperation in oceanography was not merely the search for basic knowledge, but the prediction of environmental conditions. Combining effort and coverage promised to cut costs and increase data production at the same time. The two fields most intensely explored during World War II were marine acoustics (because of the submarine threat) and ocean wave research (for planning amphibious warfare). Athelstan Spilhaus, of the Massachusetts Institute of Technology, who constructed the bathythermograph used in the late 1930s to measure thermoclines, leaked the design of his instrument to the British to help them fight German submarines, even before the United States was involved in the war. He later recalled that he "could have been shot for a spy," but his actions anticipated more intensive cooperation with the British in oceanography during the war.[41] American and British scientists working for the British Admiralty's Swell Forecasting

Section, led by British oceanographer George Deacon, made weather forecasts based upon the coordination of meteorological data gathered in both countries. Using principles developed largely by the Norwegian Harald Sverdrup and the Austrian Walter Munk (both at Scripps), the Section set up wave reporting stations on the southern coasts of England that were used to make forecasts for the invasion of Normandy in June 1944. Later the Section was moved to the Pacific to help prevent repetitions of the disastrous 1943 Tarawa landing, when ignorance of local tide conditions contributed to the death and wounding of more than three thousand American marines.[42]

At war's end, oceanographers witnessed mixed portents for international cooperation. One was the establishment of a division of oceanography within the Navy's Hydrographic Office. The director of Scripps, Harald Sverdrup, saw it as a great stride forward for international cooperation, because the Navy might pursue even larger cooperative data networks than the weather forecasting system it created with Britain during the war. Granted, the division would be charged only with the responsibility of carrying out research based upon the interests of the Navy and the Merchant Marine. But for Sverdrup, the military's new special interest in oceanography was no burden upon science. As the director of one of the two American oceanographic institutions, he had been concerned about his discipline, "by far the youngest member in the family of sciences." Previously, its greatest problem had been recruitment; however, the war revealed myriad uses for knowledge of the sea and created a sudden demand for oceanographers. Now he expected more men to work on not only military problems but also fisheries research, geology, navigation, erosion studies, and even civilian construction problems. That expansion, Sverdrup felt, should be encouraged not only in the United States but also abroad, so that everyone could benefit from the oceanographic work done in different parts of the world. It could lead to an unprecedented level of peacetime international cooperation, enabling global scientific portraits of the oceans.[43]

In an effort to assess the potential of cooperation, the National Academy of Sciences in 1946 had organized a symposium, "Problems of International Cooperation in Science." Here Sverdrup and other renowned scientists highlighted a few areas that they perceived as the best avenues for cooperation. Through international scientific unions, research within specific disciplines could be coordinated in order to prevent duplication of effort and to ensure that all work would be of value.[44] International scientific congresses similarly allowed contacts between individuals across geopolitical lines; because they were voluntary congregations of individuals within a given discipline, they should be free from "government control, propaganda, national prestige, or perfunctory representation by officials."[45] Many participants in the symposium, including the California Institute of Technology's longtime chief Robert Millikan, felt that activities such as international exchanges in scholars were important not merely for science but also for the maintenance of peace in the world.[46] All these concepts were predicated on the principle of sharing data and coordinating work on an international scale.

Despite these hopes, little was accomplished to make oceanography more international immediately after the war. Exceptions such as Operation Cabot, which employed six research vessels from both the United States and Canada to take bathythermograph readings and current measurements of the Gulf Stream, reveal the limitations on cooperation early on.[47] This expedition was coordinated between national navies for a common defense and was barely international. Sverdrup ultimately was dissatisfied with the ability of leading American scientists to make more robust cooperative efforts a reality. When he departed Scripps in 1948 to return to his native Norway, he reasoned that he could be more influential in the international arena if he operated from a smaller country.[48] Columbus Iselin stepped down temporarily as director of Woods Hole in 1950, and replacing him was Rear Admiral Edward Hanson "Iceberg" Smith. The trustees' choice of a military man (U.S. Coast Guard) symbolized a break with the informality of oceanographic science and an embrace of efficiency and professionalism. Under Smith, the trustees hoped that Woods Hole would be able to attract and handle increasing amounts of money from military sponsors.[49] On the whole, the course of oceanographic research fell into the Navy's defense-oriented hands, and it had little interest in sharing its data.

Whether the American naval establishment could properly support international activity was still open to question. It was receptive to some activities, such as an information exchange in 1951 with Norway's Defense Research Establishment.[50] No doubt Sverdrup felt vindicated by this program of oceanographic data sharing. But to some scientists' irritation, the Navy interfered with international scientific ventures that did not explicitly address security issues. As already noted, scientists at Scripps hoped in 1953 to use a Navy-sponsored expedition across the Pacific as a means to transport scientists to the Pacific Science Congress held in the Philippines that year. However, the Navy instead insisted that the visits be limited to Japan, where the scientists might concentrate on the militarily crucial North Pacific and tease any scientific information about the region from the Japanese. The National Academy of Sciences broached the idea of letting Japanese scientists come aboard the research vessel on the 1953 expedition. Although eventually the scientists were allowed on board for a brief leg of the expedition (named TRANSPAC), the Navy initially resisted the idea because some of the information or instruments on board were classified. Scientists could claim a free hand to conduct their own research aboard military expeditions, but red tape impeded the expansion of research to include partners who had not proven themselves reliable allies.[51]

The Navy was uncomfortable with the idea of scientific internationalism, but it did come to support limited cooperation during the 1950s. Its need to subdue the ocean environment, and to make it a manageable and even advantageous battleground, drove the Navy to seek ways to increase knowledge of strategic areas. One way to do this was to steal data from others. Leading scientists were highly critical of this kind of scientific appropriation, calling it something that the *other* side did. They critiqued it not for ethical reasons but because it was inefficient; compared to the expansion of research and the continuous sharing of information, onetime grabs of information did not provide a sustainable way to promote scientific growth. For example, Vannevar Bush looked disdainfully on the Soviet Union's faster-than-expected development of the atomic bomb, which he felt was due largely to the appropriation of resources and technology from other countries that it took over after the war. Such behavior, he felt, could work only in the short term.[52] The Navy did come to believe that, as a strategy to ensure a steady growth of oceanographic knowledge, continuous data exchange was preferable to simple appropriation.

The strongest military motivation for collecting data, evident by the early 1950s, was the fact that the most northerly regions, with their vast packs of ice, no longer were natural defense barriers; in light of the constantly improving long-range submarine technology, Navy planners named the northern ice packs as the most likely early battlegrounds in a general war with the Soviet Union.[53] The task of addressing this situation fell to the Navy's Strategic Plans Division, whose duty was to ensure that the Navy's plans were sound and to determine the manner in which the Navy would fight the next war. One of its more daunting tasks was to determine what specific problems had to be overcome in attacking or defending geographical areas throughout the world.[54] In April 1953, the director of Strategic Plans, Admiral Arleigh Burke, complained to the deputy chief of Naval Operations (for Operations) that in the critical Norwegian Sea and Barents Sea region, the Navy could not expect to conduct successful operations. Contrary to "an optimistic trend—unsupported by factual data or experience," he wrote, American fast carrier task groups had little or no knowledge of the ice conditions, sea state, and temperature of the region required to conduct antisubmarine and antiaircraft activities in a time of war. Burke claimed that the entire area east of Greenland and north of 70° north was relatively unknown; the Navy lacked records of marine observations, and the few bathythermograph readings available could not ensure an accurate prediction of sonar conditions. He suggested that the Hydrographic Office undertake a comprehensive oceanographic program, with the assistance of Woods Hole or Scripps.[55]

Immediately the Navy stepped up the oceanographic and meteorological research program of its submarine expedition to the area while it brainstormed possible data sources to exploit. Although the most effective way to obtain information would be through the use of an American oceanographic vessel, the Navy reasoned, "the purpose would be too obvious, and clandestine operations practically impossible." Perhaps, instead, a British ship "could traverse the area throughout the year collecting needed information under the guise of a routine research expedition."[56] This plan, one of the Navy's top secret sham operations, became known as Project Ice Pick. Its purpose would be to pretend to conduct scientific research for a civilian institution while collecting data of paramount importance to any action against the Soviet Union in a time of war. Burke, whose involvement in the "Revolt of the Admirals" and appreciation of the Hartwell study made him a believer in the centrality of science and technology to the Navy's mission, saw the project as critical. In February 1954 he wrote to the director of Fleet Operations that current war plans envisaged fast carrier task groups conducting operations in the Norwegian and Barents seas in the early phase of a general war. He was "still concerned that insufficient steps are being taken to correct the complete lack of representative weather and hydrographic information upon which plans for these naval operations should be based."[57] He cited a disastrous 1951 supply run in Arctic waters, when inaccurate information and inexperience with the weather had compelled the crews to force their ships through the ice, resulting in millions of dollars in damage. He insisted that a full year's worth of oceanographic observations should be obtained to provide operational knowledge of the Norwegian and Barents seas.[58]

The Navy never found an adequate way to implement Ice Pick, and it even considered having a foreign vessel do the work. The easiest and most effective method of obtaining the information would have been to send the Navy oceanographic survey unit into the area. However, that unit was already booked for the next few years putting the SOSUS into place under

Project Caesar, the implementation stage of the long-range detection research conducted under Project Jezebel.[59] The next option was to find a foreign research vessel to do the job discreetly. Conveniently, the two leading countries in oceanography in that region had institutes headed by old friends of the United States Navy: George Deacon of Britain's National Institute of Oceanography and Harald Sverdrup of the Norwegian Polar Institute. Even more conveniently, it seemed that they might have ships available for charter.[60] About this time, the Navy was finishing construction of an additional oceanographic center at Woods Hole. The purpose of the facility was to allow the Navy better control and coordination of oceanographic research, while simultaneously expanding the scope of inquiry and thus the creative freedom of scientists.[61] Scientists from several countries attended the dedication ceremonies in June 1954, including Harald Sverdrup, and it was there that the subject of international collaboration in the Barents and Norwegian seas was broached. Because it was necessary to conduct the research under the aegis of a civilian scientific institution, the foreign scientists were to be advised that it would be a cooperative expedition with Woods Hole.[62] But the planning of the project dragged on for several months. Burke was replaced as director of Strategic Plans and sent to sea.[63] No expedition materialized.

Although the expedition never happened, it was not due to lack of trying; both the Navy and civilian scientists expended a great deal of effort in securing a ship. Neither the British nor the Norwegian ship was available for the amount of time required. In August ONR had its liaison officer in London—at that time it was marine geologist Robert S. Dietz—gather information about ship availability. There were three Danish ships, but he found that the *Kista Dan* would be available only for short periods, the *Dana*'s use by the United States Navy would have to be approved by the Danish Parliament, and the *Galathea* would require an expensive complete overhaul of machinery.[64] At last they were left with the Canadian *Arctic Sealer,* whose small size promised little in scientific dividends. After some preliminary negotiations, the Ice Pick Planning Group decided that, because of its high cost, the *Arctic Sealer* simply was not worth the trouble of dealing with security classification, installing research equipment, and potentially provoking international discord. Constructing such an operation in secret opened up so many difficulties that gathering data in this way lost its appeal. The Navy would have to find a different means to obtain it. Project Ice Pick, in its original incarnation, was canceled.[65]

The preliminary phase of Ice Pick revealed how much the Navy could gain from the free flow of information from foreign sources. Navy leaders started to understand that they might not, after all, need to rely on costly American efforts. Up to this point, the only reliable information possessed by the United States was in the logs of ships having made the "Murmansk run" during World War II.[66] As Ice Pick was being formulated, American scientists and naval officers began to gather as much meteorological and oceanographic data as they could from their foreign colleagues. The logs of all the ships engaged in those runs between 1941 and 1943 were examined and all the relevant weather observations were extracted. Still, the information was not only incomplete but also inadequate, as it said little about the limits of ice at any given time of year, critical data for planning military operations. Soon more information was obtained from the British, including data captured from the Germans during the war, although Burke suspected that the British held back a great deal.[67] But when the Navy continued to seek information from foreign sources, it was surprised to find that the amount of available data was immense. One Navy captain estimated in October 1954 that a particular haul of data from "a foreign source" yielded one hundred fifty thousand additional meteorological observations, the equivalent of ten ships taking four observations per day continuously for more than ten years. Other lodes of data followed, including some synoptic observations and microfilmed raw data.[68] The most significant new sources were British, Norwegian, and German. The British information included meteorological observations conducted at Lowestoft and Harrow, oceanographic observations made by the vessel *Ernest Holt*, a new ice atlas of the Arctic compiled at Cambridge, and other information gathered by scientists and the Royal Navy in Liverpool and London. Data were exchanged between the British Meteorological Office and the U.S. Weather Bureau. Information from Norwegian oceanographic ships such as the *G. O. Sars* conveniently was obtained through this same agreement because of an established British-Norwegian exchange. From the Federal Republic of Germany, the Navy acquired about two hundred thousand observations and worked to establish continuous exchange.[69] The extant information was remarkably useful to the Navy, and it filled most of the gaps that existed. The volume of information was so much more than expected that the Ice Pick Planning Group estimated that scientists would need to spend several months reducing it to a usable form for operations. The Navy realized that, because of the efforts of non-Americans, a major expedition to the region was not an immediate necessity after all. The Ice Pick experience generated an increased interest and effort to begin generally to seek data from all sources—including foreign—for operational use.[70] In short, the failure to implement Ice Pick yielded a success of an entirely unexpected character: the Navy began to appreciate the value of sharing data.

Capitalizing on this realization by Navy leaders, ONR scientist Gordon Lill warned against complacency in the future. Clearly, sending a secret oceanographic cruise was inferior to sharing data over an extended period. Although oceanographic surveys were necessary both for scientific knowledge and for general Navy information, the most crucial information that was essential in conducting operations could be made available only by ongoing research and continuous exchange. "Securing 6 months more survey type data in the ICEPICK area will add practically nothing to the aerologists [sic] ability to forecast the weather or oceanographic conditions," he declared. "Current data is absolutely essential for successful prediction."[71] Lill was not criticizing the value of expeditions but rather recommending a broader policy of coordination and exchange. Oceanic conditions were not static; knowledge of the environment—its processes, its patterns, and its anomalies—required continuous observation and continuous cooperation. It could not be accomplished by a short-term appropriation of others' work. Oceanographic research, particularly of the operational kind required by the Navy, could not boil down to a tangible commodity, a secret formula that could be held in a secure vault, hidden from the enemy. Instead, the research required the kind of international cooperation that Harald Sverdrup had advocated while still director of Scripps in 1947, namely, a mutually beneficial information network.

Lill's point was well taken, and the Navy would be a great supporter of international cooperation in the coming years. It would not, however, completely abandon its practice of withholding some of its data, particularly when they appeared (as they often did) to be of an operational nature of potential use by the enemy. The specific data of concern during Project Ice Pick, weather and sea-state in particular, were primarily of an operational nature. They had immediate value distinct from being the foundation upon which new technology would be based. Yet this latter value was the leading justification that scientists gave when advocating the funding of "basic" research. In most instances, the difference was trivial. The scientists received money, and the Navy received the data it wanted. But the different outlooks on research's utility did have ramifications, particularly when the Navy insisted on strict control of information.

CLASSIFICATION VERSUS OPENNESS

The Navy came to accept that from time to time the free flow of information, at an international level, was a good thing both for the growth of science and for the pursuit of technological superiority on the part of the United States. It supported international expeditions throughout the 1950s, and during the International Geophysical Year of 1957–58, it even allowed the dissemination of all the data for use by enemy as well as ally. But the Navy knew that cooperation was a two-way street, and it resisted sharing its own data. Aside from the IGY, the Navy and the scientists did clash throughout the 1950s over the issue of total internationalism in scientific research. It may have been beneficial to share data among allies, but consistently sharing with the entire international community seemed a different matter. As historian Bruce Hevly has argued, the Navy's attitude toward research was to design or support institutions that conducted research in pursuit of national power. Not only did the Navy's influence constrain research in general through the technological commitments inherent in that pursuit, as Hevly makes clear, but also it constrained the ways in which international cooperation could be pursued because of the need for secrecy.[72] Public disclosure of research was the most significant source of friction between the Navy and oceanographers. It certainly made international cooperation attractive, because it offered opportunities to circumvent the problem.

Many scientists felt that concerns about security, particularly the fear of spies, were highly exaggerated. Vannevar Bush decried America's fears, saying, "We're really a bunch of nuts in this country; the Soviets don't really need a spy system to find out all they need to know about us." In a society as open as the United States, a spy simply needed to peruse the engineering and technical magazines and study the advertisements.[73] The military's attitude that it was better safe than sorry actually hindered rather than helped national security. Even the Hartwell scientists had criticized the notion that strict secrecy contributed significantly to security. The criticism hearkened back to the days of the Manhattan Project, when General Leslie Groves had tried to keep each scientist on a "need-to-know" basis. The scientists believed that such attitudes prohibited the more effective, system-oriented, approaches to problems. Secrecy prevented the dissemination of ideas and stifled scientific growth. Overspecialization, the Hartwell scientists felt, had blinded Navy leaders to problems within the Navy's entire organization.[74] They, like many other scientists, believed that the scientific benefits of openness far outweighed the dangers of revealing sensitive data to the Soviets.

But secrecy in science was not uncommon. Science's new position as the centerpiece of national security, despite its positive effects on government support for research, had made scientists vulnerable to the unsavory stew of vigilance and paranoia in American politics. The need for secrecy had brought prominent scientists, most notably atomic physicists E. U. Condon and J. Robert Oppenheimer, under suspicion for their alleged "anti-American" ties. Eventually both were removed from the government's corridors of power, by having their security clearances revoked. The attack on Condon focused on his internationalism; his belief that the United States should cooperate more directly with the Soviet Union struck the House Committee on Un-American Activities as naïve and even subversive. His opposition to secrecy and his support for civilian control of atomic energy only exacerbated doubts about him.[75] The same aims made Oppenheimer a target; his opposition to pursuing the hydrogen bomb struck many politicians as a sign of bad judgment, and thus he seemed a security risk. In general, scientists during the 1950s found themselves both respected and suspected. Scientists with backgrounds connected to communism routinely were denied security clearances and government funding. The facility of scientists to travel abroad to conferences was limited and controlled through legislation and the active intervention of consular officials and Department of State bureaucrats.[76] All these strictures infringed on scientific activity.

The strictures necessitated by classification struck some as more damaging than visa denials, passport restrictions, or even the blacklisting of specific scientists from government work, all of which prevented scientists from pursuing their research. Classification was particularly harmful because scientists were unable to discuss, publish, or circulate their work for the benefit of other interested scientists at home and abroad.[77] Dissemination of information was a significant problem. The scientists' argument was exemplified, Vannevar Bush later recalled, by the publication of the Smyth report shortly after the Second World War. Manhattan Project physicist Henry Smyth wrote a report about atomic energy under specific instructions that he should not reveal anything about the project that the Soviets did not already know. The result was a report that, in Bush's view, told no more than what the Soviets knew, but that told a great deal more than what most American scientists had known about the project as a whole.[78] Neither scientific discovery nor engineering development could flourish, Louis Ridenour argued after the publication of the Smyth report, "except under conditions which allow all competent men to be fully informed, and thus able to contribute to progress."[79] Bush, Ridenour, and many others felt that excessive secrecy meant compartmentalization of knowledge, which reduced the efficiency of scientific growth and thus could weaken national security in the long term.

Although scientists could point to the Navy's farsighted approach to basic research, the same could not be said about its classification policies. The Navy strictly controlled access to the oceanic charts produced from data collected during expeditions. Publication in any publicly accessible journal was impossible unless approved by the Navy, which maintained proprietorship over ocean floor soundings and other data. The soundings data came from measurements of ocean depth, using an instrument called the echo sounder, and they were the basic components of any bathymetric map (like a topographic map, but showing depth contours instead of elevation contours). When the Navy approved publication of data, it did so very slowly. H. W. Menard, who compiled bathymetric charts for the Naval Electronics Laboratory during the early 1950s, later estimated that a few scientists with the proper security clearances had access to oceanographic data about five years before anyone else did.[80] The Navy typically refused to disclose data to scientists outside the United States. For example, when British Museum scientist John D. H. Wiseman asked Menard to provide corrections to a British-made bathymetric chart of the Pacific basins and trenches, Menard responded only to say that many of them were wrong, but he could be no more specific.[81] Menard retained his security clearance when he moved to take a position at Scripps in 1956, and his colleagues there benefited from his access to the best bathymetric maps of the Pacific Ocean. The same was true of Lamont director W. Maurice Ewing and others regarding data on the Atlantic Ocean. Pockets of scientific knowledge emerged, informally connected and isolated not only from the public but also from other scientists. Menard likened it to an elite "invisible college."[82] But this somewhat positive appellation, used in the seventeenth century by Robert Boyle to describe an informal network of learned men of science, applied only when those with access could visit each other's institutions, which was not always possible. There was no substitute for publication.

Leading scientists recognized the weaknesses and inconveniences of secrecy, but also sympathized with the Navy's wish to control data. Scientists did not want to antagonize their patrons. Confronted by the specter of a closed scientific community, many scientists attempted to strike a reasonable balance between total secrecy and total openness. In September 1951, Princeton geologist Harry Hess wrote to the Navy's Hydrographic Office

Oceanography's Greatest Patron 53

about the problems of classification, acknowledging that in some strategic areas—particularly the mid-Atlantic ridge, the Alaskan coast, and the Aleutian Islands—the oceanographic data should be kept classified. But in other areas, such as the continental shelf and slope of both coasts of the United States and parts of the Caribbean, a great deal of information already was publicly available; the Navy's release of information would constitute no threat to national security.[83] Revelle supported this view and opposed "blanket" classifications of whole regions, such as the one he believed was in effect for everything north of 20° south. He felt that the Navy should make an effort to classify only areas of specific strategic significance; all other classifications were excessive and even counterproductive.[84] Hess complained in April 1952 to Revelle that he had attempted before and would attempt again to present his argument in terms that were readily understandable to the Navy, but as yet he had failed to make any progress. He cared less about winning the argument on principle than he did for seeing a more reasonable policy put into place, even if it meant that a great deal of data had to remain classified.[85] In July 1953 he wrote to the Navy hydrographer and insisted, "there is only one issue as I see it and that is what policy would most benefit the United States and the Navy." Hess pulled out the old arguments: without the free flow of information, new principles that might otherwise have been discovered may go unnoticed, delaying new weapons technology. The damage was done, Hess said, not because the country's top oceanographers did not obtain the data, but because "it never comes to the cognizance of scientists working in other fields."[86]

Valid reasons existed for this seemingly excessive control of data. One reason was the lack of attention to policy, which resulted in a tendency to err on the safe side. At that time, there was no coordinating body that addressed the needs of both the scientists and the Navy—this would come later, with the formation of the Interagency Committee on Oceanography in 1960. As Earl Droessler pointed out to Revelle in 1952, the idea of a "blanket" classification itself was erroneous; the Navy at that time had no explicit policy at all.[87] But that was precisely the problem: because the various policies of the Hydrographic Office were not negotiated, compiled, or disseminated, it was even more likely that junior officers would place restrictions upon data. Lloyd Berkner, who led the international committee responsible for organizing the International Geophysical Year (1957–58), complained that the judgment often rested with a military officer who might be censured (or worse) if his superiors suspected that he had released information improperly. That same officer "is under no penalty for improperly keeping information from the public that might have the most beneficial influence on the future of our affairs."[88]

Top Navy leaders, though sympathetic to the needs of scientists, did not make an effort to reverse the practice of classification. Although in theory the scientists' idea of openness was sound, in practice it appeared suicidal. Many naval officers were convinced that scientists' attitudes toward the Soviets were misguided and that openness was an invitation to trouble. Rear Admiral Felix Johnson, director of Naval Intelligence, rather unambiguously labeled any Soviet propaganda of peace as an attempt to exploit American ethical standards. He wrote in 1950 that "the soul-searching and expressions of fear which arose in the U. S. after the President's decision to proceed with the attempt to produce an H-bomb were interpreted by the Kremlin as signs of decay and weakness in the U.S. to be exploited at an opportune moment."[89] This sense of distrust permeated the Navy, whose leaders had to plan for the possibility of imminent war with the Soviet Union. In 1952, the chief of Naval Research recognized that de facto blanket classifications in theory might stifle the growth of the embryonic field of oceanography, which in the long term could in turn possibly reduce the Navy's military readiness. He even stated that the policy on oceanic soundings should be improved. In effect, he agreed with the scientists. He could appreciate, from the lessons of the Hartwell report, that one must think at a broad level and follow policies that benefited scientific and technological growth as a whole. Yet he also cautioned that "the classification at any given time depends on several technical and military considerations." By 1953, strategic planners made a conscious decision to withhold soundings of the ocean floor.[90]

The Navy's operational requirements prohibited it from pursuing the scientists' ideal. Navy leaders felt they could not consent to handing such valuable information to the Soviets. If the United States published soundings of the ocean floor, the Soviet Union would have more detailed knowledge of the strengths and weaknesses of any LOFAR network the United States might install. This was the "technical" consideration to which the chief of Naval Research referred. It was impossible for the Navy always to think of the long term, when in the short term the effectiveness of existing technology depended on the Navy's control of the data. The Soviets would need only to locate the stations, through intelligence operations, in order to know precisely what the Americans could see and not see inside the oceans. To supply conscientiously such information to the Soviet Union would be a service to an enemy in time of war. Burke acknowledged that "although the Soviet Union could obtain this information if sufficient effort were expended, she does not now have it nor does it appear that she can expend the necessary effort to obtain it."[91] The scientists viewed soundings as basic research; the Navy viewed them as operational information that could just as easily be useful to Soviet submarines as to American ones. The same could be said of virtually all oceanographic data that reported environmental conditions; these were the data that navies would use in wartime.

Classification proved embarrassing to scientists who found themselves unable to discuss with colleagues their most recent work or results, because of the Navy's fear that the Soviets might find it useful. Hess tried to influence the Navy by arguing that the Soviets could acquire the data easily by assigning twelve ships to the task for three years.[92] The Navy's attitude was that if it was so easy, let them expend the time and resources to do it, not give it to them for free. In 1954 the National Academy of Sciences asked Hess to give a speech to an international audience at the opening of the new Navy oceanographic laboratory at Woods Hole, on the subject of geology and geophysics in oceanography. Hess agreed to do it but protested that it would be difficult to do it properly because "the high classification placed on much of this field in recent months makes it extremely difficult to deal with the subject as a whole."[93] Not only would it be difficult for Hess himself to deal with the subject, but also it would be difficult to share what he knew with an international audience. Because of military classification, scientists increasingly perceived a threat to science that was, for marine scientists, worse than what had sparked the debate over support for basic research. They would have to adopt strategies to circumvent this problem, as their powers of argument were lost on the Navy. Their lament was nicely captured by the social scientist Edward Shils in his 1956 book *The Torment of Secrecy:*

> No professionals have been asked to sacrifice so much of their own tradition as the scientists.... They have suffered from misunderstanding, often well-intentioned, but sometimes suspicious, by their military supervisors and their security officers, who have not always understood their nature and their needs.[94]

Shils pointed to the ignorance of the military despite constant arguments on the part of scientists against the myth that anyone can keep science a secret.

In a more charitable view, one might say that priorities diverged on this particular point. Only as an afterthought did the Navy's dearest concerns enter the proclamations of scientists over the issue of classification. Spokespersons of the scientific community such as Vannevar Bush reasoned that with the amount of scientific exchange going on, in the form of cooperation and publication, scientific advances made in one country would soon seep into other countries, regardless of secrecy. He publicly declared that such secrecy made no sense. But in an oral history interview recounting the early 1950s, Bush added that secrecy in military plans was far more sacred, since "the Russians would like to know what we would do in certain eventualities."[95] Lloyd Berkner acknowledged to an audience at Dartmouth College in 1954 that obviously it would be foolhardy to completely abandon technological secrecy in military matters such as the design of specific weapons, as long as such secrecy yielded more advantages than disadvantages. "In the long view," he said, "the public interest requires the freedom of substantially all scientific and technological information as well as the potential nature and implications of its application."[96] Both these men acknowledged the logic of secrecy in extreme cases, but they focused on the positive long-term effects of openness on both science and technological growth. One is struck by the narrowness of the scientists' argument, with its unswerving advocacy of a linear progression, from basic science to applied technology, as the sole view of how science related to national security. Scientists were trying to dispel the myth that there were any real secrets in basic science, insisting that only the applications of it should be kept secret. Yet oceanographic data presented a case in which basic science itself was a commodity of extreme importance to the Navy's operations.

Despite its explicit and straightforward explanation of the danger of releasing bathymetric data, few in the Eisenhower administration were sympathetic to the Navy's needs. Instead, the Navy often was perceived as a stubborn adversary not only to scientists but also to the interests of the nation, as the administration sought to improve efficiency and cut costs. Proving that scientists often had more friends in Washington than did the Navy, Eisenhower in 1953 abolished the whole "Restricted" category, under which the Navy had classified oceanic soundings. Having experienced firsthand the tendency to overclassify military documents, Eisenhower felt that the time had come to free this particular facet of science from its Navy-inflicted constraints. But faced with the knowledge that the Navy would be forced to give a massive quantity of largely nonreviewed information to the enemy, the Chief of Naval Operations irked scientists and the administration by promptly upgrading all soundings to the category "Confidential."[97] Not long after, the Navy's plans to conduct independent expeditions in Antarctica, while scientists participated in the International Geophysical Year, were cut completely out of the budget by Eisenhower. For the Navy these expeditions had been the coat of sugar on the bitter pill of supporting such extensive unclassifiable scientific research during the IGY. But when Eisenhower quizzed National Science Foundation director Alan Waterman, formerly chief scientist for ONR, about whether the Navy's own expeditions were worth the money, Waterman did not speak up for the Navy, and the projects were immediately canceled. Such efforts to undermine their plans severely demoralized Navy leaders, including the famous naval aviator Admiral Richard E. Byrd, who hoped that the expeditions would help the United States establish strategic footholds in Antarctica.[98] It was symptomatic of a period in which the Navy, because of its attempts to place itself at the forefront of national strategy, worked at cross-purposes with an administration whose goals often kept the Navy on the periphery.[99]

The debate over classification was a conflict between strategic outlooks, on the one hand based on the free and unfettered growth of science, and on the other hand based upon the need to protect the operational effectiveness of existing technologies. Yet the conflict was not irreconcilable. Virtually all scientists recognized that the Navy should control some of its materials, especially those related to certain areas of critical importance to projected wartime submarine activities. After all, scientists too were in the business of national security. The Navy, for its part, made an effort to declassify soundings on a case-by-case basis if a scientist needed them for publication. It realized the necessity of supporting oceanography, and even of promoting international cooperation, for the health of the scientific disciplines and the probability of future technology. By way of compromise, it declassified a great deal of low precision information for wide distribution. Still, the number of civilian scientists who had access to all of the information, even at the end of the 1950s, was only about twenty, whereas scientists would have preferred to extend that number into the thousands.[100] To scientists, this was a deplorable situation, and most leaders of the scientific community spoke out against it to some degree. Scientists such as Hess and Revelle, by advocating only a limited dissemination of data, nevertheless tacitly acknowledged that the ideal of scientific growth as a means to ensure national supremacy perhaps was too narrowly defined. In the short term, the Navy by necessity had to bow to operational needs. In preserving the effectiveness of existing technology, it often refused to disclose its own data to the international community, taking a stand that was extremely unpopular among oceanographers. Even pro-Navy scientists, which Revelle certainly was, looked more and more to solving this problem through increased dialogue with Navy contacts and advocacy of projects, especially international ones, that would not leave scientists to flounder in the wake of interminable bureaucratic resistance to openness.

Oceanographers rarely complained about the Navy's interfering with basic research, though it is often identified as the leading conflict between scientists and their military sponsors. They were able to conduct enough of it, certainly more than they ever could without Navy funding, on deep-sea expeditions and in other work that the Navy viewed as essential to its own mission. The fact that the Navy was enabling the most cutting-edge work in numerous fields (the interdisciplinary MIDPAC expedition and work on marine acoustics, just to name two examples) only reinforced this belief among scientists. This was why Revelle was forging strong links between the Navy and Scripps, against the wishes of some of his colleagues there; scientists at Woods Hole, already closely tied to the Navy, kept up the relationship with enthusiasm. But oceanographic research entailed more than just providing basic research; it was also an environmental science. Its basic research was valuable not merely for ascertaining how the laws of physics functioned in the sea, but also for determining the ocean's basic topographic features and the oceanic and meteorological conditions at any given time. These facets of oceanography were specific to time and place, and the Navy understandably went to great lengths to learn them and, to the scientists' chagrin, to keep some of them secret. Was it better, in terms of both science and national security, to keep data secret or to make them public? One might claim that oceanography has always been international, because information must be coordinated throughout the whole world. At the same time, oceanography inherently challenged the concept that science was international because basic science—knowledge of environmental conditions at any given time—was a precious commodity for nations at war. In this light the successes of international cooperation in oceanography, endorsed and supported by the United States Navy, appear all the more noteworthy. Oceanographers had already made the Navy a disciple of marine science. Now they had to find a way to preserve their own interests, and they made great headway promoting their international vision of research with openness and sharing as the basic elements. During the mid-1950s, even the Navy came to accept that data sharing could be a wise means to accrue both environmental data and basic research for the benefit of American scientific, technological, and military supremacy. The most significant effort to put that vision into practice, the International Geophysical Year of 1957–58, is the subject of the next chapter.

3 THE INTERNATIONAL GEOPHYSICAL YEAR, 1957–1958

Through its efforts with Japan and its work with the Navy, the oceanographic community had promoted cooperation as a means to cultivate science in the name of American strength. But the question of the realistic extent of international cooperation still lingered, and the idea of total openness seemed a great gamble. Still, scientists' desire for cooperation had an impact upon their patrons, revealing the formidable rhetorical power of international cooperation. Cooperation was promoted by American scientists in numerous disciplines, including oceanography, with the premise that such sharing would benefit science as a whole and that the United States was in the best position to transform science into useful technology. With sponsors at least somewhat sympathetic to an international outlook for conducting research, scientists had an opportunity to attempt cooperative projects on unprecedented scales, with partners beyond North America, Scandinavia, and the United Kingdom. Already, American and Japanese scientists had worked together on expeditions such as TRANSPAC and NORPAC, and regional bodies such as the International Council for the Exploration of the Sea (ICES) (of which the United States was not a member) continued to pursue cooperative projects in the northern Atlantic Ocean. The kinds of projects to which Revelle, Sverdrup, and others aspired were those that combined the efforts of many countries to provide a synoptic (or nearly so) study of the earth's oceans. Oceanographers had an opportunity to pursue this during the International Geophysical Year (IGY) of 1957–58, a project invented by elite American and British scientists in other disciplines to make a major effort at synoptic investigations.

For the IGY, even the U.S. president, Dwight Eisenhower, provided his endorsement of international cooperation. In a letter to the National Science Foundation, he wrote that the IGY was "a striking example of the opportunities which exist for cooperative action among the peoples of the world."

Eisenhower took this even further, adding his recognition of the concept of fundamental scientific research, which the United States had been funding on unprecedented levels over the past decade:

> The United States has become strong through its diligence in expanding the frontiers of scientific knowledge. Our technology is built upon a solid foundation of basic scientific inquiry, which must be continuously enriched if we are to make further progress.

The president explained further that a project like the IGY not only would advance science but also held great promise for technological gains both for the United States and for cooperating countries.[1] This outlook, which saw international cooperation closely related to national security, was first set forth by the National Science Foundation (NSF) in a 1955 report titled "Preliminary Report on Role of the Federal Government in International Science." It called for minimal secrecy and maximal collaboration with scientists from all nations as the best means of sustaining American strength in the Cold War. Although the White House made no effort to endorse the policy officially, the NSF proceeded with its tacit approval while it planned and implemented the IGY of 1957–58. The Soviets' launch of *Sputnik* during the IGY soon challenged both America's scientific leadership and its attitude toward cooperation. Afterward, policymakers would make a greater effort to articulate a role for scientific cooperation in foreign policy more in line with Cold War necessities.[2]

The rhetoric of American scientists took the ideal of scientific cooperation to an extreme, in ways that many nonscientists were not prepared to accept; they wished to include even America's enemies within an interdependent network of knowledge production. After the death of Stalin in 1953, Soviet scientists began to participate in the international arena in great numbers for the first time since prior to the Second World War. The ideal of all-inclusive international cooperation seemed poised to become a reality. The forum in which the Soviets chose to make this leap into the international arena was the IGY itself. The Soviet delegation officially began to participate in 1954 by attending preparatory meetings and soon devoted considerable resources to it, including an ambitious oceanographic program. Because it was the first effort at cooperation—and often the first point of contact—between Soviet and American scientists during the Cold War, the IGY became a trial by fire for the scientists' rhetoric of openness and all-inclusive international cooperation. Since then, the IGY has been hailed as an example of what scientists could achieve even in the midst of intense geopolitical rivalries. This view emphasizes the creditable international agreements that eventually sprang from the kind of science-related diplomacy fostered by the IGY, namely, the 1961 Antarctic Treaty, the Limited Test Ban Treaty of 1963, and even the Space Treaty of 1967. Some commentators noted that, in light of the pattern of cooperation set by the IGY, the Department of State should take a leaf out of the book of the IGY scientists.[3] George Kistiakowsky, one of Eisenhower's chief scientific advisors, later wrote that international science had proven itself to be a salve to the world's ills, because science was one of the few common languages of mankind, independent of political boundaries and ideologies.[4] These sentiments, however, were not altogether true. One could just as easily argue that politics were a hallmark of the IGY and that world tensions, particularly over the Antarctic, rocketry, and oceanography programs, were only exacerbated by these large-scale scientific projects.

THE IGY: POLITICAL FROM THE START

The IGY did not begin as an oceanographic project at all. Most agree that the idea was born in 1950, at the home of American geophysicist James Van Allen, through discussions between Lloyd Berkner, British geophysicist Sydney Chapman, and others. These scientists wanted a third "Polar Year," which had brought scientists throughout the world together in 1882–83 and 1932–33 to coordinate scientific investigations in the high latitudes.[5] Although 1957 would be only twenty-five years after the previous Polar Year (as opposed to fifty years between the first two), Berkner felt that scientific ideas and technological capabilities were sufficiently improved to warrant an early "Year." Berkner in fact was becoming a strong advocate of expanding American efforts in international cooperation. The same year, 1950, a government committee that he chaired issued a report that emphasized the need to expand the science liaison system throughout the world to improve America's understanding of the status of scientific development. As historian Allan Needell has demonstrated, Berkner himself often was a liaison between the State Department and American scientists, balancing the promotion of national interests with the pursuit of science.[6] He and Chapman were to become members of the Special Committee for the IGY (or CSAGI, taking the acronym from the French) that officially launched the project in 1957. The "Polar Year" idea died early on, however, because they realized that the project needed to be a bit more broadly defined to arouse the interest of national academies of science throughout the world. Consequently, in 1952 they expanded the scope to include a wide array of geophysical phenomena, including oceanography.[7]

The IGY was promoted as a purely international pursuit, devoid of politics. The Cold War was not supposed to become involved; this was an arena for science to be the common language of mankind. It was an opportunity to prove that relations between individuals, citizens of all nations, even those at odds with one another, could contribute to productive activities of mutual benefit. According to Berkner, it was not even supposed to be a cooperative effort between nations. The programs were the result of individual scientists working together on an international scale, "with the consent, cooperation, and aid, but not the direction, of governments."[8] This distinction favoring "international" over "intergovernmental" would resurface constantly for oceanographers over the next decade. One American geologist wrote that this explicit policy of never discussing politics was the "guiding light" of the IGY, and understandably so. In 1955, as planning got under way, there were two Chinas, two Koreas, and two Germanys, each wishing to send representatives to represent the whole. To welcome politics into the IGY would have been to cripple it from the beginning.[9]

Nevertheless, despite widespread insistence that the IGY was nonpolitical, it was laden with politics. Many nations believed that the scientific activities planned in Antarctica, for example, would establish precedents for future territorial claims. Although eventually the 1961 Antarctic Treaty made the continent an international site for scientific purposes, prior to the IGY the political fate of Antarctica was far from certain. Also, the launching of satellites, despite the ostensibly peaceful forum in which *Sputnik* was to appear, was particularly haunting, given the evident potential to use rockets to launch intercontinental ballistic missiles and to use satellites for spying. Looking back on the era, *Sputnik* is typically invoked to illustrate the danger and hostility of the Cold War, exemplified by the space race and arms race that ensued after its launch. In fact, it is rarely remembered that *Sputnik* was part of the Soviet IGY program and that Soviet scientists had announced it beforehand. Another sticky political issue was that of radioactive waste disposal in the deep sea. One aspect of the IGY was to find deep portions of the ocean where water was relatively stagnant, so that radioactive waste could be deposited without harmful effects. One possible location was the Aleutian trench in the North Pacific, but international fisheries groups protested even investigating the possibility, because of potential damage to salmon and other fish life that might frequent nearby waters.[10] Soviet scientists likewise protested when such potentially stagnant areas as the deep Black Sea were suggested. The long-term ramifications of such studies, despite their supposed nonpolitical character, were abhorrent to countries trying to keep radioactive wastes out of their backyards.

The mere presence of the Soviet Union in the IGY was an unexpected political hazard. After Stalin's death in 1953, Soviet scientists asserted themselves more and more in the international arena, and the IGY provided a kind of debutante ball for Soviet scientists to "come out," and they clearly expected to play a major part. Geophysicist Vladimir Beloussov, who attended the 1954 IGY planning meeting in Rome, complained that many of the world's largest nations had not been asked to participate in the CSAGI. At the time, there were three CSAGI officers—Sydney Chapman, Lloyd Berkner, and Marcel Nicolet. The Soviets claimed that this provided representation for the United Kingdom, the United States, and Belgium, respectively, to oversee all other scientists in the CSAGI. Trying to maintain the nonpolitical spirit of the IGY, the three officers countered that these members represented scientific unions, not nations. But in light of the Soviet complaint, the CSAGI officials had to confront the question of whether it would be more "political" to reject the Soviet demand or to include a Soviet in order to provide more equitable national representation. The issue was somewhat resolved when the International Union of Geodesy and Geophysics appointed Beloussov as its representative to CSAGI. Soon Jean Coulomb of France was made an officer as well, making CSAGI look like Berlin in 1945.[11]

The steady evolution of Soviet participation in the IGY and the geopolitical unease it created centered largely on the Antarctic, which opened new questions about the long-term political ramifications of the IGY. Although the Soviets announced their participation in the IGY during the 1954 Rome meeting, it was not clear whether their work would include an Antarctic component. At once diplomats began to consider the implications for territorial claims in Antarctica, where already more than a half dozen countries had made claims and the United States seemed ready to make claims as well. According to the British Foreign Office, if the Soviets wanted to establish a strategic foothold on the continent too, the IGY "would provide a plausible opportunity to do so with 'international blessing.'" To prevent this from happening, the United States and Britain urged other countries to take up the observation locations on the continent that no one yet intended to occupy (referred to as "gap locations"), particularly one near Vahsel Bay in the Weddell Sea. France, which already had difficulties enough financing its existing base, was not interested, and neither Norway nor Sweden expressed any desire to occupy it.[12] So by late 1954, both the Americans and British made preparations to fill the "gap location" themselves. The diplomats "had no doubt that some kind of accommodation could be worked out between our two Governments, and that the important thing was to keep out the Communists."[13]

Concerns that the Soviets themselves would fill the "gaps" were well founded. It soon became clear that they would go to Antarctica regardless of the scientific need for them to do so, and a number of signs from within the Soviet Union pointed to a major orientation toward the southern continent. Intelligence reports indicated that news media and other literature inside the Soviet Union speculated about a Soviet expedition to Antarctica equipped with helicopters, jeeps, and tractors. In addition, the Soviet government broadcast a special television program marking the 135th anniversary of the "discovery" of Antarctica by Russian explorer F. G. Bellingshausen, and the naval museum in Leningrad opened a special exhibit commemorating the event. To the American State Department and British Foreign Office, this escalation of interest strongly suggested that the Soviets intended to make a major commitment to Antarctica during the IGY.[14]

The politicization of IGY activities was only to escalate. At a planning meeting in Paris in 1955, scientists tried to reaffirm their desire to leave such issues behind and just stick to science. "Early resolutions that the proceedings would be entirely scientific and non-political," one British official nevertheless noted, "had little effect." Both Argentina and Chile sent large delegations from their embassies to discuss the political factors in conducting IGY work on the continent of Antarctica, and the Americans and Soviets had to negotiate over who should have the privilege of setting up a station at the South Pole. Ultimately, the Soviets consented to set up theirs some distance away, at the Geomagnetic Pole.[15] But toward the end of 1955, a British diplomat noted that it was "becoming more and more apparent that the Russian and United States expeditions are being 'stepped up,'" not only for prestige value, but because neither seemed likely to leave the continent after the close of the IGY.[16]

The Americans had a major military presence in the region. This was largely because of the logistical support for science provided by the Navy. The funding for American participation in the IGY was allocated by Congress and given to the National Science Foundation, but the Navy was heavily involved not only in logistics but also in its own exploratory work prior to the IGY (many of its expensive plans, as noted in the previous chapter, eventually were terminated by President Eisenhower). The presence of so many men in uniform did little to allay concerns that the United States saw the IGY as a major opportunity to extend its power. The Soviets persistently drew attention to this point while noting their own peaceful scientific activities. Navy leaders certainly treated Soviet participation in the IGY as a threat, and their activities reflected it. The commander of the Navy's Task Force 43, Rear Admiral George Dufek, later wrote that he saw himself racing against the Soviets on the continent, not cooperating with them. In particular, there was the symbolic achievement of landing a plane at the South Pole, to scout it out for the station to be placed there for the duration of the IGY. "In the back of my mind," he wrote in his journal on the night before the successful American landing in October 1956, "is the haunting concern that the Russians will beat us to the South Pole."[17] Capitalizing on the huge American military presence, the Soviets emphasized their own peaceful intentions. In contrast to the American base "Little America," a *Pravda* article of November 1955 announced, the main coastal Soviet base would be named "Mirny" (peaceful) in honor of one of Bellingshausen's ships from the 1819–21 expedition. Around the same time, an article in *Soviet Fleet* repeatedly emphasized that the organization in charge of the American Antarctic expeditions was the Department of Defense. It quoted from Admiral Byrd, who openly had emphasized the strategic value of a permanent base in Antarctica, and from Admiral Dufek, who mentioned the possibility of uranium and other materials to be exploited in Antarctica or using the region for weapons testing.[18] These issues, and the Cold War anxieties of the participants, clouded the cooperative spirit of the IGY long before *Sputnik* became involved.

OCEANOGRAPHERS IN THE IGY

For oceanographers, the IGY provided an opportunity to expand the scope of international cooperation in oceanography beyond the traditional partners in North America, the United Kingdom, and Scandinavia. In 1956 Robert Dietz wrote to Woods Hole director Edward Smith that American oceanographers needed more direct liaison with scientists in Western Europe, particularly France and Germany. These countries should be invited to participate in conferences and their rising stars (Dietz identified Henri Lacombe at France's National Museum of Natural History and Günther Dietrich at the German Hydrographic Institute) should be cultivated as active partners in oceanographic projects.[19] The IGY presented an opportunity to achieve this broadening of activity and liaison, but unfortunately those who were planning the IGY had not intended to depend very heavily upon oceanographers. Much to their chagrin, there was no seat in the CSAGI to represent oceanography at all, despite the fact that the committee had recently expanded to accommodate more members. The highest international planning organization for the oceans would be an international "working group" on oceanography led by British oceanographer George Deacon, who directed Britain's National Institute of Oceanography (NIO). Deacon and others resented the lack of representation for oceanography, especially given the fact that marine research was going to be more active in the IGY than many other aspects of geophysics. He protested that a huge field of inquiry was being given short shrift for political reasons, namely the CSAGI's need to give the IGY a sufficiently "international" character. "If the Committee can be expanded to include two Russian scientists," he complained to Smith, "it can be expanded a bit more to include one representative for the oceans."[20] Berkner had hoped to avoid this predicament by having the International Union of Biological Sciences elect as its representative Anton Bruun, a noted Danish marine biologist. Unfortunately this did not happen, so the scientist who ultimately represented the oceans was Georges Laclavère, who described himself as "a very humble geodesist." Laclavère wrote that despite rumors that he was not up to the task, he took his position very seriously and he intended to take his cues from the oceanographers themselves, provided they all could agree. He announced that oceanographers wishing to participate in the IGY should convene for planning purposes at the next meeting of the International Union of Geodesy and Geophysics in Brussels in September 1955.[21] It was there that oceanographers of East and West would meet for the first time.

Although atmospheric physics and research in Antarctica were at the core of the IGY's research agenda, and there was no oceanography seat in the CSAGI, oceanographers ultimately took the international project as an opportunity to coordinate their work as well. One of the originators of the IGY, Sydney Chapman, helped to create the British National Committee for the IGY, comprising not only atmospheric physicists but also oceanographers. The primary object of the IGY was to collect synoptic data in many fields. To take synoptic data meant to observe the conditions of the earth by taking specific measurements at the same (or very near the same) moment at different places throughout the globe. One could not accomplish this very well with expeditions, because no matter how fast a ship traveled, the conditions of the sea would change from one observation to the next. Synoptic observations provided a "snapshot" of the earth; it was only possible by standardizing the kinds of measurements and coordinating observation intervals at an international level. British scientists George Deacon and

Edward Bullard both saw the IGY as a tremendous opportunity for work at sea. In Deacon's view, the IGY promised to give physical oceanographers a chance to make a major synoptic study of ocean circulation. Robert Dietz, the U.S. Navy's science liaison officer in London, reported that the flow of deepwater masses was one of NIO's most promising lines of research, because it attacked a specific problem. It would address a discrepancy among oceanographers, who calculated the movement from high latitudes to the tropics in terms of decades, and carbon-14 measurements, which put the figure as high as eighteen hundred years.[22] Attempts to measure deep currents on a large scale could be made during the IGY, using buoys recently designed by the NIO's John Swallow. For Bullard, these investigations would give geophysicists an opportunity to come along for the ride. "The things I am interested in," he wrote to the Royal Society, "do not really fall directly within the scope of the International Geophysical Year, since it is not essential that they be done simultaneously in different places." Nevertheless, if ships were going to sea to make synoptic observations, then they could certainly do other things too.[23]

Other European scientists took an early lead in conceptualizing oceanographic projects for the IGY. For Harald Sverdrup, now head of the Norwegian Polar Institute, it was the time to pursue the kind of cooperation that he had advocated while director of Scripps in the 1940s. He was also president of ICES, a body created early in the century for research in the service of fisheries exploitation, and he was well positioned to create a coordinated plan for physical oceanography. The attention of ICES was concentrated on the North Atlantic. The United States had been a member prior to the First World War, but had never rejoined, and Canada was not a member either. However, the Federal Republic of Germany joined it in 1952 (Germany had been a member before World War II) and the Soviet Union in 1955, and coordination with the North Americans did not pose any serious problems. A German, Günther Böhnecke, took over the coordination of ICES's contributions to the IGY. Through ICES and also the International Commission of Northwest Atlantic Fisheries (ICNAF), eight countries, including Canada, would conduct the Atlantic Polar Front survey using twenty-two vessels.[24]

The extent of American oceanographers' participation in the IGY initially was uncertain. At Woods Hole, few scientists were spearheading international efforts, and projects were not particularly ambitious. The Air Force Cambridge Research Center thought of sponsoring some of its scientists to observe the electric potential of submarine cables throughout the world during the IGY. Henry Stommel wrote to Deacon in 1953 to sound him out, "before some official inquiries or invitations came to you through the long, tortuous, devious route of international scientific organizations." It was clear that Woods Hole had no definite plans to participate in the IGY to do strictly oceanographic work. Deacon's attitude was quite different. He responded, listing three oceanographic projects that the British National Committee already had recommended: studies on surges and long waves (especially those of meteorological origin), slow deepwater movement, and the distribution of temperature, salinity, and direction of drift in the northern part of the Gulf Stream. He added, "Why don't you get an oceanographer on to the US National Committee for the International Geophysical Year?"[25]

Oceanographers in the United States soon began to view the IGY similarly, and by mid-1954, they were planning for a major effort. One project, similar to that suggested by British scientists, was to set up a number of "island stations" to record mean sea level using tide gauges. This would not necessitate any expensive ship time, as the stations would simply record the data and send them to a central data bank. Ships would be dedicated to the study of oceanic circulation, taking measurements of subsurface currents in the Atlantic, Indian, and Pacific oceans. The decisions to pursue these lines of investigation were not due to international coordination; they were entirely of American design. When Deacon heard about them, he wrote to a colleague in the Admiralty that "[i]t was impossible to avoid the impression that they would organize it all themselves if no-one else was prepared to help," but that it was probably a good plan for everyone. The British had planned to set up stations along similar lines, and the German, Finnish, and Swedish proposals likely could all be covered by what the Americans wanted to do.[26]

Competition with Scripps was a major factor in getting Woods Hole more interested in jumping into the international fray. Woods Hole's director, Edward Smith, attended the Brussels conference in 1955 and discovered that Revelle, Deacon, Bruun, and others not only were hoping for a major oceanographic program in the IGY but also were planning to meet in Tokyo to discuss the development of marine science under the auspices of Unesco (this is discussed in greater detail in the next chapter). Upon his return, Smith wrote to Athelstan Spilhaus about Scripps's orientation toward international cooperation and noted "the feeling I have that Woods Hole is not tied into this movement so far as I know, and it is my desire that we should be." Although Woods Hole was making a strong effort in the IGY (Smith chaired the working group at Brussels, and Columbus Iselin represented

Woods Hole), Smith felt that Revelle was going further and that he was trying to define the parameters of future cooperation in oceanography. He sensed that the IGY was going to give a major stimulus to internationalizing oceanography, and earth science in general, possibly in a permanent collaboration between Unesco and the International Council of Scientific Unions. "Revelle realizes the advantages of the early bird in such matters and is seeing to it that Scripps is properly represented," he wrote, adding that it was his desire to see Woods Hole equally involved.[27] By the end of 1955, Smith was actively cultivating transatlantic relationships. "It would seem that Roger [Revelle] has done much to enthuse his Pacific colleagues with the ambitious ship program of the IGY," he wrote to Deacon, enjoining him to help "keep the fires stoked during the next year" with Atlantic-based organizations such as ICES and ICNAF.[28]

Most scientists welcomed the large effort by the United States, despite its unilateral approach to cooperation. Although there would be some difficulties in arranging ship cruises to coincide with the American ones, the benefits outweighed the inconveniences. For example, Deacon felt that his efforts to have Britain's oceanographic plans approved would be greatly strengthened by the fact that he could point to a much larger effort being made by the Americans. He wrote to Böhnecke that "we ought to be able to argue that the advantages likely to be gained by cooperation will justify some extra work and expenditures."[29] By the same token, American oceanographers could point to increased activity by other countries to justify more funding. Smith wrote to National Science Foundation director Alan Waterman in 1956 urging him to allocate money for building new ships, specifically designed for oceanographic research (not simply overhauled Navy vessels), to take the United States into the next decade with the best capabilities in oceanography. Other nations, he noted, had demonstrated a recognition of this need: Japan built the *Umitaka Maru* for training and research purposes; the Federal Republic of Germany built a new fisheries research vessel, the *Anton Dohrn;* and of course the Soviets had been operating the huge 5,600-ton *Vityaz* for many years and had several other large ships (more than 1,000 tons displacement) dedicated to research. In the United States, he noted, such modernization through building new vessels to replace old ones "is seriously lagging." The only U.S. vessel specifically designed for research was Woods Hole's *Atlantis,* built in 1930, which Smith and others judged to be obsolete and badly in need of replacement. Smith also argued that the Navy's interest satisfied only a small part of the nation's broad needs related to oceanography. He pointed to the Atomic Energy Commission, the Departments of Interior, Commerce, and State, and Health, Education, and Welfare as agencies whose immediate problems all called for increased knowledge of the sea. These problems included radioactive waste disposal at sea and biological effects of nuclear weapons testing, fisheries resources and conservation, the study of hurricanes, relations between maritime nations, and general research and training.[30] Already, leading oceanographers were seeing the government's enthusiasm for a major international scientific endeavor as a means to give their own science a boost at home.

The international working group that convened at Brussels in 1955 hinted at the new international face of oceanography. Smith, Ewing, and Revelle, were members, and Deacon chaired it. Other members included the German Günther Böhnecke, the Norwegian Håkon Mosby, and the Japanese Koji Hidaka. It was at this meeting that these scientists realized that the Soviets actually meant business. They increasingly felt that within the Soviet Union a new policy had been instigated, which stressed international friendliness and cooperation among scientists. Americans in other disciplines had perceived this only recently at other international meetings, such as the Geneva Atoms for Peace conference in August 1955 and the International Astronomical Union meeting in Dublin in September 1955. This Brussels meeting was the first time oceanography experienced it. Seventeen delegates from the Soviet Union attended the CSAGI meeting in Brussels, with two of them dedicated to oceanography. One was Vladimir Kort, head of the Institute of Oceanology in Moscow. Kort would become known during the ensuing years as one of the leading Soviet marine scientists. However, in 1955, none of the other delegates knew him at all. Indeed, this was one of the first appearances of a Soviet oceanographer in a noncommunist country since before the Second World War. The other delegate, Mikhail Somov, had been deputy head of the Arctic Institute in Leningrad, but now he was the leader of the Antarctic expeditions planned by the Soviet Union. These scientists made an effort to express a new spirit of cooperation, claiming their government's willingness to release data from some Soviet national expeditions and to make available books by Soviet authors on the field of oceanography. They also suggested formal visits to the Soviet Union by scientists and students, even proposing the exchange of scientists aboard oceanographic vessels during the IGY.[31] All of these suggestions seemed to represent complete reversals of previous Soviet science policy.

The Americans were cautiously receptive to the Soviets' overtures. Although they were skeptical about whether such attitudes would last very long, they felt that the Soviets had revealed some fairly detailed information about their intentions, and it would be awkward for them to back away from their new attitude in the immediate future. They seemed to be pressing forward energetically, having planned an extensive oceanographic program. They expected to utilize at least seven oceanographic vessels, all varying in tonnage from 1,200 to 5,000 tons. This contrasted sharply with the relatively tiny American vessels, which all displaced fewer than 1,000 tons. The Soviets' 5,000-ton ship was the *Vityaz* (actually around 5,600 tons), well known to the Americans because it had been operating since 1949 in the Pacific Ocean, in the Kuriles and Kamchatka area, and had made four large-scale expeditions. The *Vityaz* was Soviet oceanology's flagship, with accommodations for eighty scientists and elaborate equipment for deep-sea coring and trawling.[32] With all these plans, the Soviets were committed to an ambitious oceanographic program at least for the duration of the IGY, and they seemed perfectly comfortable with sharing their data.

Despite weighing in strongly in terms of size of ships and numbers of scientists, the Soviets were not in a position to dominate the IGY. Their plans only moderately impressed colleagues in other countries. Soviet instrumentation, judging from a handbook passed around during the conference, appeared rather far behind Western techniques. One observer noted that its overall appearance was that of a 1940 text. Although the Soviets did use the all-important bathythermograph, they had achieved no great advances in instrumentation. The best echo sounders they possessed had been manufactured by the British during the Second World War, so the accuracy of bathymetric mapping was certainly no better than that accomplished by Americans. According to Robert Dietz, the overall vision of the Soviets' plan seemed uninspired; it was more an exercise in exploration than a program designed to address specific problems.[33] Although perceptions of Soviet science would soon change, the notion that Soviets were more concerned with exploration and data collection, rather than scientific problems, would last a long time.

Like other countries, the United States set up a national committee for the IGY, and within it was the Technical Panel on Oceanography (TPO), which initially was headed by Edward Smith. But as director of Woods Hole, Smith felt that handling the expected large sums should be done by an agency other than one of the principal institutions to conduct the work.[34] Scripps director Roger Revelle suggested that Gordon Lill, head of the Geophysics Branch within ONR, take up the post. Because the Navy was expected to play a major role in carrying out the oceanographic program, regardless of any formal arrangement, Revelle hoped Lill's appointment would expedite such involvement (and he was correct).³⁵ Meanwhile Smith, Revelle, and Lamont director W. Maurice Ewing, appropriate to their roles as directors of the leading American institutions that pursued oceanographic research, became the general spokesmen for American oceanography. Smith and Ewing were responsible for facilitating contacts with international partners in the Atlantic region, while Revelle took responsibility for the Pacific.³⁶

The IGY planning helped bring about a stronger national effort by the United States. Although the American TPO and Deacon's international working group functioned as appropriate national and international bodies during the IGY, many oceanographers strove for a permanent setup. The chief of Naval Research, Rear Admiral Rawson Bennett, wrote to the National Academy of Sciences in August 1956 that it needed to address the fact that there was no established means through which oceanographers could exercise concerted action on national and international issues. As only one example, no organization existed that might help integrate important issues of radioactive waste disposal, the responsibility of the Atomic Energy Commission, with the expertise of oceanographers in major academic institutions. The director of the U.S. Fish and Wildlife Service agreed that a permanent oceanographic committee might be more effective in eliminating duplication of effort in several areas and in ensuring that scientists outside physical oceanography had access to enough of the Navy's precise hydrographic data.³⁷ Bennett summarized the effects of IGY planning:

> In recent months there has been an increasing demand for advice on oceanographic problems of great magnitude. Many of these questions are of broad scope and long range, having far-reaching effects on the safety and benefit of mankind as well as considerable influence upon the foreign policy of the United States. They often require the concerted action of oceanographers and scientists in related fields.³⁸

The general consensus of representatives from the Office of Naval Research, the Atomic Energy Commission, the National Science Foundation, and the Fish and Wildlife Service was that the United States needed a group for advisory purposes and for planning, coordinating, and directing oceanographic research across a broad range of disciplines.³⁹

As one step in the direction of creating such a group, in October 1956 the governing board of the National Academy of Sciences Division of Earth Science approved the concept of creating a committee on oceanography. With the addition of the committee, scientists in the United States had to confront some discipline-related issues, including the ever-present and not totally trivial one of whether to call the science "oceanography" or "oceanology." Americans typically felt that the law of inertia compelled them to stay with the term oceanography, although they admitted that "oceanology," a term used by Soviet scientists to differentiate themselves from their naval colleagues, was a more proper name for a science.[40] But at the same time, American oceanographers were accustomed to blurring the lines between naval work and fundamental research, which may explain why so few pushed for a change in name. More critical than changing the name was the need to address squabbles between disciplinary branches that thus far had crippled oceanographic committees under ICSU. Many felt that an impartial committee under the National Academy of Sciences, adopting oceanography broadly conceived but with a "clear vision and perspective[,] could render a real service and see this steadily growing field through some of its current infant diseases of internal dissension."[41] Lloyd Berkner hoped that such a committee might assume the function of the TPO at the close of the IGY, while the newly founded Special Committee for Oceanic Research (SCOR; the S later stood for "Scientific" instead of "Special") would take the place of Deacon's working group as the permanent international organization.[42] By forming such a committee under the National Academy of Sciences, oceanographers hoped to have a permanent mechanism to place their discipline at the forefront of the international scientific community, a role that planning for the IGY helped to solidify.

SCIENTIFIC EFFORTS

There were numerous logistical difficulties in securing cooperation in oceanography by nations other than the United States, Soviet Union, and United Kingdom, all of which were committed to oceanic studies that were global in scope. Some problems were organizational. Various cartographic institutions in Central America offered to participate, but this only meant that they were willing to request that their governments establish national committees. Guatemala actually had a national committee organized and it tried to take the lead in organizing a general plan for Central America. However, Guatemalan scientists could promise no support from their government unless the focus of research was in local waters such as Amatique Bay; as was true for many nations, any synoptic data collection of the entire earth needed to be supplemented by more useful local surveys.[43] Most problems stemmed from financial constraints, and the Americans attempted to alleviate them. Supply shortages at Bolivia's La Paz Observatory were happily addressed by the United States Coast and Geodetic Survey, since that organization received data from the observatory already and had a vested interest in keeping the data flowing during the IGY and beyond.[44] One Japanese scientist reflected that he and his colleagues wished to attend international meetings, "but it seems to me very difficult because the Japanese Government is very poor." The Americans tried to assist in such cases when they could, often by arranging for transportation through the Military Air Transportation Service (MATS).[45] In general, the American scientists made sincere efforts to include as many participants as possible in oceanographic studies.

Despite these difficulties, scientists managed to conduct a well-coordinated plan of action. Scientists in all fields recognized "World Days," during which a predetermined number of stations would conduct the same kinds of observations. The stations would then submit the data to several World Data Centers, located in various locations throughout the world, which acted as clearinghouses of scientific data for dissemination to any of the IGY participants. One such World Data Center, for oceanography, was located at Texas A&M University, which along with the University of Washington occupied a place behind Woods Hole, Scripps, and Lamont as a significant American center of oceanography. The concept of World Data Centers was in fact one of the most important legacies of the IGY, resulting in two major centers for the dissemination of data. World Data Center A was located in Washington, D.C., while World Data Center B was located in Moscow. The choice of these locations is obvious, given the political climate of the era. Scientists would write to whichever location would cause them the least difficulties in acquiring the data. Technically, both A and B were supposed to have the same data, drawn from the other centers; in practice, however, this was difficult to achieve. One example was the participation of mainland China in data exchanges with the Soviet Union but not with the rest of the world. In any event, there was also a World Data Center C, which was dispersed throughout the world. The first two World Data Centers would survive the IGY and become major vehicles of data exchange programs in oceanography during the 1960s.

The most important synoptic oceanographic project of the IGY was the Island Observatories Project. By observing tide gauges, oceanographers hoped to make a more precise correlation between changes in sea level and ocean circulation; knowledge of circulation would, it was supposed, lead to a better understanding of weather and climate. By placing tide gauges on islands, the oceanographers hoped to avoid coastal influences. Specifically, the project followed up on findings by scientists at Scripps who discovered that the sea level of the oceans depended upon the season. The Island Observatories Project might, at the very least, help discern what happened to a huge volume of water between seasons. The methods of the project were to measure sea-level oscillations having periods anywhere from five to fifteen minutes of duration and also to study a longer oscillation (with a period of 3.84 days) that seemed to correspond to the equatorial wind system. Also important were the seasonal changes, which obviously had much longer periods. In addition to these oscillation records, scientists at each of the island observatories would make meteorological observations.[46]

Scientists at Lamont and Scripps carried out the American portion of the Island Observatories Project. At Lamont, scientists used microbarovariographs to observe slight changes in barometric pressure, which eventually allowed them to perceive atmospheric waves that corresponded closely to oceanic waves with the same periods and moved in the same direction. Such studies reinforced perceptions of the interconnections of geophysical circulation patterns. In the Pacific, Scripps scientists cooperated with the Dutch, Japanese, French, Chileans, Ecuadorians, British, and Mexicans at various islands. In the Arctic Ocean, stations were manned by American, Canadian, Finnish, and Soviet scientists. All of the stations in the Pacific Ocean encountered difficulties due to storm damage, and the Arctic stations endured freezing temperatures and ice movement along the coast.[47] These observations, which promised to unveil a great deal about the circulation of both air and water, could not have been accomplished without the cooperation of scientists and the coordination of information on an international scale.

Also crucial were the American maritime agencies such as the Coast and Geodetic Survey, the Coast Guard, and the Navy, all of which helped cut costs of the largest IGY effort, namely, the American one. The Coast and Geodetic Survey made many of its own observation stations a part of the Island Observatories program, as did the Coast Guard, which coordinated data from its own weather ship program. The American oceanographers asked the Navy's Bureau of Ships to act as the purchasing agent for their equipment; much of this equipment could simply be lent to the scientists for the duration of the IGY. Gordon Lill, in seeking such support from the Bureau of Ships, admitted that "the benefits to the navy of broad oceanographic studies such as those presented here [for the Island Observatories Project] are somewhat obscure," but that such general understanding would benefit the Navy in the long term.[48] The Navy generally was cooperative throughout the IGY, and its various agencies took on logistical tasks such as processing bathythermograph slides, lending equipment, and providing broad support for IGY activities both within and external to budgets allocated officially to the IGY through the National Science Foundation.

The Americans also planned expeditions, the bread and butter of oceanographic research. Although the Island Observatories Project addressed specific questions, expeditions covered a broad swath of observations that often had no specific problem to solve. Scripps director Roger Revelle noted that expeditions, as obscure as their value might seem to the laboratory scientist, were necessary for such a youthful science. He commented on two specific constraints on oceanic science:

> We tend to look at the ocean rather than ask questions of it. Because our ignorance is so great, descriptive exploration is necessary, but I believe that we ought to use it primarily as a means of formulating problems—as an aid in thinking of questions to ask. In the second place, the size and intractability of the ocean makes it hard to answer such questions as we can ask, because it is difficult to carry out controlled experiments, those peerless tools of the classical sciences.[49]

Indeed it was this difficulty in attacking oceanography with the traditional tools of scientists that tainted the discipline's image. Geologist Harry Hess, for example, felt that marine geology would not be helped by the kind of activities engendered by the IGY, which as a whole probably would not produce much of interest to the scientist. "Fifty six million dollars," Hess wrote to then-controversial writer Immanuel Velikovsky, "will produce a lot of scurrying back and forth to the South Pole and an indigestible mass of random observations on everything."[50]

Observational expeditions nevertheless constituted a significant portion of the American oceanographic program. The Navy's Task Force 43, led by Rear Admiral George Dufek, was charged with supporting the Antarctic scientific work throughout the IGY. The icebreaker vessels he commanded were also suitable for oceanographic work, and when Navy commanders felt it was practicable, such work was done in Antarctic waters. The most significant work done by oceanographers aboard Navy vessels in this region was on the Antarctic Convergence, a line around the continent approximately a thousand miles from its shores, where northern warm water met southern cold water. The properties of the two relatively unmixed regions were distinct in terms not only of temperature but also of sea life and meteorology, making the convergence an area of special interest to those investigating the dynamics of the sea. In addition to these observations on Navy ships, more formal scientific expeditions were conducted by academic institutions. Texas A&M sent its ship, the *Hidalgo,* to study the waters connecting the Atlantic and Caribbean to the Gulf of Mexico. The University of Washington sent the *Brown Bear* along the western coast of the United States and northward to the Aleutians, where it studied the composition of water at various depths.[51]

Woods Hole and Scripps also conducted expeditions. Scientists at Woods Hole sought to test Henry Stommel's recent hypothesis of oceanic circulation, which stated that surface water subsides to the bottom in the subarctic region and rises from the bottom again in the subantarctic, creating a global circulation pattern that reached the oceanic depths. Stommel's notion conflicted with many scientists' expectation and even the hope, for those who wished to dump radioactive wastes in the deep sea, that waters in the great depths were relatively stagnant. Coordinating closely with British scientists at the National Institute of Oceanography, oceanographers aboard the *Atlantis, Crawford, Chain,* and the British vessel *Discovery II* made east-west intensive studies across most of the Atlantic Ocean searching for deep currents and making other measurements of the deep water. One of the many scientific results of this cooperative work was the clear evidence of deep currents, which added fuel to the argument against deep-sea radioactive waste disposal. Scripps conducted similar work. Its 1958 Dolphin expedition traced a deep current for 3,500 miles under the equatorial Pacific (this was named the Cromwell Current, after Fish and Wildlife scientist Townsend Cromwell, who first recognized it). The subsequent Doldrums expedition showed that another deep current, the Equatorial Countercurrent, was much larger than previously believed.[52]

Lamont, although a relative newcomer to the field of oceanic research (it was founded in 1949), also conducted work throughout the IGY. Its first director, W. Maurice Ewing, had initiated a period of intensive deep-sea investigations, and Lamont's significance as a center for studying the oceans was exceeded only by Woods Hole and Scripps. Ewing was once asked by a British colleague about where Lamont, a part of Columbia University and located inland, kept its oceanographic vessels. Ewing answered, "I keep my ships at sea."[53] Certainly this was true during the IGY, as the schooner *Vema* was at sea nearly continuously throughout the year conducting surveys, among other things, of the Mid-Atlantic Ridge. Not all the work was purely exploratory, as Lamont scientists had specific questions to answer. Ewing and Bruce Heezen in 1956 had revealed that earthquake activity was strong not only on the Mid-Atlantic Ridge but also along other ridges throughout the globe. Heezen and Marie Tharp, both Lamont scientists, revealed the existence of a rift valley along what they believed to be a "world-girdling" ridge and rift system, where seismic activity was concentrated. Ewing and Heezen published a paper, "Some Problems in Antarctic Submarine Geology," just in time to have their research problems explored during the geophysical year. Their work posed questions that addressed not only undersea topography but also the origin and processes of the seafloor and the earth's crust as a whole.[54]

The IGY provided a means to collect even more crucial earthquake data to be analyzed, and the Lamont scientists' work prior to the IGY posed ample problems to be investigated on scientific expeditions. One of the principal aims of the Scripps Downwind expedition, for example, was to investigate Tharp and Heezen's rift. H. William Menard indicated to Heezen that he had found no evidence for a rift in the Pacific or the Indian oceans. Thus, he had been "increasingly distressed to read one account after another in the press and magazines of this fabulous rift 2 miles deep, 20 miles wide and 40,000 miles long." At the time, Menard felt that perhaps Ewing and Heezen were conflating the ridge and rifts with the fracture zones that Menard himself had discovered and that their analysis had misread the seafloor's basic topographical features. He expected to use the IGY to confirm or deny the existence of the rift. In the end, Downwind did neither. Upon his return from the expedition, Menard wrote to Heezen, "I have the mournful honor of informing you that the median trough of recent fame does not exist along the crest of the East Pacific Rise between 48° south and 43° south." But this proved only a temporary blow to the idea of a world-girdling rift that appeared in many other places throughout the globe.[55] Still, this was a case in which expedition research was problem based, not merely exploration based.

American oceanographers were also committed to solidifying the international basis of the IGY, with planning input from scientists in other countries. In October 1956 the American TPO recommended participation in an Arctic Ice Reconnaissance Project, because of the unique opportunity offered by the IGY to have an international cooperative effort to cover the Arctic Ocean. The Americans appropriated some preexisting Navy programs to make up the American side of the effort. One was the measurement of water transport across the eastern part of the Bering Strait, already planned by scientists from the Naval Electronics Laboratory. Japan and the Soviet Union would be invited to conduct similar measurements in the central and western portions of the strait. Also, because underwater sound propagation studies might provide a means for communication between ice pack stations, undersea acoustics (traditionally the province of the military) would also be pursued.[56] The Navy thus entered the IGY as a partner on the American side, with the understanding that all data taken in the program would be made available for international exchange. The Navy agreed to this arrangement; the chief of Naval Research argued that such a project in the Bering Strait would serve as a means to obtain information from the Soviet side of the strait where American scientists were not permitted to work. However, the Navy did feel that the operation of American ships in Soviet waters, and more important, vice versa, was not acceptable. Data exchanges would have to suffice.[57]

AN ALTERNATIVE VISION: SOVIET IGY PLANS

Despite American criticisms of Soviet oceanology as strictly observational and not problem based, the Soviets were eager to assert their own role in the international effort to solve specific problems of the deep ocean. In a *Deep-Sea Research* letter to the editor, Soviet oceanologist Lev Zenkevich stressed the importance of systematic worldwide research work in the deep ocean, exemplified by the Swedish *Albatross* expedition, the Danish *Galathea* expedition, and the British *Challenger II* expedition. He also applauded the various intensive regional expeditions of Woods Hole's *Atlantis,* active since the 1930s, and the Soviet Union's *Vityaz,* which had operated since 1949 in the Okhotsk, Bering, and Japan seas. Zenkevich complained that previous scientific work too often had been confined to mapping. Soviet scientists were ready, he claimed, to move forward and eliminate the poverty of knowledge of the oceanic depths that impeded progress in so many physical and biological disciplines. More studies of the depths themselves, such as bottom relief, bottom sediments, and composition and distribution of abyssal fauna, were required to develop a more complete understanding of the general geological and oceanic processes over time.[58] Thus, work aboard the *Vityaz* would continue during the IGY, exploring deep trenches with its sixty-plus complement of scientists, with the aim of determining the properties (such as suitability for sustaining life) of such deep areas.[59]

For the most part the United States and Soviet Union each planned their programs separately and implemented them independently. It would go too far to say that cooperation was in form and not in substance. Certainly principles such as data sharing and even some scientist exchanges were respected.

Throughout the planning stages there were efforts, on both sides, to conduct scientific research worthy of such massive financial and time commitments, while respecting the necessity of trying to work together. Still, the Soviet Union had its own oceanographic agenda, independent of American plans. Its Marine Antarctic Expedition was organized in 1955, and it did not wait for the IGY officially to begin before starting its preliminary scientific work. For two years, Soviet scientists traversed the waters between the Antarctic and Indian oceans aboard the *Ob,* a 130-meter-long diesel-electric ship of 12,600 tons displacement. Equipped as both an icebreaker and a scientific vessel, the *Ob* contained laboratories for meteorology, hydrology, chemistry, biology, marine geology, and geophysics. Enjoying four such laboratories, the marine geologists used echo sounders to determine the topography of the seafloor, taking samples of bottom sediments using various coring methods, analyzing the sediments using seismic methods, and collecting suspended matter by using filters and pumping water. From this work the Soviet scientists developed typical profiles of the continental shelf and continental slope of the eastern portion of the Antarctic continent.[60]

Soviet scientists sought to alter the IGY agenda to bring other nations' work closer to their plans. At the 1955 Brussels meeting, Vladimir Kort proposed that all nations should join them in devoting their energies to the oceans around Antarctica, which seemed natural given the IGY's focus on that continent. Lloyd Berkner warned American oceanographers at Woods Hole that the Soviets would be attending conferences and trying to convince everyone to change their plans and follow the Soviets' lead. Woods Hole's Edward Smith wrote to George Deacon in England that there would be as many ideas about surveys as there were oceanographers; they should try to coordinate, "but we will not be able to satisfy everyone." Deacon responded that they should not feel obligated to accommodate the Soviets, even if their desire to participate was such a surprise. "I hope we shall treat the Russians like anyone else and not let the novelty of their attendance influence our arrangements." He doubted that it would be possible to achieve genuine collaboration with them anyway.[61]

By the end of 1955, scientists were firming up a program of action during the IGY, and the Soviet proposals had little impact. After a Copenhagen conference of ICES members, the West German Günther Böhnecke was confident that all of the big gaps in their program could be closed by negotiation over the next several months, particularly if they could convince the Soviets to conform to the others' plans: "The discussion with [the Soviets]

The International Geophysical Year, 1957–1958

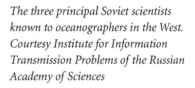

The three principal Soviet scientists known to oceanographers in the West. Courtesy Institute for Information Transmission Problems of the Russian Academy of Sciences

Marine biologist Lev Zenkevich

Oceanologist Vladimir Kort

Geophysicist Vladimir Beloussov

was similarly tiresome to that in Brussels. Their attitude was in general, however, positive. Apparently they have to ask at first their superiors at home."[62] Smith doubted whether the Soviets would persuade any ships from other countries to participate in their plans, but publicly he and others generally applauded the Soviet plan for its ambitious scope, which at least showed a determination to make a major effort. Writing to Laclavère, Smith called it "very desirable field work."[63] However, diverting efforts to the waters surrounding Antarctica was anathema to the Americans and others who already had spent a great deal of time coordinating the existing program.

But the Soviets were persistent in pushing their plan. Kort's 1955 proposal had met a cool reception among oceanographers, but the members of the Third Antarctic Conference had lauded the plan as a nice complement to the work to be done on the continent. Because Antarctica was the focus of the IGY, they reasoned that it would be appropriate to conduct extensive investigations of the surrounding waters. The Soviets, meanwhile, tried to demonstrate their willingness to cooperate in other projects. During a Tokyo meeting in 1955, Soviet (Lev Zenkevich and N. N. Sysoev) and Japanese (Koji Hidaka and Kanji Suda) scientists coordinated their plans for work in the Pacific Ocean.[64] Bolstered by these cooperative plans and by the endorsement of the Antarctic Conference, Kort proposed in 1956 a Soviet program that could be supplemented by the participation of scientists from Argentina, South Africa, Australia, France, and the United States. This proposal suggested that the Americans work in the South Pacific, particularly along the New Zealand shelf and between the Ross and Bellingshausen seas (off Antarctica).[65]

Despite general agreement that oceanographic work in the Antarctic was desirable, few were willing to support the Soviet-backed proposal. The Working Group on Oceanography, led by Deacon, had already rejected the plan in 1955, and only a few countries—Argentina, Chile, and France—seemed willing to participate in the Soviet plan. Now that the proposal had resurfaced through a different channel (a conference on Antarctica), Deacon felt obliged to stand behind his previous rejection. He believed that the Soviets, although taking up oceanographic research with an admirable enthusiasm, were getting ahead of themselves. Most of the ships visiting Antarctica to land scientific parties were not equipped for extensive oceanographic research but rather would conduct such work only when it did not interfere with other logistical business. Such work would be relatively simple and was likely to be confined to studying the balance of carbon dioxide between the ocean and atmosphere. Deacon also felt it was unlikely that such an expedition would accomplish substantially more than what already had been done by the extensive expeditions conducted by his own National Institute of Oceanography between 1924 and 1951; most of this information had yet to be analyzed. What the Soviets truly had in mind was a project that would require "a dense coverage of marine meteorological observations to be made by many ships over many years," which was beyond the resources currently allocated for the IGY.[66]

The rejection of Soviet plans did not encompass all IGY projects. At the Barcelona conference in July 1956 (at which Deacon was not present), the Soviets introduced their plan for a unique worldwide expedition aboard the nonmagnetic schooner *Zarya*. The schooner was equipped to allow continuous observations of the geomagnetic field. The vessel could visit many different areas and obtain data that, when coordinated with data from magnetic observatories set up for the IGY, would provide a more precise picture of the distribution of magnetic properties over the oceans. The *Zarya* planned to cover about 45,000 miles in the North Atlantic Ocean; the Black, Mediterranean, and Arabian seas; and along coasts as far apart as those of Brazil and Western Australia.[67] The Americans had no such schooner, and they affirmed that this was a valuable contribution that the Soviets could make. The American nonmagnetic ship *Carnegie* had done similar work before being destroyed by an explosion in 1929, and the Americans meanwhile had taken to making such surveys by air, although the ship measurements were slightly more accurate. American scientists applauded the cruise of the *Zarya* as a genuine, unique contribution to the IGY.[68]

Although the Soviet Union intended to carry out its work regardless of whether other nations joined it, it did make an effort to develop studies that were coordinated at an international level. Adhering to the recommendations of the 1955 Brussels meeting to attempt conformity of oceanographic observations, the Soviets disclosed most of their plans to the West. N. N. Sysoev, who handled instrumentation aspects of the Soviet projects, described the Soviets' methods of measuring currents both from anchored ships and drifting ships and their methods for chemical analysis. He also voiced the Soviets' wish to procure better equipment from Western countries.[69] Both American and Soviet scientists were committed to the idea of exchanges. After discussing the international resolutions made at Barcelona, the American TPO endorsed them and particularly stressed the importance of exchanging scientists on oceanographic vessels. By putting scientists aboard foreign vessels, they reasoned, the international scientific community would take another step forward in establishing uniformity of methods.[70] Also they advocated exchanges of data, whether more direct cooperation was possible or not. Soviet research on deep oceanic circulation, however, was hardly coordinated at all with the American efforts. Yet the Americans and the Soviets were seeking the same answers, and both even began with Henry Stommel's work on oceanic circulation (though G. B. Zaklinskii, director of the Murmansk Hydrometeorological Service, credited both Stommel and the Soviet scientist A. M. Muromtsev with the portrait of oceanic circulation that provided the framework for IGY research and the subsequent International Indian Ocean Expedition).[71] The results of the expeditions, despite the lack of close agreement in planning, were shared for the common benefit. The IGY was a promising step in cooperation, even more than previous efforts because of its explicit crossing of Cold War boundaries and particularly because of its inclusion of Soviets in planning and implementation.

Despite publicly noting how the IGY helped transcend political differences, however, American scientists used the entry of the Soviet Union in the IGY to solicit money from foreign policy arms of government. During the first couple of years of his administration, President Dwight Eisenhower had emphasized the importance of trade instead of aid—reciprocal arrangements rather than give-away programs. To ensure that state-sponsored foreign activity all conformed to his executive policies, in 1953 Eisenhower placed scientific and technical assistance under the jurisdiction of the Foreign Operations Administration (FOA), which looked to the Department of State for guidance. Cooperative scientific programs were to be explicit in their mutual beneficence and their alignment with Eisenhower's commitment to open trade and free enterprise.[72] Dissatisfied with director Harold Stassen's ability to achieve these ends—Stassen had in mind programs more akin to the Marshall Plan—in 1955 the administration allowed FOA to be abolished. Its functions were divided and assumed by the Department of Defense and the newly created International Cooperation Administration (ICA) under former congressman John B. Hollister, with an even more subservient relationship to the Department of State. The year 1955 nevertheless was one of soul-searching in terms of foreign aid, even among Republicans. On the one hand was the driving "trade not aid" philosophy of the Eisenhower administration and on the other hand was the realization that, without spending more on direct assistance, many developing countries—particularly those in Asia—might soon fall under the communist sphere.[73] The Soviets in 1955 conducted an arms deal with Egypt, and Premier Nikita Khrushchev visited Afghanistan, India, and Burma at the end of the year to emphasize the Soviet Union's desire to help these nations, all of which challenged the United States to compete for their friendship.[74]

American scientists tried to capitalize on these anxieties. They themselves politicized the IGY by repeatedly seeking support from the ICA and using national security as the basis for it. But when in 1956 the United States National Committee for the IGY asked the ICA for funds to support the scientific work of other nations during the IGY, the response was far from receptive. The ICA felt that if certain Latin American countries, particularly Brazil, wished to acquire and finance personnel and equipment to carry out the IGY, they would have to seek it elsewhere. Hollister acknowledged that indeed "it might well be in the national interest to assist Brazil and other countries of the free world in their participation in Geophysical Year activities." However, he did not believe that such financing fell properly within the purposes of the Mutual Security Program. He suggested instead that perhaps the National Committee should call this to the attention of the National Science Foundation, which could in turn request additional funds from Congress to finance other countries' projects.[75] Knowing this was not likely, National Committee chairman Joseph Kaplan continued to pester Hollister, claiming in July 1956 that the Brazilian Air Force was providing many of the scientific delegates to planning conferences and that the Brazilian government felt that the IGY would be important for the future development of Latin America. At one conference to be held in Rio de Janeiro, ten Soviet scientists were expected to attend, he said, "at which time they can be expected to learn of the US intentions with respect to support to Latin America for its IGY program." What should the Americans tell the Latin American countries, Kaplan asked, when they ask about American assistance?[76]

Publicly, American scientists emphasized the nonpolitical aspects of the IGY; behind the scenes, however, they tried to gain support by capitalizing on Cold War fears and sensibilities. At the tenth meeting of the National Committee on July 13, 1956, members discussed the fact that some countries—Egypt, India, Indonesia, Israel, and several in Latin America—had requested that some of the equipment for the IGY be furnished by the United States. Knowing that the Indonesians and the Egyptians had already asked the Soviet Union for assistance, they decided to make another plea to the ICA.[77] Lloyd Berkner wrote to Herbert Hoover, Jr., the undersecretary of state, imploring him to help Latin America take part in the IGY. This was an opportunity, he noted, to conduct scientific research in areas of critical importance without dealing with problems of national sovereignty. He added other benefits:

> It would generate additional interest and expertise on the part of the Latin American nations involved, it would lead to advancement in science and technology, and it would stimulate the intimate relationship of the US and the Latin American nations—conditions that I believe are in accordance with our present foreign policy.[78]

Berkner tried to make an appointment to see the undersecretary, but Hoover was too busy with other matters to see him, and ICA's acting director, D. A. Fitzgerald, wrote with the same negative response. Not only was it inappropriate to do this with funds made available under the National Security Act, he claimed, it would have involved the diversion of funds from other projects. Besides, the costs for individual countries were low enough that if the countries really considered it important, they could do it themselves. If the United States decided to pay for them all, the cost would be rather high.[79] Using such reasoning, perhaps the ICA felt that these other nations were simply playing one superpower against the other in order to participate in the IGY without paying for it. Although it was concerned with science, the ICA was more worried about the Soviet economic offensive, which consisted of technical assistance, trade, and the extension of credits to nations such as Egypt, Syria, India, Afghanistan, Indonesia, and Yugoslavia. The ICA was particularly concerned about losing India's huge population to the charms of communism. In 1956 Eisenhower had announced a change in the nature of the Cold War, sparked by the post-Stalin Soviet foreign policy. He recognized that the Soviet Union had shifted its emphasis from the threat of military force to the threat of economic dominance throughout the world, and this became the focus of ICA's concern.[80] The IGY's importance in these areas, at the time, seemed minimal.

American and British scientists and officials nonetheless were deeply suspicious of Soviet intentions in the IGY, beyond the territorial mess looming in Antarctica. In 1956, the British Ministry of Defence even considered appointing physicist Edward Bullard as a science liaison officer in Moscow to help develop a clearer picture of what sorts of activities the Soviets were focusing on. Bullard himself thought it was a terrible idea, given his knowledge of American rocket programs; people connected to any classified projects, he felt, ought to avoid going to the Soviet Union altogether.[81] But contacts with Soviet scientists, when made in the West, were routinely reported to government and military officials. Bullard himself reported an early encounter with Soviet geophysicist Vladimir Beloussov, who traveled to Cambridge University in 1956 to familiarize himself with the developments of its Department of Geodesy and Geophysics in the field of marine geophysics. The Soviets were beginning to pursue the kind of seismic studies that Britain and the United States had been pursuing since the end of the war. In Cambridge, Beloussov told Bullard that it would be the first work of its kind by Soviets, although Bullard later noted that this contradicted what a Soviet had told him at the recent Unesco meeting in Tokyo. In the course of the conversation with Beloussov, Bullard mentioned that he was interested in the fact that the depths of trenches in the Pacific Ocean were all nearly equal, and he mentioned a sounding by the *Challenger II* (a British vessel) that was supposed to be the deepest. This elicited a strange reaction from Beloussov, which Bullard recorded in a letter to a contact at the Ministry of Defence:

> Beloussov became very heated, and said that this was untrue, and that the deepest sounding was a Russian one in the Kurile Trench. I was a bit surprised about his enthusiasm over the trivial question of which sounding happened to be the deepest, but [American marine geologist Robert S.] Dietz tells me that this has been made a great propaganda story on the superiority of Russian scientists!

Bullard was astonished that the eminent Soviet geophysicist would equate measuring a deep trench with scientific prowess, as if the location of the trench and the nationality of the scientists who happened to measure it had something to do with the quality of scientific activity. Baffled by the encounter, he concluded of Soviet scientists: "They really are odd people."[82]

Americans also reported their impressions through official channels. The U.S. Navy had its liaison system, through which Robert Dietz in London routinely sent intelligence of European scientific activities, touching on the Soviets as often as possible on even the most trivial of matters. At the Thirteenth International Limnological Congress in Helsinki, he was able to make many observations of Soviet limnologists and oceanographers, including some on their typical dress:

> Russian scientists can be instantly spotted in gatherings by their clothing, as follows: conservative styling with poor materials; invariably a double-breasted suit; baggy, almost bell-bottom pants with pajama-like fullness; almost never a white shirt but generally a blue one with a conservative overhand big knot tie; simple black shoes.

Dietz noted that most scientists were of the "old school," fitting the common Hollywood caricatures of professors, although this "single-suited group" did not appear to be paid particularly well. It was clear, however, that their activities were very well supported by the state.[83]

Two Soviet oceanographers also went to Cambridge, where they told visiting American scientist Russell Raitt of their plans to do some observations off the coast of California, although they did not know quite how to integrate the work with the IGY program. This elicited some suspicion by the British Ministry of Defence, which suspected they might be doing work of a military nature there. The pair also visited Britain's National Institute of Oceanography, where George Deacon judged only one of them to be a real scientist. The other, who expressed little interest in scientific matters, Deacon presumed was watching over the Soviet marine scientist. It was obvious to him that the value of oceanic research had reached high levels in Moscow. "There is no doubt," he wrote, "that Russia and some other countries are using the IGY oceanography programme to show what a lot of science they are doing. If one has the money ocean exploration is easy, relatively useful, and a good way of showing the flag, and showing how keen you are on science." There was no cause for alarm, however, because the Soviets were not doing much creative or original work. They just wanted to send out a lot of ships and record data. The Soviets' style of exploration, in Deacon's view, was reaching the point of diminishing returns. He believed that British and American scientists were fast coming to the point at which it would be more profitable to attack and resolve basic problems rather than simply accumulate more data on the distribution of physical and chemical properties.[84]

SOVIET OCEANOGRAPHY AND WESTERN DISTRUST

The Soviet oceanographic program began its major international phase in 1955. Robert Dietz was able to gather information on the two main research vessels for Soviet oceanography, the *Vityaz* and the *Ob*. The *Vityaz* belonged to the Oceanological Institute, located in Moscow (the vessel itself was not based in land-locked Moscow, of course, but far to the east in Vladivostok). In addition, the institute planned to build two new vessels for oceanic research. The *Ob* belonged to the Arctic Institute in Leningrad, which had access to nearby port facilities. The ship itself, even prior to the IGY, operated in Antarctic waters under Kort's scientific leadership. During 1955 and 1956, the *Ob* obtained a considerable amount of data, including a detailed section along a line of longitude from the southern tip of India to Antarctica.⁸⁵ In geographic scope, Soviet oceanography was just as ambitious as the programs offered by Americans. The same year, the *Ob* visited New Zealand, and Kort gave a lecture in Wellington to the Royal Society of New Zealand on recent activities of Soviet oceanography. Scientists aboard the *Ob* left the Soviet Union in November 1955 bound for Antarctica, and in early 1956 they began to organize and construct their station on the coast of the Davis Sea. A month later, a second ship, the *Lena*, came into the region to help, and by mid-March the Soviets had essentially finished the construction of their base, Mirny, named after the ship that had transported the Russian explorer Bellingshausen to the region in 1820.⁸⁶

Western oceanographers did react to the escalation of Soviet scientific activities. George Deacon had rejected the Soviet plan to conduct coordinated research in the Antarctic because, he said, it could not add significantly to the material already collected in previous years. As secretary of the Working Group on Oceanography, Deacon's position represented the majority of the non-Soviet contingent of oceanographers. Nevertheless, soon after Kort's proposal, Americans began to formulate their own improvisational Antarctic plans. Joseph Kaplan wrote to Chief of Naval Operations Arleigh Burke that, although the primary mission of the Navy's presence in Antarctica was the establishment and logistical support of stations on the continent itself, the Navy should carry out oceanographic work as well. The Navy already had agreed to conduct such work when practicable, but the scientists hoped to make the oceanographic work a formal part of the Navy's role. Sympathetic, Burke replied that it might be possible to do so in the coming months, but that the oceanographers presently would have to "operate on a catch-as-catch-can basis, taking advantage of what opportunities are offered by circumstance."⁸⁷ This was hardly a methodical way to investigate the sea, but Burke seemed to suggest the possibility of greater flexibility once the IGY got under way. The American TPO decided to change its own position vis-à-vis Antarctic oceanography and to give more explicit attention to such a program, going against the position of the international Working Group on Oceanography. But it did not jump into the camp of the Soviets, who were proceeding with Kort's plan; rather, the Americans instigated a separate plan and made some arrangements to collaborate with New Zealand.⁸⁸

Prior to *Sputnik*'s launch, American scientists and policymakers remained strongly in favor of international cooperation, despite the rising challenge of the Soviet Union. Nelson A. Rockefeller, who was assisting the president in formulating strategy for government organization, had requested that the National Science Foundation prepare a report laying out its position on the federal government's role in international science. Reflecting the beliefs of many American scientists, the 1955 report emphasized the importance of international research as a means to use science to build up "free world" strength and to ease international tensions at the same time. Rockefeller made no specific comments on the report over the next few years, indicating at worst ambivalence toward science as a means to achieve foreign policy objectives and at best a tacit agreement with the report's general claims.[89] But during the first months of 1957, as the world prepared to kick off the IGY at midyear, Congress's praise for the IGY "wizards" seemed limitless. At early 1957 congressional hearings regarding the NSF's appropriations, congressmen repeatedly congratulated NSF director Alan Waterman for his work on the IGY, expressing their high regard for the top American scientists, particularly Lloyd Berkner, Antarctic geologist Laurence Gould, and Scripps director Roger Revelle. These men, in Waterman's words, "have succeeded in 'selling' the IGY program" to Congress.[90] In a March 1957 speech titled "The Scientist and the Politician," Revelle reciprocated Congress's sentiments of goodwill by claiming that the words "politician" and "scientist" easily could be combined. "Neither commands much respect or understanding in our society," he claimed, "yet between them they hold in their hands the future of mankind." The IGY reflected a recognition that the future would be shaped both by scientific knowledge and by technological developments and that scientists and politicians would have to cooperate to help the world adjust.[91]

This warmth of feeling toward international cooperation changed dramatically as the IGY got under way, and international tensions caused some difficulties. One of the most unfortunate political incidents of the IGY was the 1957 withdrawal of the People's Republic of China, which, in addition to having many interested scientists, had a great deal of land area that might have contributed considerably to the geophysical observations. The Chinese withdrew because of the last-minute inclusion of Taiwan in the scientific program. Mainland China had stipulated as early as 1955 that it would refuse to participate if this occurred, and at the time Taiwan had expressed no interest in the IGY. But, in the end, Taiwan decided to take part, and its inclusion alienated the People's Republic. It continued, nevertheless, to conduct the scientific research that it had planned, but not as part of the cooperative enterprise. The "two Chinas" problem was a great disappointment of the IGY.[92]

The International Geophysical Year, 1957–1958

Most of the political difficulties were related to the Soviet Union and the uneasiness brought about by its participation. After the Soviet nonmagnetic research vessel, *Zarya,* traveled south from Puerto Rico in early 1958, it received a cold reception in Brazilian ports. At the port of Belém, the ship's crew was forbidden to leave the ship, and local authorities insisted that it remain anchored well offshore. Wary Brazilian police even came aboard the ship to ensure that no spies or propaganda material could find their way to land.[93] American scientists tried to ensure that no similar embarrassing situation would arise when the *Zarya* visited American territories. Hugh Odishaw, secretary of the U.S. National Committee for the IGY, wrote directly to the Soviet Desk Officer at the Department of State to assure him that the ship's visit to Koror Island in the Pacific not only was relevant to the Soviet plans but was essential to the overall success the IGY's scientific program.[94] Canadian authorities acted with similar deference when the *Vityaz* visited Vancouver in November 1958; they accorded it the status of government vessel and received it with due ceremony. When American oceanographer Richard Fleming visited the vessel, he reported that there were about sixty-five scientists aboard, including about thirty-five women (the prevalence of women in Soviet oceanology was remarkable given the dearth of them in American oceanography during the same period). The work all appeared to be of high quality, and although it was not very original, "they are doing so much in so many different fields that it makes the U. S. efforts look very puny."[95]

The Soviet Union's October 1957 launch of the first artificial earth-orbiting satellite, *Sputnik,* did the most to shatter the spirit of the IGY. Although the Soviets launched it as part of their planned scientific program, the shock caused by *Sputnik* suggested that it was perceived less as an innocuous international scientific undertaking and more as a direct challenge to the technological supremacy of the United States. Because its successful launch meant a great stride in Soviet ballistic missile technology, no country could consider itself safe from nuclear attack. It also showed that the Soviet Union's ability to develop new technology certainly was not inferior to that of the United States.[96] Most American leaders played the blame game in some form or another with regard to this Soviet "first." Eisenhower put himself in NSF's camp, blaming Congress for making cuts in the American satellite program, "bent as it was at that moment on 'economy.'" The military took its share of abuse, because missile and satellite programs had been divided, perhaps inefficiently, between the Army and Navy. Despite these frustrations, Deputy Secretary of Defense Donald Quarles found a silver lining in *Sput-*

nik, namely, that the Soviets had unintentionally "done us a good turn" by establishing the principle of freedom of international space.[97] Only three months after it began, the IGY's appearance had taken on a Cold War hue that its planners had hoped to avoid.

The Soviets understandably relished their successes. Halfway through the IGY, the Soviet program's chairman, Ivan P. Bardin, announced his country's great strides not only in scientific cooperation but also in scientific leadership. Nearly a hundred scientific institutions had made preparations, he claimed, in a very short period of time; in addition to its astounding satellite program, the Soviets had organized a "mighty Antarctic expedition" and assigned twelve large ships to conduct oceanographic work in all the world's oceans. Bardin pointed to the "daily aid from the [Communist] party and government" as the source of the Soviet Union's sudden world leadership in many scientific pursuits.[98] Although it was clearly propagandistic, the fundamental points of Bardin's message could not be denied. The Soviets had succeeded without much help from scientists of other countries, despite having offered to coordinate their work in Antarctic waters and thus make their successes more international in character. Chagrined by Soviet activities during the IGY, Canadian geophysicist J. Tuzo Wilson later wrote:

> The Soviet State may be likened to an adolescent . . . sloppy in appearance, secretive in behavior, unpredictable, sensitive, and brash. It is also inclined to show off, which is annoying when, on occasion, it happens to be right.[99]

These sentiments perhaps encapsulated many of Western scientists' feelings about the strong showing of the Soviet Union in the IGY.

Many Americans felt that the Soviets could not be trusted to fulfill their obligations during the IGY. After *Sputnik*'s launch, the Central Intelligence Agency (CIA) made repeated inquiries to the National Academy of Sciences regarding the status of Soviet data sharing. The Soviets initially had listed the termination of their information sharing to be the first week of January 1959, officially beyond the timeframe of the IGY. Of course, this would not allow enough time to process data collected during the year. The CIA was concerned that the Soviets would attempt to find ways around the sharing aspect of the IGY, particularly with regard to information on rockets and satellites. The CIA also suspected that the Soviets were trying to continue something similar to the IGY during 1959, but confined to countries within the communist orbit.[100] One particularly distressing detail involved countries with which the Soviets were sharing data. Alan H. Shapley, a scientist in the IGY's Sun-Earth Relationships Section, wrote to an IGY official:

> I notice, as possibly have you, page 14 of PB131632-7 Soviet Bloc International Geophysical Year information, March 28, 1958, that in a December 1957 article China is still listed as one of the countries in the European-Asian region of the IGY program for World Days of Communications. Interesting, n'est-ce pas?[101]

This suggested that, despite the fact that communist China had dropped out of the IGY, the Soviet Union still considered it a partner. In other words, the Chinese were receiving all the information from the IGY, and their own research was being communicated only to the Soviets. In reality, Soviet scientists wanted to extend the IGY for another year for everyone. But British scientists objected, feeling that the eighteen months already planned were going to produce a great deal of data and that time and efforts ought to be directed toward processing and studying it. After some negotiation, scientists adopted a compromise that "saves Russia's face and commits no country to continue extra observations unless it wants to do so." This gave each national committee for the IGY the discretion to judge at which level they would participate in a continuation, and the name for the duration of 1959 was changed to "International Geophysical Cooperation."[102]

Soviet enthusiasm again raised the question, who truly benefited from scientific cooperation? *Sputnik* was an embarrassment to the National Science Foundation, which still awaited the judgment by the government on its 1955 report regarding federal support for international science. That report stated that even with internationally coordinated programs such as the IGY, during which all scientific data would be shared, the United States was by far the best equipped to transform the basic research data into tangible results. Despite the free access to IGY data for all nations, the Americans would "continue to maintain a leading position relative to the speed and efficiency with which the results of basic research are converted into technological advancement."[103] Recent Soviet activities had undermined the entire concept, unleashing a wave of resentment against international science. *Sputnik*'s launch moved Rear Admiral John E. Clark, deputy director of the Advanced Research Projects Agency, to describe the Soviet Union as "coy and superior" during the IGY while the United States "lived in a glass house." In December 1958, the Associated Press reported his saying about the Soviet Union that "[the IGY] has meant a great benefit to itself with little return to others in kind. We should ask ourselves why and we should keep putting this question to ourselves until we make the honest answer."[104]

In some newspapers, including the *Washington Post*, the statement by

Clark appeared next to a report of an exchange of IGY data between the United States and Soviet Union, which seemed to insinuate that American participation in the IGY was part of an unwise exchange that provided more benefits to the enemy than it did to the Americans. Joseph Kaplan was outraged at this kind of bad press, because it undermined future cooperation by staining America's good faith and inviting like charges by the Soviets.[105] Further, some scientists felt that roadblocks to cooperation were being thrown up merely by the anxiety brought about by *Sputnik*'s launch. Soviet scientist Nikita Zenkevich, as late as May 1959, wrote to American marine geologist H. W. Menard that he was having some difficulty acquiring the data from the American Downwind expedition. Menard, writing that American authorities had no right to withhold any IGY data, simply had to send the data to Zenkevich himself instead of waiting for them to clear through proper channels, knowing quite well that the data might otherwise be tied up interminably.[106] Such attacks on the principle of data sharing, brought about by anxiety over *Sputnik*'s launch, threatened to discredit international cooperation.

Despite the doubts raised over international science, the launch of *Sputnik* also ushered in a wave of science advocacy that would last well into the 1960s. This resulted in an extraordinary increase in funds for the National Science Foundation, which was responsible for disbursing money for scientific projects, including the IGY. When it was established in 1950, its budget had a ceiling of $15 million, though it never achieved that maximum before 1956. During that year, the budget more than doubled to accommodate the expenses required for the IGY, and the budgets for both 1957 and 1958 reflected a steady rise in overall budget, from $30 million for 1956, to $53 million for 1957, to $69 million for 1958. After the launch of *Sputnik*, however, Congress allocated an unprecedented $138 million to the National Science Foundation for 1959, which was even more remarkable because the percentage of this budget allocated to the IGY itself dropped from approximately 28 percent to 4 percent. That level of funding for the National Science Foundation continued to rise every year until the mid-1960s. The ramifications of *Sputnik*'s launch on American science and society in general are too many to mention in detail here; the establishment of the President's Science Advisory Committee, the beginnings of the National Aeronautics and Space Administration, and the growth in science education are just a few.[107]

The tenor of the IGY scientists' pleas for continued support of science became far more nation oriented than international. Lloyd Berkner called the rise in funding for the National Science Foundation after *Sputnik*'s launch a "radical recognition of the neglected importance of science and technology."[108] In early 1958, the House Committee on Interstate and Foreign Commerce presided over hearings regarding the extension of scientific activity beyond the IGY. The committee echoed the sentiments of many Americans who perceived Soviet involvement in the IGY as a major challenge, not a testimony to the power of cooperation. Committee chairman Oren Harris noted that "if we are going to be derelict in doing what we ought to do and let some other country step in and take over important work that we have started, then I think it would be a terrible situation."[109] Most of the concern, of course, was directed at rocketry and satellites. But when geologist Laurence Gould and meteorologist Harry Wexler testified, they drew the committee's attention to comparable deficits in other fields, notably in Antarctic science. They hinted that the United States needed to maintain a strong physical presence on the continent to prevent the Soviets from dominating it in the next few years. This was not merely a symbolic problem of having the Soviets take leadership over Antarctic science, but a political one as well, because the territorial questions of the continent had not yet been settled.

In their congressional testimony, Gould and Wexler also raised the issue of Antarctic oceanography, which as yet was an informal activity conducted when possible aboard Navy vessels doing other work. The scientists repeated earlier pleas to have vessels whose primary task was oceanography, so that the scientific work would not suffer at the whims of ship commanders.[110] The Soviets, of course, had been trying to get the Americans and others to join their work there since 1955. Now Gould and Wexler emphasized the threat to American leadership in oceanography. They presented in detail the capabilities of Soviet oceanographic vessels, particularly the *Ob*, which had been operating in the region for some time. They emphasized the efficiency of Soviet work aboard these vessels:

Mr. Macdonald: ... Am I correct in my understanding of both you and Dr. Wexler that the Russians have the single ship that is both an icebreaker and a floating laboratory?
Dr. Gould: A magnificent one.
Dr. Wexler: At least two, Mr. Macdonald. The *Ob* and the *Lena*. Now they have this atomic icebreaker launched a few months ago. I do not know if it is operational but there may be more icebreakers.
Dr. Gould: There are few areas of science in which we have lagged further behind than oceanography.

Mr. Macdonald: But it is a combined icebreaker plus a floating laboratory? *Dr. Wexler:* Also a cargo carrying vessel. The *Ob* and *Lena* went down to the Antarctic a few years ago. Then they pulled into an Australian port and took back 6,000 tons of grain to Hamburg, Germany, which helped pay part of their expenses.[111]

The two scientists painted a portrait of Soviet oceanography that focused on its broad scope, its efficiency, and the level of government commitment. S. D. Cornell, executive officer of the National Academy of Sciences, testified that one of the most pressing subjects confronting American science was the need to fund more facilities for oceanographic research, either private or governmental, which simply "do not exist today in the elaborateness and size that are needed."[112] Such advocacy was widespread among oceanographers. When the *Vityaz* visited Vancouver just prior to the close of the IGY, the University of Washington's Richard Fleming mentioned that if the ship were to visit San Francisco—as was planned during 1959—the event should receive wide press coverage. He hoped it not only would promote good relations with the Soviets but also would "provide strong argument for developing the oceanographic efforts in this country."[113] Oceanographers were finding their own silver lining in the launch of *Sputnik,* namely, that it might provide an impetus for supporting more of their own research.

The most significant legacy of the IGY was not cooperation but, rather, renewed competition with Soviet science. It is easy to forget that it was under this "cooperative" effort that the Soviets launched *Sputnik,* because the space race that ensued appeared to be the opposite of cooperation. At the very least, the IGY warranted a reappraisal of the outworn perception of Soviet science as backward. Even before the launch of the artificial satellite, psychologist Ivan D. London wrote in the *Bulletin of the Atomic Scientists* of the perilous American tendency to assume that Soviet science was too laden with ideology to produce anything valuable. He likened it to Americans reading Norman Mailer's *The Naked and the Dead* and asking, "If that book is a picture of the American Army, how come we won the war?" The same could be said of Soviet science—if it was so bad, why were Soviet scientists challenging American leadership in so many disciplines?[114] Oceanographers needed only compare the size of ships and the extent of national programs in both countries. The comparative weakness of American oceanography was amplified in a 1958 report of the President's Science Advisory Committee noting that a ship designed explicitly for oceanographic research had not been constructed in the United States since 1930.[115] These indications of the state of science, and oceanography in particular, focused upon the dangers of Soviet science and the threat of Soviet leadership in ostensibly international activities.

Although internationalism did not disappear after the IGY, its popularity was constrained by global tensions. The international scientific committees that emerged directly from the IGY (SCAR, COSPAR, and SCOR, devoted to Antarctic, space, and oceanic research, respectively) may seem to have attested to the enduring ideal of cooperation. Some scientists even went so far as to say that the Antarctic Treaty was the first treaty designed to protect a scientific program.[116] But examples such as the (eventual) internationalization of Antarctica were not merely instances of the power of international cooperation. One just as easily could argue that scientific activity in Antarctica had stimulated a new source of tension between the United States and Soviet Union, making the issue of territorial claims an immediate one. At a 1958 meeting between the Department of State and the Joint Chiefs of Staff, military leaders urged that the United States lay claim to all areas in the Antarctic to which it might conceivably have a legitimate claim. One official even suggested invoking the Monroe Doctrine to prevent Soviet bases in parts of the continent in the Western Hemisphere. The chief of Naval Operations, Admiral Arleigh Burke, was concerned that the Soviets had more oceanographic vessels devoted to the area than any other country, including the United States. Internationalizing the whole area was only one of the many options discussed, and it was far from the most popular at the time.[117]

As a result of Soviet activities, international cooperation had lost some of its luster. As most countries were curtailing their activities in Antarctica toward the end of the IGY, the Soviets stepped up theirs. The Central Intelligence Agency reported in December 1958 that the Soviets were sending a submarine (equipped for scientific research), a nuclear-powered icebreaker, and a second whaling fleet to the Antarctic in an effort to consolidate their presence. The Soviets planned at least three more stations on the continent; in addition, they were about to turn over one of their stations to Poland, thus reinforcing the political position of the communist bloc.[118] The submarine alone had clear propagandistic shock value on an international scale, particularly for the Australians, who feared the Soviets were establishing a submarine base in Antarctica. The whaling fleet stirred up alarm among Norwegians, who feared that Soviet economic exploitation of the area would result in Soviet domination of the industry. These whaling vessels could be used to expand scientific research or, worse, to extend military power secretly into the Southern Hemisphere, by using them to transport satellites or missiles. In addition to these potential threats, there was evidence that the Soviets had been withholding some scientific results of their Antarctic activities, particularly hydrographic charts, echo soundings, and other oceanographic data. At the very least, many Americans felt, the U.S. scientific program had to be stepped up to compete with the Soviet one, which according to a 1959 State Department report presented "an immediate and long-range challenge to the US scientific and political position."[119] If the Soviets indeed were using the IGY to initiate such activities, the idea of international cooperation, despite the success of the IGY, was even more open to criticism than before.

4 THE NEW FACE OF INTERNATIONAL OCEANOGRAPHY

Despite its successes, there were many challenges to cooperation during the IGY, the most obvious being the launch of *Sputnik* only three months after it began. This momentous event cast doubts upon the conceptual foundations of scientific cooperation as a whole, in particular the assumption that the gains of cooperation outweighed the risks. But the IGY presented opportunities as well. In the United States, various government agencies sensed that the IGY was a chance to develop not only a unified policy for cooperation but also a common forum for the discussion of America's national efforts in oceanography. This led to the creation of the National Academy of Sciences Committee on Oceanography (NASCO). In the wake of *Sputnik*'s launch, this organization smoothed the efforts of American scientists, military leaders, and policymakers to develop a coherent national plan to compete with the Soviet Union. In addition, the IGY provided a vehicle for establishing permanent international bodies devoted to oceanography. Prior to the IGY, no organization crossed political boundaries and disciplinary boundaries to encompass oceanography broadly conceived with membership broadly inclusive. There was no forum, for example, in which a Soviet marine zoologist could debate with a British physical oceanographer or an American marine geologist. When the IGY officially ended after December 1958, many American scientists would withdraw from the international arena, directing their oceanographic plans toward national goals that virtually ignored cooperation. However, a number of prominent Americans continued to promote international science. They joined with other world leaders of marine science, such as George Deacon (United Kingdom), Håkon Mosby (Norway), Carl-Gustaf Rossby (Sweden), Günther Böhnecke (Federal Republic of Germany), Vladimir Kort (Soviet Union), and Anton Bruun (Denmark), wishing to establish a more permanent working group or committee to plan cooperative ventures beyond the IGY. These scientists helped to define the parameters of international cooperation in oceanography over the next several years.

Oceanographers began to promote cooperation not merely as a way to ensure scientific growth, or even to ease international tensions, but also as a means to address problems of society. This was a crucial step, as it signaled the decline of the rhetoric of "easing tensions" and it institutionalized a development-oriented agenda for international oceanographic research. In the IGY, easing tensions had never been the object; scientific research always had been, and the good relations were to be a possible byproduct. No one gave support to the IGY on the *condition* that tensions had to be eased in the process. In adopting a more development-oriented approach, scientists were crossing a boundary, consciously or not, that had the potential to compromise basic research more than ever. In addition, in courting governments, oceanographers moved toward *intergovernmental* cooperation, which stood in stark contrast to the international cooperation between individual scientists that prevailed in existing international scientific unions and that the IGY was supposed to maintain.

The present chapter traces the evolution of oceanographic cooperation from the close of the IGY to the implementation of the International Indian Ocean Expedition (IIOE) in the early 1960s. Initially American scientists sought to bring men and women of several nations together for a common cause, following the rhetoric of the IGY, by establishing the Scientific Committee on Oceanic Research (SCOR) and by sponsoring the First Oceanographic Congress of 1959. In planning its first major project, SCOR expanded upon existing cooperative relationships by including some developing nations. In handing over the coordination of the IIOE to the Intergovernmental Oceanographic Commission (IOC), SCOR allowed both oceanographic cooperation and the development-oriented agenda itself to be institutionalized. By moving the IIOE and all future projects to an intergovernmental forum, SCOR succeeded in establishing a permanent body to promote and manage the world's oceanographic programs, but at the same time abandoned hope of having a scientific body detached from politics.

UNESCO AND MARINE SCIENCE IN THE 1950S

One legacy of the IGY was the boost that it gave, at the international level, to establishing a semipermanent oceanographic organization for coordinating research—the Special Committee for Oceanic Research ("Special" soon became "Scientific"). Such efforts had faced several starts and stops since the late 1940s. The first Joint Commission on Oceanography under ICSU was formed in 1948, to plan cooperative research and establish common nomenclature, under the guidance of scientists in the International Association of Physical Oceanography (IAPO). It had been a failure, largely because of opposition to its wide scope, and a new commission was formed in 1951 with almost exclusive attention focused upon research in the deep ocean. One of its great successes was the establishment of the international journal *Deep-Sea Research,* first published in 1953.[1] Two British scientists, John D. H. Wiseman and C. D. Ovey, led that commission, but it included in its membership the American scientists Roger Revelle, Mary Sears, and Harold Urey. Revelle's efforts at Scripps were an object of special praise from Wiseman, who felt that including British, Australian, Swedish, and French scientists on Pacific Ocean expeditions during 1952 had established a positive precedent for international cooperation.[2] Personal contacts between scientists of different nations, he felt, were crucial in order to use science to foster broader international understanding. It appeared that with Revelle at the helm, at least one of the two leading American oceanographic institutions would be committed to promoting international activity despite global tensions.

The path to establishing a lasting international body for oceanography amid Cold War tensions was circuitous; perhaps surprisingly, the decisive conflicts preventing consensus were among Western scientists, not between Americans and Soviets. For example, the Joint Commission on Oceanography in 1953 outlined an ambitious role for itself in international oceanographic research, seeing its role as encouraging cross-disciplinary participation, organizing symposia, promoting publications, and making efficient use of existing research vessels and laboratories. But many physical oceanographers urged ICSU to disband the body in October 1953. It did so, leaving only the journal to survive. The friction among scientists was strongest between the Americans (Revelle and Urey) and the British (George Deacon and IAPO president Joseph Proudman). The latter two felt that the commission had overstepped its boundaries in making extensive recommendations on international cooperation, without due consideration of the role of IAPO. To replace the commission, in 1954 ICSU appointed a smaller committee to advise on problems for which two international unions (International Union of Geodesy and Geophysics and International Union of Biological Sciences) might cooperate. Revelle was invited to join this committee, and eventually it issued a report describing itself along the lines of the recently killed joint commission, but making sure this time to include a caveat that it would address only problems that could not be addressed adequately by IAPO.³ This territorial conflict, between physical oceanographers and others who wanted a body defining the field more broadly, would plague international marine science for years to come.

With these experiences behind him, by the time Revelle in 1955 became involved in planning and implementing the oceanographic program of the IGY, he had firsthand knowledge of the difficulties in trying to start up a standing international body. Many prominent oceanographers embraced the idea of cooperation only to the extent that it facilitated scientific activity and made it more efficient. They were wary of a broadly defined "marine science" stepping in to confuse boundaries and diffuse effort. What Revelle and others ultimately did with SCOR was equally threatening, because SCOR would define its jurisdiction widely *and* try to sell science for its social ramifications. Many prominent physical oceanographers feared that neither strategy would serve their interests.

To appreciate fully the process of resistance to "marine science," one must look to the short-lived existence of another body, formed under Unesco. During the planning stages of the IGY, which was conceived only as a "one-shot deal," the foundations for a long-term organization were being laid under the auspices of Unesco, by many of the same actors who were spearheading the IGY. A 1955 meeting in Tokyo provided a unique opportunity for scientists from Europe, the United States, and Japan to meet among themselves and with scientists from other marine institutions in the region. It was a forum to make new contacts and for the well-supported scientists to commiserate with each other while offering encouragement to their colleagues and suggesting ways to improve and increase their efforts. For example, contacts made at the meeting encouraged Klaus Wyrtki, a scientist working for the government of Indonesia, to file a request of funds for an oceanographic survey of the Java and China seas. Many of the scientists at the meeting concluded that a permanent body could do a great deal to help scientists like Wyrtki promote oceanography in their own countries. Roger Revelle, Lev Zenkevich (Soviet Union), George Deacon, Koji Hidaka, Marc Eyriès (France), and Anton Bruun all were among the first members of the International Advisory Committee on Marine Science (IACOMS), formed under Unesco toward the end of 1955. Officially, each was chosen to represent a particular branch of science, not his own nation; however, given the distribution of representation, clearly IACOMS gave some consideration to national origin.⁴

This body managed to avoid jurisdictional problems initially because its entire purpose fell outside the purview of existing organizations. The purpose of IACOMS was not to coordinate scientific action among those most capable of doing it, as in the case of IAPO or the IGY programs, but rather to find ways to cultivate interest in marine science in countries where it was weak and to make such activities part of Unesco's Marine Sciences Programme. Deacon wrote that there was a vast difference between Unesco and the International Council of Scientific Unions. One body had to worry about development and regional representation, while the other "ought to have the most active and stimulating men from each aspect of the subject wherever they live."[5] Thus Unesco's cooperative programs would be devoted explicitly to raising the standards of oceanographic work in less developed countries, with the advice of organizations such as the Food and Agriculture Organization (FAO), the International Council for the Exploration of the Sea, and scientific bodies such as IAPO.

Initially, Unesco's Marine Sciences Programme did provide funds outside this purpose. For example, it provided funds for a current-measuring program in the North Atlantic by the United States, the United Kingdom, and Norway. It also gave a grant to ICES, vaguely rationalized because of its liaison with FAO, and a grant to an IAPO symposium on the basis that it might be "helpful in attracting fresh interest and support to marine research," both of which technically fell outside the parameters that IACOMS had created for itself.[6] These actions no doubt resulted from the views of scientists like George Deacon, who felt that Unesco also should lend support to existing organizations and help the best scientists conduct their work. But here Deacon was on shaky ground, because in truth he did not want to mix the purpose of Unesco with that of top scientists. Anticipating the formation of IACOMS, Deacon expressed his reservations about the body's ability to ensure the health of the science as a whole. He urged his colleagues not to forget the needs of top oceanographers, "lest the admirable work which is being done to widen interest in oceanography should prevent concentrated attacks on some of the outstanding problems." He felt that scientific progress depended upon a far better understanding of oceanic processes, and only the existing scientific communities were poised to do the necessary experimental and theoretical work, not scientists from developing countries. Such work would require a concentration of effort by the best scientists, "since one expert among a lot of newcomers is not sufficient to build up a science."[7]

In Deacon's view of the special mission of Unesco was the seed of future discontent among scientists about the mode of support for international research. He feared that a new organization, by focusing on development, might blend physical oceanographers with marine biologists. At that time, IAPO was part of the International Union of Geodesy and Geophysics, while marine biologists were more closely related to the International Union of Biological Sciences. Deacon, and many other oceanographers wanting to avoid the dominance of fisheries research, wished to keep it that way.[8] He stated that he would like to see the science itself move forward and that the best way to achieve this would be to let competent scientists attack specific problems facing the discipline. This point would come to a head in Deacon's general frustration with intergovernmental oceanography during the 1960s.

One of Deacon's like-minded colleagues was the Norwegian Håkon Mosby, also a physical oceanographer. Like Deacon, he perceived the risks from the start, and in 1956 he complained about this new mode of support for developing countries. He wrote to Unesco in preparation for the first meeting of IACOMS, insisting that "the simplest, the most effective and therefore also the cheapest way for Unesco to increase our knowledge of the sea must be to further oceanographic research within countries where such research does already exist." If Unesco must help scientists from other areas, then those scientists should come and study in advanced universities. The Norwegians even offered to create a training center at the Biological Station at the University of Bergen. One of its professors, Hans Brattström, felt Bergen was the natural setting for such ventures, as it was "of old a well-known international centre of oceanography," due to the large number of institutions attached to the university with the ocean as the focus of research. However, the views of both Mosby and Brattström were outweighed by those of Harald Sverdrup, former director of Scripps, who now headed the Norwegian Polar Institute. Sverdrup took a line quite similar to his American colleague Revelle that Unesco should use its resources "for encouraging and assisting the development of the Marine Sciences in the countries in which they now are receiving only small attention."[9]

Despite the reservations of a number of oceanographers, IACOMS focused on less developed scientific communities in an effort to widen oceanographic research throughout the world, finding disciples of marine science where there had been few or none before. One way it did so was to hold professional meetings in developing countries in which there existed some body of potentially interested people, so that the meeting itself might help stimulate action in support of work in those countries. Its first meeting was held in Lima, Peru, in 1956, providing opportunities for scientists throughout

South America to interact with their colleagues from countries with better facilities, better funding, and higher levels of commitment from their governments. More important, it sought to find ways to improve opportunities through cooperation. With limited resources, less developed countries could only hope to achieve something significant if they acted in concert. To Klaus Wyrtki, this was evident in the mentality of future IACOMS members during the 1955 Tokyo meeting. He wrote, "We have heard from Unesco, that only plans on a super-national level have a chance of support." It would not be enough for Wyrtki to do good work within Indonesia; IACOMS's goal was not really to aid specific scientists or countries, but to help them to pool their resources, to accomplish something that would have been impossible to do alone.[10]

The Danish marine biologist Anton Bruun was sent by IACOMS on a journey to Southeast Asia to survey its capabilities in marine research and to make recommendations on cooperative efforts. Bruun, who had traveled to the region in 1929 as part of an expedition aboard the *Dana,* now characterized it as "the most challenging in the whole field of oceanography." Its area was vast, from Taiwan to Australia and from Sumatra to New Guinea. The region was more complicated than any Bruun knew either at sea or on land, in terms of its geology and its variety of flora and fauna, the extent of which was completely unknown. "This challenge to all branches of oceanology," Bruun wrote, "becomes still more pressing when one thinks of the hundreds of millions of people of the adjacent countries who could get a better living if the resources could be explored, evaluated and utilized." Here was where Unesco would be of service, to help these people build up their scientific strength in order to help themselves.[11]

After visiting the Philippines, Hong Kong, South Vietnam, Singapore, Malaya, Indonesia, and Thailand, Bruun was distressed by what he felt was the general inefficiency of scientific activity in the region. Bruun looked to Indonesia and Hong Kong as its potential backbone of marine research. The University of Hong Kong was a possible font of scientific advice for training programs. Marine research had been pursued in Indonesia for more than a century, with a good amount of it published. These observations, Bruun noted, "really form the most important founding-stone of our knowledge of the whole huge region." The facilities were fine, and there was at least one geophysicist doing research there (Klaus Wyrtki). Like other countries in the region, however, Indonesia and Hong Kong did not make efficient use of their resources because of the lack of trained personnel. In the Philippines, the level of training was generally insufficient for advanced scientific activity. Bruun was exasperated to find one physical oceanographer who had been trained in the United States now working in an agricultural bureau. So even though someone had the appropriate training, he was not doing physical oceanography. As for marine biologists, the damage done to the collections in museums and libraries during World War II had never been addressed. In general, the staff concerned themselves primarily with fisheries research, and scientific exploration of nearby waters had not been done in earnest since before the war. In South Vietnam, the facilities were good, "among the very best in the whole region," but there was no ship regularly available for research. In Singapore, there were good facilities and a ship for research, but the government professed little interest in anything but fisheries research. In Thailand, there were willing scientists and two small research vessels, but the scientists were poorly trained. In all these countries, there were the ingredients of successful marine research, but no country by itself possessed all the ingredients. To sum up his assessment, Bruun proposed a "utopian institute of oceanology for the Far East." It would pool the qualified personnel from all countries, put them in a location with the best facilities, and provide them with the region's best research vessels. Acting in concert, the scientists of the region could vastly improve their efficiency and do a lot more science.[12]

Bruun's assessment strongly favored the biological sciences. The study of the region's seas, Bruun felt, was still "in the exploratory stage, particularly in biology" (although as a marine biologist, Bruun was likely more sensitive to this subject). He insisted that the fisheries research must be accompanied by studies of more complicated scientific problems of hydrography and biology. He suggested that the fisheries departments of each country loosen their "too-firm grip" on marine scientists, to allow them to conduct basic research. Also, the countries should be encouraged to expand their collections. "Every single country must have a National Museum of Natural History; this is a need too often overlooked." One needed to remember that unlike physicists and chemists, who could apply their knowledge universally, biologists depended heavily on guides of local flora and fauna, as well as on collections. In addition, more professors were needed; only Singapore and Hong Kong had the people to provide the kind of training necessary for productive research. With all these elements, a proper atmosphere for marine science could be created. Bruun complained that "far too often the young students, who have been abroad and enjoyed such facilities, give up in despair when they face the difficulties in these matters in their home country." Research had to be expanded, particularly in coun-

The New Face of International Oceanography

tries such as the Philippines and Indonesia, whose home waters were vast. "This may sound utopian, but if these countries want the same level of knowledge of their own natural surroundings as advanced modern civilization it must be done."[13]

Bruun's trip had an immediate effect on the mentality of IACOMS in its support for research. One of his suggestions, to equip an international vessel for oceanographic work, was particularly intriguing. Such a vessel would have the benefit of being both a research and a training vessel, and there would be no question of its work being relevant to Unesco's purpose, because it would be a Unesco ship. The idea remained high on Unesco's agenda as a potential avenue for scientific cooperation, even beyond the life of IACOMS and into the early 1960s (ultimately, the idea would be abandoned). For the time being, the most direct way to cultivate scientific development, IACOMS reasoned, would be with training courses. Already there were a number of fellowships available, but these would take students out of the region to study in advanced institutions. The members of IACOMS felt that "more effective results can be obtained by having the students trained within the region, where the conditions of work and the problems faced are sometimes different from those that the workers would be encountering in well established laboratories." Thus they hoped to address one of Bruun's observations of the people he encountered on his visit, namely, that they often were overwhelmed by the multitude of obstacles in their work environment, which typically posed problems (poor work facilities, few library or specimen resources, lack of ships, no political support) that did not exist in the laboratories where they were trained. The psychological factor, of returning to such an environment after experiencing a far more positive situation, might be eliminated if training was kept within the region.[14]

Over the next couple of years, Unesco put these recommendations into effect. Unesco's Masao Yoshida praised Bruun for sowing seeds of interest that no doubt would last for years to come, but he cautioned that Unesco itself "should try not to let their hopes die out" and should actively nurture its contacts with them.[15] It sponsored scientists from advanced institutions to visit locations in South and Southeast Asia, and it convinced a number of countries to participate in regional training courses. But IACOMS did not limit its activities to Asia. Its first significant training courses, one in Bombay in 1958 and the next in Nhatrang in 1959, were followed by a 1961 course held in Morocco. The course was attended by twenty individuals from Kuwait, Lebanon, Libya, Egypt, Sudan, Syria, Tunisia, Turkey, and Morocco. In all the courses, the object was to teach things that the participants could then go and teach in their own countries. They focused on instruments that could be acquired at low cost, so that countries with small budgets could begin rudimentary oceanographic activities. Their main preparatory task was to identify the simple instruments, gather the most straightforward handbooks and charts, and find good films to show the students. The director of the Moroccan course, British oceanographer J. N. Carruthers, later reported that their intention was not necessarily to instruct the principles of oceanography but rather "to arouse interest enough in things which can be done with very modest resources, to give grounds for the hope that some attention would be paid to the marine sciences in the countries from which the students came."

Unesco embraced marine science broadly defined. Its training courses did not emphasize any particular discipline such as physical or biological oceanography, and indeed they did not have a fixed curriculum at all, given the need for flexibility with the varying range of student backgrounds. Carruthers summed up the views of the course's tutors after its completion: "If Unesco can continue to stage such training programmes aimed at producing interested young men and women who will be disciples of the marine sciences in their own countries, much of value will be achieved." Ambitious university teaching was not their objective; instead, they focused upon stirring up interest in marine science and promoting its value to these countries.[16] Unesco hoped that training courses might help identify candidates for fellowships and thus contribute to building a cadre of marine science experts in Asia. In 1958, it set up a "refresher course" in marine biology at the University of Bombay.[17] The following year, it convened a meeting with representatives of nine Asian nations to discuss ways to expand marine science in Southeast Asia through international cooperation. The immediate prospect was a six-month (later shortened to about four) training course at Vietnam's Institut Océanographique in Nhatrang, in which scientists could familiarize themselves with techniques of marine research.

Despite these actions, IACOMS was looking to do something even more ambitious. At IACOMS's second session, in 1957, the members resolved to expand Unesco's Marine Sciences Programme to include a major project, to begin by 1961. Exactly what form this project would take, however, was unclear.[18] The international vessel appeared to be an ideal way to combine resources and manpower, while attracting attention and support by flying the flag of the United Nations. Revelle particularly emphasized the "attention" aspect, which he felt would come naturally from the ship visiting various ports along its journey and showing the United Nations flag, a symbol of international cooperation. The ship could travel to areas that were not already being explored for commercial exploitation or to areas where deep-sea scientific expeditions were not being organized by countries with suitable vessels. This would be no ordinary ship, but an oceanographic vessel built for research purposes, which was very uncommon at the time. Members of IACOMS judged that there were only five such vessels "in the true sense of the word, fully engaged in transoceanic scientific expeditions." France had one, the United Kingdom had one, the Soviet Union had one, and the United States had two. The Unesco vessel would belong to the world in general and could conduct work in service to mankind. Henry Stommel suggested putting it to work on monitoring radioactive waste disposal in the oceans or on the contamination effects of nuclear weapons testing, both "of such obvious and deadly importance that it obviously should not be entrusted to any nation individually."[19]

In addition to serving such worldwide needs, the vessel would serve the scientific interests of the developing world in particular. It would give its scientists an opportunity to conduct world-class scientific work by pooling their resources and to do it in a region that was most relevant to them. Members of IACOMS were thinking specifically of countries in Southeast Asia. The vessel, the committee believed, "would without doubt contribute to rectifying the state of our knowledge of such areas as the Indian Ocean which, according to Dr. R. Revelle . . . 'is less than what we know about the surface of the moon.'"

Revelle's reasoning touched on the reality that oceanographers were fast learning, that they had to compete with more glamorous scientific pursuits that held the public's attention far more than oceanography did. Also cognizant of this problem, Harald Sverdrup warned that the proposed ship, despite being a good idea, might run into financial difficulties. He pointed to the experience of CERN, the European Center for Nuclear Research, which had cost many of its participating countries even more individually than they had spent prior to its creation. Sverdrup doubted that countries would go along with that sort of increase for something like oceanography. "Marine research," he warned, "does not have an appeal to the public—or Governments—in a manner comparable to that of nuclear research." Sverdrup's compatriot, Håkon Mosby, again expressed his dismay at the ever-increasing interest in diverting interest and funding toward helping scientists in the developing world, when clearly the industrialized countries had a large enough task trying to find support for their own research. He repeated his previous sentiment that scientific support was best given to countries that already had well-developed research communities. He also expressed his displeasure that people tended to assume that all countries with advanced scientific institutions were as well off as the United States or the Soviet Union. "It must be remembered," he wrote regarding the proposed international vessel, "that the marine scientists in countries where oceanography is highly advanced are not necessarily spared the difficulties in financing their national research projects."[20]

With the support of members such as Revelle and Bruun, IACOMS nevertheless helped to organize the first major oceanographic project in the Southeast Asian region, namely, the Naga expedition. Marine biologists identified the area between the Indian and Pacific oceans as containing a greater variety of marine species than any other region in the world. This, combined with the seasonal current variations due to the monsoons, and the area's complex geology, made an oceanographic expedition particularly enticing for scientists. A number of deep-sea expeditions had taken place there in the past, notably by Denmark, the Netherlands, the United Kingdom, and more recently France, Japan, and the United States. "These efforts," IACOMS determined, "have only served to bring into relief the vastness and complexity of the scientific information remaining to be assembled." A systematic scientific investigation, moreover, was required prior to any rational program of fisheries exploitation.[21] The United States had planned a two-year series of cruises, conducted by Scripps to collect data in the South China Sea and Gulf of Thailand. Bruun, a Dane, led the expedition. Both Bruun and Revelle agreed that this would be an ideal opportunity to bring scientists in Asia aboard American ships. They planned to cooperate with local institutes and coordinate the expedition with the Nhatrang training course, to give the students practical research experience at sea. In return, the Asian representatives agreed to seek out ways to equip and staff their own laboratories, in order to make full use of the cooperative project, and to help defray the expenses of having a joint research ship. Cooperation in the name of economic development was under way.[22]

THE CREATION OF SCOR

Despite these successes in providing training courses, conferences, and even the Naga expedition, IACOMS fell on the periphery of most scientists' interests. Cooperative programs typically ran into difficulties with men such as Deacon and Mosby, who wished to use funds to tackle scientific problems and who felt that diverting money to the developing world would not accom-

The Scientific Committee on Oceanic Research (SCOR), 1957. From left, first row: Lev Zenkevich, N. Marshall, Günther Böhnecke, R. Fraser, Anton Bruun, Y. Miyake, Norris Rakestraw. Second row: P. Galtsoff, Columbus Iselin, Roger Revelle, Maurice Hill, Håkon Mosby, Luis Capurro, George Deacon, E. Nielsen. Courtesy Woods Hole Oceanographic Institution Archives

plish this. At the same time, there seemed to be great financial opportunities for scientists who could demonstrate the social relevance of their research, and helping developing countries had great potential in this regard. No one knew this better than Roger Revelle, who dealt with problems of funding not only in IACOMS but also in the highly successful international program of the IGY and at his own institution, which had been balancing the needs of science, the Navy, and fisheries research since he took over its directorship in 1950.[23] Given his experience in the international arena, Lloyd Berkner approached Revelle about setting up a body to continue international oceanographic investigations beyond the IGY. Because of the presence of Soviet scientists, designing a new organization promised to be even more daunting than his past experience. He and other prominent oceanographers met in Göteborg, Sweden, in January 1957, to assess the future of international cooperation in oceanography.

By insisting on a strategy of framing SCOR's agenda in terms of practical problems shared by all humanity, Revelle was instrumental in shaping the underlying theme of SCOR's work. Only with this in mind can one understand why the initial problems of SCOR were conflicts among Western scientists, not between the Americans and Soviets. The Soviet Union sided strongly with the United States in supporting SCOR. Vladimir Kort wanted the organization to agree upon basic problems, to coordinate work, and to standardize research methods. Revelle complimented the Soviet delegates at the Göteborg meeting for their effort to outline specific problems in a report that they had brought with them. Working closely together, he said, scientists from many countries could come to a consensus. Many of the other delegates adopted a similar attitude, emphasizing the importance of formulating specific problems or tasks for oceanographers to address over a five-year period. For Soviet biologist Lev Zenkevich, SCOR seemed the perfect body to investigate a host of problems in marine science, a term to be understood broadly. He suggested emphasizing the importance of general scientific problems, such as developing a picture of oceanic history, determining the circulation of water masses, and studying the oceans' biological structure, all general problems in marine science. Revelle agreed that a broad interdisciplinary approach to defining oceanographic problems was wise; in addition, he suggested that SCOR explicitly incorporate social benefits into the problems, to emphasize that cooperation addressed the larger problems facing not just science but all mankind. For example, Zenkevich's research agenda could be rephrased as developing a picture of climate change over time, studying the effects of radioactive wastes in the oceans, and assessing the productivity of food resources in the sea.[24] SCOR's role could be to help scientists bring new techniques to bear on these practical questions. For example, physical oceanographers could use natural radioactive substances, such as radiocarbon, to observe mixing of water masses and to track current flows. Also, scientists were developing free-floating buoys and new current meters to study deepwater motion. Other techniques included better salinity measuring methods, new uses of heat flow calculations, new ideas for sediment studies, and even the possibility of introducing large amounts of artificially radioactive substances, on the order of tens of thousands of curies, into the deep ocean to see what would happen to them.[25] All of these promised to address scientific problems, but in SCOR they were framed to provide as much social relevance as possible.

The New Face of International Oceanography

The early meetings of SCOR sought to forge a sustainable organization that could point to an overall direction for oceanic research—broadly defined to include both physical and biological studies—on which the leading scientists could agree and then return home with the imprimatur of the international scientific community, to bolster their cause in gaining support from governments. The organization could "serve as a sounding board to emphasize the economic and social importance of greater knowledge of the oceans and thereby assist marine scientists in different countries to obtain support for their work." It could also help coordinate cruises, point out areas where work needed to be done, and facilitate the exchange of techniques, samples, data, and scientists themselves, all of which could drastically improve the efficiency of marine science as a whole. This was especially important for deep-sea research, the most expensive form of marine science. Already SCOR envisioned its first major project, a combined scientific assault on the Indian Ocean, which would help scientists answer their pressing questions (this was particularly true of physical oceanography, because of the seasonal variations in current), and also it "would have a lasting effect in encouraging and developing the marine sciences and fisheries in those countries." As a means to promote oceanographic research with greater success, SCOR combined the goals of organizations that hitherto had been separate.[26]

At the first meeting in Göteborg, the attendees were far from united on the purpose of the new body. Because the scientists had spent the previous few days attending meetings of IAPO's Working Group on Oceanography to plan for the IGY, many participants were unsure what purpose SCOR might serve, beyond that already covered by IAPO. Once again, old jurisdictional disputes arose. Deacon was skeptical about the organization, disliking the idea of creating a group to deal with oceanography broadly conceived, which threatened to make the work of physical oceanographers more difficult. He wrote to IAPO colleague Börje Kullenberg that Unesco was harmless because it dealt with attracting interest where it previously had not existed. "I am not so sure about SCOR," he wrote. "It will probably do its best to make all the oceanographic observations among the stars above, or in the earth beneath, and none in the waters in the midst of the firmament." The IAPO, he added, "surely must speak up for the water."[27] He later wrote that he feared the post-IGY plans "might allow a small group of people to use the prestige of an International Organization to browbeat the rest of us into doing things we are not very keen on." The IGY, he felt, was fine because it did a lot to publicize the subject, not because it was really the best use of funds for advancing science. Senior British oceanographer Arthur T.

Doodson shared Deacon's sentiments, feeling that SCOR's terms of reference appeared to go far beyond their appropriate boundaries, to interfere with the work of existing bodies. This body of course was IAPO, which Deacon admitted perhaps was not "sufficiently enterprising to satisfy present needs" but which certainly could rise to the task and do the job as well as any new organization.[28]

The ambitious scope of SCOR made it the first major international organization committed to understanding the oceans as a whole, defining the subject more broadly than many were prepared to do. It sparked anxiety that such a reorganization of scientific activity might mean that the support and coordination for physical oceanography would diminish in favor of biology, simply because of the fisheries aspects. Geophysicist Jean Coulomb, director of France's Centre National de la Recherche Scientifique (CNRS), repeatedly stated during international meetings that the main gaps needing to be filled in marine science were in physical oceanography and that a lot of work already had been done in marine biology. Coulomb insisted, for example, that Unesco not support science simply for its fisheries applications. When the development-oriented approach was appropriated by SCOR under the banner of the International Council of Scientific Unions, a body typically unconnected to such an outlook, objections from likeminded scientists only increased.[29]

The genuine roadblocks to SCOR came from the scientists who saw no need for a new organization incorporating a broader vision. Primarily concerned with physical oceanography, the objecting British scientists were quite satisfied with the international arrangements already provided by IAPO. At the Göteborg meeting, Danish oceanographer Anton Bruun complained that these physical oceanographers were sluggish in supporting the creation of SCOR, whereas the marine biologists and meteorologists were anxious to establish a body that would improve the as yet inadequate cooperation between physicists and biologists.[30] The culprits were Proudman, Doodson, and Deacon, who all initially were unenthusiastic about the formation of SCOR. The ICSU's secretary-general, Harold Spencer Jones (also British), speculated that this was because of their narrow vision. They did not appreciate oceanography as a science beyond its physical aspects, and perhaps they took the creation of SCOR as a "reflection on the IAPO . . . which has been ineffective in achieving anything of importance." Jones expected that these men would expend considerable energy stirring up dissent against SCOR, and he urged ICSU president Lloyd Berkner to stimulate enthusiasm for it

The New Face of International Oceanography

in the United States so that the British view might be overcome. Wanting to create SCOR, the ICSU leadership criticized the views of British oceanographers as being "narrow and parochial and not representative of oceanographers as a whole."[31]

Berkner was not surprised that the British oceanographers opposed SCOR. Agreeing with Jones, he wrote home to the National Research Council:

> This is part and parcel of the old professionals who insist on dragging their heels. The general state of the science of oceanography at the present time is simply a scandal, and the persistent conservatism of the British group in the face of this situation seems to me to border on the outrageous.[32]

Berkner wrote back to Jones and insisted that ICSU not be influenced by the reactionaries, who he believed already bore a great deal of responsibility for the neglect that oceanography had suffered to date. He emphasized the need to support the actions of others, particularly Revelle, who were trying to improve the dismal state of oceanography at the international level and to establish a body to steer international cooperation to be more inclusive, in terms both of disciplines and of nations. He stressed that if ICSU did not press forward with SCOR and take leadership in oceanography, a livelier group would do so; he did not want ICSU to let SCOR go.[33]

Revelle, not wishing to alienate his British colleagues permanently, moved to ameliorate these problems. First of all, he explained that there was no conspiracy; the actions to define SCOR's role were not yet a fait accompli and would require input from all participants. In addition, he assuaged IAPO's fears by claiming his own hope to strengthen and extend IAPO's work by using SCOR to coordinate more closely with other organizations and to gather financial assistance from national oceanographic committees. This naturally would provide even more funding for IAPO, not replace it. Writing to Maurice Hill, one of Deacon's colleagues at the National Institute of Oceanography, Revelle explained that "our experience with the IGY and our work with the Japanese in the Pacific demonstrate the importance of international co-operation on a broad basis." The ability to connect projects to interdisciplinary bodies and to practical aims such as fisheries research had enabled a large amount of research in various fields. Only a broad-minded approach made it possible. Future success would hinge upon the enthusiasm, sympathy, and effort of scientists of all backgrounds and outlooks. This was what SCOR would represent. Encouraging Hill to show his comments to Deacon, Revelle injected some flattery into the argument. "George Deacon," he wrote, "is the most effective and dedicated promoter of the marine sciences I know. I hope with all earnestness he can be persuaded that SCOR will help in winning his own lifelong crusade."[34]

Although Deacon ultimately conceded the need for SCOR for precisely the reasons delineated by Revelle, he remained skeptical. Such international cooperation was at least ten years off, he believed. It might be useful for gathering data on such subjects as mean sea level, but to attempt to coordinate the study of ocean circulation or other big problems was simply ridiculous. Advances in theories and methods would be too rapid, and scientists "should not be tied down to programmes arranged long beforehand and selected because they are within everyone's capabilities." Organization should be more lax, not more strict, to provide flexibility. "No-one can organize oceanographic research," he wrote to Maurice Hill a month after Revelle wrote his fawning letter, because "what we do next is largely determined by the interests of the men we can get and by unpredictable advances in methods and techniques from many sciences." Deacon did not care that ICSU could point to American approval. He argued that ICSU wanted an administrative organization simply because it had one during the IGY, but the only reason it had worked during the IGY was because the scientists liked the science, not because they liked the organization. "If you go on upsetting the scientists," he wrote, "you may have an administration which satisfies ICSU and science which satisfies no one." Deacon eventually supported SCOR, but insisted that it not be overly bureaucratic and that scientists themselves should be given the responsibility of coming up with their own plans. If SCOR wanted to try to coordinate them, that would be fine. These sentiments, strongly worded in SCOR's infancy, would resurface consistently during the life of the Intergovernmental Oceanographic Commission during the 1960s.[35]

THE FIRST INTERNATIONAL CONGRESS ON OCEANOGRAPHY, NEW YORK, 1959

An important step in creating an atmosphere of acceptance for a broadly conceived discipline of oceanography was the First International Congress on Oceanography, sponsored by SCOR and IACOMS to take place in New York in 1959. Both bodies were interested in more than coordinating research by individual nations. Personal interaction still was an important aspect of international cooperation. Together they helped bring about the first major post-IGY event that brought oceanographers together to discuss the state of the

Scientists gather at the First International Oceanographic Congress in New York, 1959. From left: Roger Revelle, W. Maurice Ewing, Edward Bullard, Edwin Hamilton. United Nations photograph

field, and Unesco convinced the United Nations to hold it in its headquarters. Although organized by the American Association for the Advancement of Science (AAAS), the congress would not have occurred without Unesco cooperation and financial backing. The AAAS initially planned a modest meeting, but after oceanographers led by Woods Hole scientist Mary Sears took over the planning, the congress blossomed into a full-scale meeting to last two weeks. The congress not only included scientific sessions but also had field trips, receptions, and other events. This first congress was unprecedented in scope, with a budget of more than one hundred thousand dollars. The IGY had stimulated interest, there was increasing attention to oceanic science by government, new discoveries in marine geology were challenging old ideas, and IACOMS and SCOR were preparing to plan the future of international oceanography. The AAAS found the money without difficulty, for, as its executive officer Dael Wolfle later noted, "it was the right time for a large international meeting on oceanography . . . a time when the world's oceanographers wanted to get together to review progress and to think about what should be done next."[36] The congress planners wished to provide a major forum for individual scientists to exchange ideas—in the style of SCOR, namely, by addressing the direction of organized scientific activity as well as the substance of science.

The congress came at a time of international goodwill on a more general scale. By August 1959, events indicated that Cold War antagonism might be at a threshold of better relations. Soviet premier Nikita Khrushchev was planning a trip to the United States, and he said he wanted to leave his technical aides and military advisors at home. It was to be a peaceful journey meant for touring rather than negotiating, and he would bring along his wife and children to acquaint them with American culture.[37] On the American side, a few days earlier the State Department announced that the United States would extend its self-imposed ban on nuclear weapons tests for the rest of the year.[38] Although the situation of the Cold War remained essentially unchanged, the United States and Soviet Union both appeared to proffer measures of good faith.

One such measure of good faith for oceanographic cooperation by the Soviets was to send a shipload of scientists to the oceanographic congress at the United Nations headquarters. When the congress began on August 31, 1959, the number of scientists who attended was astounding. Revelle, who was president of the congress, recalled later that he had no idea there were so many oceanographers in the world. If the meeting had taken place twenty years earlier, he claimed, there would have been at most 50 people in the world with enough interest to attend such a meeting.[39] The AAAS was surprised too; the registered attendees numbered 1,175, whereas the expected number had been 500–800. Scientists from fifty-four countries were represented, including 63 from the Soviet Union, the largest foreign delegation (840 were American). The Soviets arrived aboard the *Mikhail Lomonosov*, an oceanographic research vessel that gave a reception one evening for the scientists. Wolfle vaguely described that reception as "memorable."[40] Revelle described those memories in more detail:

> The Soviets gave a reception onboard the ship. [Vladimir] Kort and I somehow got into a vodka drinking contest; we were drinking vodka in champagne glasses, big champagne glasses . . . I stayed on my feet, and so did he. Several other people didn't.[41]

Undoubtedly, some tensions were eased that evening. Although not all examples of Americans trying to make friendships across Cold War lines entailed the consumption of vodka, the congress itself represented a large-scale effort to consolidate the oceanographic community and establish a precedent for personal contact and cooperation in the post-IGY era.

Despite such goodwill, the efforts to preserve contacts involved political concessions on the part of the United States. The State Department initially tried to block travel of Soviet scientists and to contain them within the New York area for the duration of the conference. In response, the National Academy of Sciences Committee on Oceanography (NASCO) passed a resolution stating that it was in the best interests of the United States to lift such restrictions and to allow travel to any oceanographic institution. The chairman of NASCO, Harrison Brown, felt that in the face of such restrictions the Soviet delegation might refuse to attend.[42] A similar restriction placed on Soviet scientists at the June 1959 International Petroleum Conference (also in New York) had resulted in only one Soviet actually showing up, whereas previously they had planned a large delegation. The State Department, however, felt that the Soviets should show some sign of reciprocity before the Americans loosened their own constraints. It offered to concede to the scientists' demands only if they could prove that such reciprocity had been obtained by American oceanographers at the last IGY meeting in Moscow or that reciprocity would be obtained in the future. Alternatively, the State Department challenged NASCO to take the position that reciprocity was not necessary at all. Eventually, NASCO agreed with this last condition, deciding "that *any* person to person contact with senior Russian marine scientists will pay off to our benefit regardless of whether the contact is made here, there or in between" (emphasis in original). The committee believed that the congress could add significantly to the mere handful of contacts already made, almost all in connection to the IGY, between American and Soviet oceanographers. To scientists, jeopardizing such progress in international cooperation, over a matter of State Department policy, seemed unwarranted.[43]

The congress became the forum to discuss a new trajectory for international oceanography, one that supported the scientific and economic development of other countries. Meandering among oceanographers at the congress was a "cultural officer" from Indonesia who witnessed this change in the focus of international research. In his efforts to explain to as many people as possible that the absence of a scientist from his country was "not because of lack of interest but due to 'circumstances,'" he gathered details about SCOR's first big project. He was pleased to find that the participation of all countries bordering on the Indian Ocean was being solicited, instead of the expected cooperation between countries whose scientific communities were well developed. The cultural officer made it clear that his country had a scientific manpower problem and had no suitable vessel. Revelle and

Deacon responded that they were intending to provide nations such as Indonesia with help in training scientists and obtaining equipment, to ensure their participation in an expedition that was aimed at understanding the environment that was most crucial to them.[44] This project, the International Indian Ocean Expedition (IIOE), would be the first large-scale oceanographic enterprise of the 1960s that was modeled upon the IGY, in the sense that it included all countries who wished to participate, regardless of politics. Yet it was also a new direction for cooperation, because it solicited the participation of developing countries. It reflected Revelle's (and others') wish to frame research problems in terms of society's needs; in this case, science would be directed toward a development agenda, helping scientists in the Indian Ocean area to understand critical problems of ocean food resources. Even more than the establishment of SCOR, the IIOE, which was to extend over a number of years, would set a new standard for international cooperation.

THE INTERNATIONAL INDIAN OCEAN EXPEDITION

International cooperation in oceanography, despite initial indications that it would be modeled after the IGY, was changing. When the International Indian Ocean Expedition first was formulated, well before the 1959 congress, it mirrored more closely the pattern of the IGY. If it hoped to survive, SCOR had decided, it should attach itself to a tangible large-scale project in which the need for international cooperation would be obvious. The choice of the Indian Ocean came fairly early. One explanation of it reveals the whimsical way in which some decisions might have been made during the late 1950s. Woods Hole's Henry Stommel recalled that at SCOR's first meeting he and another scientist were gossiping about personalities, when Woods Hole director Columbus Iselin walked in. In an effort to change the subject, Stommel asked Iselin to comment on a chart of the distribution of oceanic data that Stommel had helped to create prior to the IGY. The director commented that there was not much on the Indian Ocean; he then had some coffee and walked out. Stommel later conjectured, "I have a suspicion that's how it all started."[45] In fact, interest in the Indian Ocean began earlier, and IACOMS had identified it as an area in need of a major study. In any case, SCOR's leaders—particularly Deacon, Revelle, and German oceanographer Günther Böhnecke—decided in early 1958 to make plans for an "Indian Ocean Year" in 1962–63.[46] But the conformity to the IGY model did not last long.

The most obvious change from the IGY was the inclusion of poorer countries in the program. Even during the IGY, a Unesco fellow, T. S. S. Rao, wrote to IACOMS that Southeast Asian countries were in need of food resources to feed their populations and that this might come from fisheries exploitation. He hoped that the American vessels then operating in the Indian Ocean would visit local ports to help stir up some interest in the potential of oceanographic research to solve this problem.[47] Although the lack of data on the Indian Ocean may have sparked the oceanographers' interest in the region, it was the potential to build up scientific communities and possibly even economic well-being that SCOR latched onto in promoting the expedition. This was the most significant way in which it differed markedly from the IGY: the IIOE had an expressly practical objective, aimed at economic development. The expedition's promoters often referred to a story that in June 1957, British and Soviet ships simultaneously reported seeing millions of tons of dead fish floating mysteriously in the Indian Ocean. Although no one knew the cause, the quantity of fish revealed the great store of untapped food resources that might be used to feed the starving millions in and around the Indian subcontinent.[48] This practical angle gave the expedition a means of attracting support. Certainly it was a huge area that was comparatively (to the Atlantic and Pacific regions) blank in the scientific literature, thus a major study was most likely to contribute to all aspects of marine science. That and the uniqueness of the region because of seasonal variations in wind and current patterns would be enough to spark the interest of international scientific unions, even skeptics of Unesco such as George Deacon. But because its shores were homes for heavily populated and relatively impoverished peoples, the expedition would have economic and social value as well. This could attract the interest of governments around the Indian Ocean, and they might be willing to participate in the expedition.[49]

The desire of SCOR to include countries with embryonic oceanographic communities caused the IIOE to become a long-term project for the development of science in the Indian Ocean region. The immediate effect was upon the duration of the expedition itself, which was lengthened so that poor countries could spread the costs over several years. These participants included the largest country in the region, India; it also included Pakistan, Thailand, Ceylon, Indonesia, Singapore, and a number of countries in eastern Africa. The hope of SCOR was to use the IIOE to help these other communities develop, much as cooperation with Japan in the early 1950s aided in the growth of that country's oceanographic community. This aspect of the IIOE appears to have made some success; both Thailand and Indonesia eventually set up their own marine science institutions, and Pakistan's University of Karachi created an oceanographic studies program as the expedition began in 1959. A year after the IIOE's end in 1965, one of its leading participants, N. K. Panikkar, became the first director of India's new National Institute of Oceanography.[50] To have a training program prior to the expedition was SCOR's idea and also contributed to the lengthening of the IIOE. Scientists and technicians from the Indian Ocean countries would have to spend time at intensive training sessions. Such direct tutelage was supposed to help integrate scientists from developing countries into the international scientific community.

To gather interest and begin planning, SCOR had to reach out to the developing world somehow. By the end of the 1959 congress, SCOR had a rough outline of its aims for the IIOE. The Soviets had worked in the southern part of the region during the IGY, and by October 1959 they had a ship on its way to the area to begin work ostensibly for the IIOE. The United States began its efforts in 1959 also, and both nations conducted work preliminary to the main phase of research. But the United States wanted to do more, particularly to drum up support and make contacts in other countries. Thus the Americans paid for Robert G. Snider to travel to the region in order to sell the project to the region's scientists and governments.

Not an oceanographer, Snider had worked as a Navy test officer for antisubmarine warfare projects for a number of years before his involvement in the IIOE. Revelle later described him as "an entrepreneur, a promoter, a very hard working and earnest sort of guy."[51] Snider toured the various countries and promoted the IIOE with laudable energy, drawing on his strengths as a promoter. One of his first visits was to Japan, where he solicited Japan's participation, not as a developing country, but as one of the key leaders. Snider appealed to a movement within Japan that sought to mimic the aid policies of the United States, emphasizing Japan's new role as a world leader in oceanography. American officials in Japan had informed Snider that the Japanese were "very interested in helping Southeast Asian countries on a sort of technical assistance or Japanese point #4 basis." In addition, Japan did not want to send its scientists for training in the United States or a European nation, but rather hoped to be a center for the training of other Asians. Seeking any angle to promote the expedition in Japan, Snider played up these possibilities in his presentations, emphasizing Japan's role as a scientific benefactor instead of beneficiary. The Japanese oceanographers were ready, he learned, to assume a leadership position in oceanography, particularly if that cooperation promised to expand Japan's influence in Asia.[52]

Snider could not play the prestige card with everyone, and often the IIOE was hard to sell to developing countries. In particular, he had difficulty explaining how sharing data could be beneficial. This point concerned the Ceylonese, who believed that if all the data of the expedition were openly shared, some nations (Japan was not specified, but Snider suspected) would be able to exert too much influence, particularly by dominating the fishing industry. Snider countered that Japan would acquire this information anyway, but the other countries would not have access to it unless they too participated.[53] He encountered a different view when visiting Pakistan, where he was pleased to find at least one confirmation of the positive effect international cooperation was having upon scientific practices. His primary point of contact was a naval commander, S. R. Islam, who insisted that the Pakistani Navy play the predominant role in the country's participation. Islam wished to concentrate the work of his country off Pakistan's own coasts, and he was concerned about the kind of data that his country might be expected to share with others. Snider told him of a recent declassification of virtually all American bathymetric data, by order of the chief of Naval Operations of the United States Navy. Islam took this into account and said that it was a very good precedent—a measure of good faith—for Pakistan to do the same.[54]

The extent of participation in the IIOE by local scientific communities varied. Snider told his contacts in each country that they did not need to plan extensive programs; even minimum participation would be a great help to the overall success of the IIOE. Minimum participation meant that no ship would be used, but that the country would set up tide gauge stations, much like those set up for the IGY, which could automatically record data over a long period. But although he was prepared to accept this as a minimum, Snider hoped for greater participation. In Japan there were few difficulties, because scientists and business leaders alike understood the potential economic benefits of the IIOE. The ICSU simply would write directly to Japan's prime minister, the prime minister would then contact leaders in the fishing industry, then the fishing interests would put pressure on the Diet to legislate in favor of the expedition. The channels through which to promote research were clear, as was the necessity of touting the economic implications of the project.[55] Other nations possessed no such clear channels to support research. Upon consulting the Indonesians, for example,

Snider found that the cultural officer at the 1959 congress had not overstated the difficulties that Indonesia would have in pursuing oceanographic research. In the whole country, there was only one oceanographer holding a Ph.D. This meant that control of Indonesian activity would have to lie in the hands of at least some foreign personnel and that certainly there would be a very limited pool of potential trainees to choose from. Most of Indonesia's science-oriented agencies were fisheries organizations, which expressed interest in the expedition, but its meteorological service was only a year old, and its hydrographic office was not equipped with the rudimentary tidal stations required for minimum participation.[56]

Scientists from other nearby countries faced similar obstacles to participation. In Singapore, a seven hundred thousand–dollar oceanographic research facility had been sitting virtually unused since 1957. The reason was that the Singapore government was not willing to pay for fundamental research to be conducted. There was one oceanographer there, but he refused to be restricted to applied research, so nothing at all was being done. The Singaporeans expressed their willingness to use the facility for oceanography during the IIOE, but it would expend its resources only if the results focused upon the neighboring South China Sea.[57] In Ceylon, the situation was slightly more conducive to cooperation: there were more scientists, including one who had received his Ph.D. from Carl Hubbs at Scripps. But it had no hydrographic office set up to install tide gauges. Moreover, the scientists themselves had very little power in the country. One of them advised Snider that, although there was a Ceylon Association for the Advancement of Science, it was more appropriate to talk directly with the government about the expedition, since the scientists could only offer encouragement without giving any assurances of participation. Snider also found that the Soviet Union had contacted Ceylon already about a joint Soviet-Ceylonese marine expedition in local waters. The Soviets wanted Ceylon to put up some money and to house Soviet scientists in air-conditioned quarters. The Ceylonese had refused.[58] Limitations such as these indicated that the inclusion of these countries would entail more help by the richer ones and that the relationship might more closely resemble foreign aid than international cooperation.

Despite the independent Soviet initiative and the likelihood that the United States would have to carry some of the burden of the developing countries, Snider insisted on maintaining the IIOE as an international venture, not a national one that emphasized Cold War divisions. These countries, he felt, were growing accustomed to playing the superpowers off each other, to their own benefit. Some Ceylonese in particular welcomed American financial aid, reminding Snider that a little good press could go a long way toward making the expedition a prestige plum for the United States. But Snider was not prepared to run the financial risks of competing with the Soviets for the favor of small nations. He insisted that, although various American government agencies would take on some of the costs of the IIOE—particularly the International Cooperation Administration, which had not done so during the IGY—the participating countries would not ride for free. More than once Snider had to clarify this internationalist philosophy, as when Ceylonese scientists requested that SCOR provide the plans they wished Ceylon to pursue, so they could consider them. That was precisely what SCOR did *not* want, Snider explained; each country should make its own plans, then submit them to SCOR, so the international coordinating body could try to integrate all the plans together. Snider also insisted that each country show its good faith vis-à-vis the IIOE, by making a token commitment of future payment, even if only for a few tide gauges. Only then could SCOR hope to solicit the help of outside agencies, such as the United States through its International Cooperation Administration or the United Nations through its Special Fund.[59]

The possible use of the United Nations Special Fund revealed a significant point of contention among Indian Ocean countries. Upon visiting India, Snider found a marked reluctance to participate in any joint ventures with the surrounding countries, particularly if a request for outside money was involved. Snider had suggested that India might be able to obtain some assistance for participation in the IIOE by submitting a joint proposal for the Special Fund with other nations. An officer in the Indian Navy told him that as a matter of policy, India did not like to be involved in such things and that Snider ought to drop it if he wanted his ideas to fall on receptive ears. Snider reported this opinion:

> India is by far the largest nation in the Indian Ocean, and it does not like to do things with the smaller countries because then it gets in the invidious position of being accused of hogging the whole show.

Snider eventually dropped the issue, but not before receiving a similar reaction from leading scientists who indicated that India would find its own means of participation, without applying for funds together with the surrounding countries.[60]

The Indians seemed to feel that, scientifically, they should be more closely identified with Japan rather than these other countries. The conceptualization of SCOR, that the expedition would include both "developed" scientific communities and "developing" scientific communities, could be very distasteful to Indians who knew SCOR considered them to be in the latter category. After all, the situation in India differed markedly from the other Indian Ocean countries. Snider had come to India armed with the same practical arguments he presented other countries: the Indian Ocean was 14 percent of the earth's surface; more than 25 percent of the world's population surrounded it; the Indians would work in their own backyard under expert guidance. And of course he had the story about the unexplained swath of dead fish. But India required little prodding in the area of basic research; it already had the will and some means to participate. There was an Indian Navy prepared to provide some logistical support. There were a number of scientists at Madras University and the Bombay Institute of Science who were excited about participating. India had a science advisory system set up within its cabinet, an organization (its National Committee for Oceanic Research) devoted explicitly to planning and coordinating marine research, and also private organizations, such as the Tata Foundation, which supported science. Like Japan, India had a scientific tradition that predated the economic and political turmoil that plagued it after World War II. One of its prominent scientists, K. R. Ramanatham, was the president of the International Union of Geodesy and Geophysics and was involved in ICSU affairs. In general, the interest in the IIOE among Indians was high, and those in power were willing to cooperate. Snider found that the Indians already were preparing for the IIOE; one Indian scientist had departed to do research aboard the Soviet oceanographic vessel *Vityaz*.[61]

Snider and other Americans were unsettled to learn that the Indians already had begun to work with Soviet scientists. The Soviets intended to be heavily involved in oceanographic activity in the Indian Ocean over the next few years and were pleased at the prospect of working under international auspices. Visiting the Soviet Union, Snider met with Lev Zenkevich and Vladimir Kort. They announced what by then Snider knew, that the vessel *Vityaz* already was working in the Indian Ocean, and it would be joined soon by two smaller vessels for work in 1960 and 1961. The Soviets' "minimum program," that which the scientists could confirm regardless of special funding for the IIOE, would involve these ships, although another large one would replace the *Vityaz* sometime in 1961. The large ship would carry a complement of sixty to seventy scientists who would do work in all relevant disciplines. As for helping other scientists to train for the expedition, the Soviets did not wish to participate in SCOR's plans; they already had a

The New Face of International Oceanography

couple of Indians working aboard the *Vityaz,* and they preferred to make their own arrangements anyway. In the same vein, they were displeased by the extent to which SCOR's plans for the expedition did not allow for very much flexibility on the part of the individual scientists.[62]

Complaints about the expedition's inflexibility were echoed by other scientists as well, including many in the United States. Some feared that SCOR would adopt a "grid" system to divide up the Indian Ocean and assign areas of work (as opposed to leaving it open to address particular problems at scientists' discretion). This would, according to most, invite mere surveys in conformity with a master plan. Scripps's Robert L. Fisher complained that the cost of a grid system was too high. "If you cover everything, you may get 95 per cent result for a 95 per cent effort. If you find a critical spot and concentrate time and resources on it, you will get an 85 per cent result for a 35 per cent effort."[63] Fisher's point accorded with the views of the Soviets, many of whom wished to study the Indian Ocean, not just map it. This attitude stands in sharp contrast to the spirit of the comprehensive survey plans that they would later propose on a routine basis. But at the time, the Soviets posed no particular problems to the initial conceptualizing of the IIOE. The Americans' only concern was that the Soviets were pressing on with their own collaborative relationships, particularly with India, without coordinating them with SCOR. But overall, the changing face of cooperation, characterized by the inclusion of nations with embryonic oceanographic communities and an explicitly development-oriented agenda, seemed to be fully endorsed by scientists in both the United States and the Soviet Union.

SELLING SCIENCE: NEW RHETORIC FOR OCEANOGRAPHY

By planning a long-term scientific project involving many nations and promoting a development mission for science, oceanographers were in fact helping to define national foreign policies for science. Even before the launch of *Sputnik* in 1957, American scientists had clamored for such involvement. University of Rochester chemist Albert Noyes, Jr., argued in the pages of the *Bulletin of the Atomic Scientists* that some of the nation's broad foreign policy issues could benefit from the advice of scientific experts. This advice should not be limited to nuclear matters but rather would include other issues "to ensure that science and scientists play their proper parts in world development." Such involvement, Noyes argued, should include the activities of international scientific organizations, of the United Nations in general, and of foreign policy arms of the United States.[64] One *Bulletin* author went so far as to claim that the growing popularity (among scientists) of science fiction as a genre was due largely to scientists' desire to take part in national policy decisions. Sociologist Arthur S. Barron felt that scientists wished to guide policy to serve man and that science fiction itself could be seen as a protest against the appropriation of scientific developments for ends that menaced, rather than helped, humanity. He argued that if scientists could take on a greater role in shaping policy toward the needs of humanity, they would find such alternative outlets less satisfying.[65]

The voices of scientists in politics certainly were heard by the late 1950s, but not primarily as advocates of science in service of humanity. In the wake of *Sputnik* a science advisory office was revived within the Department of State after many years of neglect. In addition, President Eisenhower created the President's Science Advisory Committee (PSAC) and made increased use of his personal science advisor. These measures were designed to keep the United States abreast of science and its potential technological applications in the struggle against the Soviet Union. Both PSAC and the science advisory position within the Department of State were integral parts of the bureaucratic decision-making process by the 1960s. As noted by *Science* journalist Daniel Greenberg, this was due to a decisive change in attitude within government. The position within the Department of State, Greenberg noted, was "here to stay. This is not so much because of its performance, which according to almost everyone involved, is much in need of improvement, but rather because political leaders today accept it as an article of faith that science is inextricably involved in public affairs."[66]

American and British oceanographers felt that they enjoyed a relatively free hand in contributing to science policy, compared to their Soviet counterparts. An illustration of this perception was SCOR's initial effort to construct a permanent administrative body to handle oceanographic cooperation. At a 1960 planning meeting in Paris, leading oceanographers Roger Revelle, George Deacon, John Lyman, and Vladimir Kort convened to work out the general plan of a new Intergovernmental Oceanographic Commission (IOC). They argued over issues such as the organization of the IOC Secretariat, the source of funding, the amount of data to be exchanged, and other details. Revelle recalled the frustrating experience of those days:

> [T]he interesting thing about it was that we'd hammer all day on Kort, and we'd finally come to an agreement by, say, five o'clock or six o'clock in the afternoon. The next morning the agreement had come completely unstuck, and we had to start all over again.[67]

Revelle assumed that Kort was not speaking for himself. Although typically the Soviet scientists were afforded some autonomy in organizations such as SCOR, which ostensibly was made up of scientists representing the scientific community (not individual nations), the IOC would be an intergovernmental organization in which each nation would have to judge the merits of projects based on a host of economic and political considerations, not just scientific ones. Revelle felt that Kort could only agree to something in principle, and then he would go back to his hotel room and phone Moscow, only to have his superiors tell him to disagree. Revelle later recalled feeling the presence of the Soviet government—not just a Soviet scientist—in those discussions. He felt that this was the crucial difference between how Americans and Soviets behaved toward science policy. American scientists were free to do what they thought was best for science; Soviets answered to government superiors.[68]

Revelle's attitude toward the personal influence (or lack thereof) of his Soviet colleague implied that American scientists had a remarkably greater influence over foreign policies related to science. Department of State science advisor Wallace Brode appreciated this perception among Americans. He said that in the Soviet Union, few scientists acted upon the political ramifications of scientific developments. In the United States, by contrast, scientists could point proudly to agencies such as the National Institutes of Health and the National Science Foundation, as testimony to the growing role of science in national policymaking. This role was also reflected in the presence of influential science bureaucrats in government such as George Kistiakowsky (special assistant to the president for science and technology), Herbert York (director of defense research and engineering in the Department of Defense), and Brode himself.

Although Americans were quick to criticize Soviet science for being controlled by government, national policies shaped scientific activities in the United States as well. Even in the creation of IOC, more was at stake than scientific considerations. As Brode noted, "While we scientists are generally willing to concede that science should influence our national policy, it is often more difficult for us to admit that our national policy should also influence our science programs."[69] The American government's approaches to cooperation in general were transforming the nature of scientific cooperation. For example, in 1961 Adlai Stevenson, former Democratic presidential candidate and the American representative to the United Nations, emphasized the need to use scientific cooperation to help developing countries, with their "almost unlimited need for education, health, industrial development, agricultural improvement, communications, and exchanges in the fields of science and culture." The path to peace, Stevenson pointed out, lay with the humanitarian aims of the United Nations.[70] Secretary of State Dean Rusk told an audience at the Massachusetts Institute of Technology that within science "we discover a world which President Kennedy recently referred to as the world which 'makes natural allies of us all.'" Looking at the world's citizens as Homo sapiens, Rusk argued, the issues between man and his environment reduced to insignificance the petty quarrels among men. The policies of the American people could be found, Rusk added, within the United Nations Charter, with its devotion to the human needs of the world. The world required "development scientists." So great was the need that "the nations and peoples of the earth seem to be pinning their hopes on the possibilities of scientific and technical development for the satisfaction of the basic human needs."[71] American oceanographers' embrace of "development" as a justification for research must be viewed in the context of Democratic leadership. Scientists were matching the administration's rhetoric.

The development agenda for science was viewed with mixed emotions within government. Brode was rather critical of this world-development objective as a science policy. Like some of SCOR's critics, he felt that the problems addressed by a humanitarian strategy were not particularly scientific ones. The scientists should instead concentrate, he said, on either scientific advancement or the exploitation of such advancement for the purposes of the United States. "One naturally expects the new advances in science and technology," Brode said to the Senate Foreign Relations Committee in January 1960, "to take place in the scientific centers of the world, and we have no unique monopoly on these centers."[72] It was far more beneficial to focus science policy on cooperative measures designed to acquire knowledge, not to disseminate it to the needy. Others saw that a strong science policy was one based upon multiple premises and that the development aspect could provide a rather useful veneer while serving national aims. It painted real social benefits onto packages that were designed as anticommunist measures or as ways to enhance the power of the United States. George Kistiakowsky, PSAC chairman, argued along such lines with regard to developing countries:

> The foreign aid programs supported by the United States are powered by a matrix of motivations made up of altruism, a belief that it is to our best interest to strengthen independent nations, and a desire to contain menacing philosophies.[73]

The other facets certainly remained, namely, the economic and political struggle against international communism, but a development agenda had insinuated itself firmly within the rhetoric of American science policy.

With political and foreign policy trends focusing on development, it was sensible for SCOR and other bodies to promote marine science along these lines; however, some scientists disliked the idea of pursuing oceanography while claiming a development agenda. Marine biologist Gilbert Voss, for example, complained that the purpose of the IIOE had been described in a deliberately vague manner. He had heard that the expedition aimed to study how the slope of the Indian Ocean was affected by the monsoon season and that such knowledge would fill many gaps in oceanography. He also heard many say that improving the world's understanding of food sources would provide material benefits for the peoples surrounding the ocean. But how were these aims connected? Voss suspected that it was all rhetoric and that the "development" aspect was just a scheme to obtain funds for ship time, a method that some physical oceanographers had learned to exploit in the years since the IGY. If the objective of the expedition was truly to contribute to the region's development, he felt, the expedition should be organized quite differently and should be directed toward fisheries exploration. "You can't eat temperatures and salinities," Voss wrote, questioning how much physical oceanography could really benefit starving people.[74]

Biologists were particularly opposed to touting the economic aspects of the IIOE, because they felt that the real substance of fisheries research, namely, biology, was being treated as a subject of secondary importance to physical oceanography. Biologists argued that their research programs had not been prioritized enough. Voss in particular felt that including the biologists on the expedition was a disingenuous effort to preserve the expedition's development-oriented image. Virtually all the scientific plans were designed around physical oceanographic aims. He complained:

> Karl Banse stated that an exciting idea originating at Copenhagen was short period transects of the Equatorial Current ... he then said that "the biologists will have some fun too" presumably by doing some biology at the same time. If we biologists are going out to the Indian Ocean just to have fun, I suggest that we stay home.[75]

Voss's comment revealed the resentment many scientists felt when they perceived that support for their research lagged behind physical oceanography (it also revealed that some biological oceanographers, such as Banse, were willing to compromise and to look for opportunities). It was worse when the physical oceanographers seemed to want the biologists to be involved only for their role in legitimizing a false humanitarian agenda.

The biologists' irritation was certainly justified, considering the attitudes of the leadership in the American IIOE program, namely, the National Academy of Science's Committee on Oceanography (NASCO). Lamont director W. Maurice Ewing strongly cautioned against giving approval to many proposals for the IIOE that were not explicitly connected to physical oceanography, even if they would result in good science. Instead of evaluating each proposal independently, he urged NASCO (of which Ewing was a member) to make "some statesmanlike decisions" and instigate policies to give preferential treatment to certain programs. Although it would have been nice to evaluate each proposal on its scientific merits alone, the United States could not afford to support the optimal development of every branch of science. Ewing at the time was criticizing a proposal to have separate ships devoted specifically to meteorological studies. He felt that it was silly to spend great amounts of money on special ships when the scientists could simply do the same work on ships that were budgeted for other purposes. The same attitude applied to biological studies.

Despite this trend among the American scientific program planners, the biologists eventually were given their own ship. This was due largely to another skeptic of American oceanographers' humanitarian intent, marine biologist Dixy Lee Ray, who later would go on to serve as chairman of the Atomic Energy Commission and then as governor of Washington. In 1960, Ray took leave from the University of Washington where she taught zoology to act as a consultant to the National Science Foundation. After realizing that IIOE biologists were essentially going along for the ride, she worked to secure funding for a ship—Harry Truman's presidential yacht, the *Williamsburg*—to be devoted to biological research. Ray reasoned that the biologists would have to separate themselves almost completely from the physical oceanographers if they wished to have control over the research program of the vessel in which they operated. The ship was rechristened *Anton Bruun* (Bruun, who died shortly after the start of the IIOE, had also complained against the preponderance of power by physical oceanographers within marine science).[76] Ray's actions worked against the hypocrisy of a program that boasted a devotion to fisheries research yet at the same time alienated the biologists.

Although being given a separate ship may appear to have been a victory for the biologists, the idea of having clear boundaries between disciplines drew strong criticism. Revelle, of course, did not accept such lines of

The New Face of International Oceanography

demarcation, nor had Bruun. On the one hand, scientists feared that one group would dominate the others; on the other hand, they wanted to maintain as broad a definition of the field as possible. Another critic of such divisions was Lev Zenkevich, the Soviet marine biologist. He wrote to Robert Snider that the Soviet National Committee on Oceanography was repulsed by some scientists' persistent separation of oceanography from marine geology and marine biology. "We are accustomed," Zenkevich wrote, "to the idea that oceanography (=oceanology) is a complex science composed of five large divisions—physics, meteorology, chemistry, geology and biology." Zenkevich noted that this attitude extended far beyond the Soviet Union and that the recent international congress in New York had carried the name of oceanography not of marine biology or marine geology. In discussing the IIOE, Snider often spoke in terms of three separate fields of study—oceanography, marine biology, and marine geology. To Zenkevich and his Soviet colleagues, this was "wrong at its very core and in the long run may lead to serious complications."[77] Zenkevich, Revelle, and Bruun all knew quite well that blending disciplines made it much easier to blend motivations, particularly the strictly scientific ones and those related to economic development.

Impressions about blending fields and motivations initially were mixed, strongly tempered by fears of biting off more than the community could chew. George Deacon, for example, reluctantly admitted that large projects addressing human issues appeared to be the only way to compete with more glamorous sciences (such as those related to space and Antarctica) for funding.[78] The public interest and support for science, certainly one of the most significant legacies of the IGY, convinced him and others who typically detested ambitious survey-type expeditions that an even bigger project was desirable. Governments, it seemed, were far more enthusiastic about "development" projects than purely scientific ones. But at the same time, he wanted to move forward cautiously, to ensure that they adopt a sound plan of action. He feared that everyone would try to get on board the project without having a clear and practical goal in mind, and it might expand to unmanageable proportions. "We cannot afford," he wrote to a colleague in SCOR, "to have a South Sea Bubble."[79] Deacon was making a historical reference to the rush of investors and overspeculation in the South Sea Company, which led to the first great stock market crash in England in 1720. When the bubble popped, fortunes disintegrated. Likewise, Deacon did not want everyone's enthusiasm for the IIOE to endanger the quality of the project.

But as the IIOE dragged on through the planning stages, the image of the

South Sea Bubble prevailed among cynics. American physical oceanographer Henry Stommel even created a newsletter called the *Indian Ocean Bubble*, which served, before the Americans had a national committee for the IIOE, as a vehicle for discussing plans for research in such areas as ocean dynamics and meteorology. As in the case of the South Sea Bubble, everyone wanted to be part of the new big project. But, as Navy scientist Eugene LaFond wrote in the *Bubble*, "the problem is not what to do, but rather, *who in the Indian Ocean region can be rounded up to do it?* Everyone should be reminded that this is the Indian Ocean, not the Woods Hole or Scripps one" (emphasis in original). LaFond called to mind the commitment to work with local scientists. If anyone hoped to attain information from the Indian Ocean after the expedition had finished, local science needed to be cultivated, and Asian scientists needed to gain some experience.[80] Selling the project to them would be Snider's task as he toured the area to stir up interest. Their participation would be a crucial step in legitimizing the development aspects of the expedition, which would help all scientists, regardless of their true motivations.

Scientists were wary of SCOR's embrace of the notion that it should "sell" science on the basis of practical applications. Unsurprisingly, George Deacon was very sympathetic to those decrying such efforts. But he also defended the IIOE as "only a trial of strength, whetstone, flag to wave and perhaps cause célèbre, as well as good science." Like most oceanographers concerned about patronage, he was annoyed at how much harder SCOR had to work than either SCAR or COSPAR. Prestige alone sent expeditions to the Antarctic and to outer space, and the scientists needed only conduct as much research as possible. By contrast, governments had to be sold on the oceans. Scientists had to connect their work to a popular political or foreign policy agenda. In addition, "SCAR and COSPAR are clubs of a few active countries— they don't have to worry about less developed countries."[81] However, SCOR was in the invidious position of having to advertise the practical benefits of oceanic research. Snider's role had made the blending of scientific and economic priorities embarrassingly apparent to many scientists. Japanese marine scientist Kazuhiko Terada urged his international colleagues to establish a firm science-based philosophy of its large-scale projects and to abandon the practice of pandering to the economic needs of individual states.[82] Concurring, the future president of SCOR, George Humphrey, was eager to relieve Snider of his duties and to transfer responsibility for the IIOE to a new organization tied more firmly to Unesco, namely, the new Intergovernmental Oceanographic Commission (IOC). In doing so, SCOR

The New Face of International Oceanography

hoped to return to more science-oriented tasks of identifying problems and advising, rather than selling and managing unwieldy intergovernmental projects.

THE INTERGOVERNMENTAL OCEANOGRAPHIC COMMISSION

At the turn of the decade, SCOR became the dominant marine science advisory committee not just for science but also for the application of oceanography to practical problems. It was officially born at a meeting in Woods Hole in August 1957, and Revelle was its first president. The following year, ICSU accepted the body as an official special committee. Although many people agreed on the need to create SCOR, some scientists were irritated at the sudden proliferation of international scientific bodies, whether they were committees, councils, intergovernmental organizations, or regional commissions. Vladimir Kort complained that there was really no proper coordination between these bodies, some of them competing with each other for influence. This was despite the fact that many of them worked toward the same ends and even comprised some of the same members. All of these bodies lacked "the necessary concentration of attention, energy and means in the study of the fundamental problems of present-day oceanology," and Kort urged a change for the better. He pointed to the World Meteorological Organization as a "much more rational organization of international effort." Kort suggested combining SCOR and IACOMS into an organization whose primary raison d'être would be to coordinate the efforts of all organizations, to plan expeditions, and to coordinate the methods of research.[83]

Kort was correct that there were too many bodies, especially now that SCOR had adopted an approach focusing on social ramifications. In particular, IACOMS seemed superfluous. Thus, SCOR tried to dissolve the older committee. No one wished to expend resources on duplicated effort, but individual scientists had invested the majority of their time and energy—not to mention egos—in one or the other body. A brief struggle took place that ended in the downfall of IACOMS in 1960 and the ascendancy of SCOR as Unesco's official advisory body for marine science.[84]

Oceanographers learned quickly that planning big projects took far too much time and attention away from discussing science itself. Not having been conceived as a managerial body, SCOR realized early on that a huge project like the IIOE would require more than just a group of advisory scientists. So in 1960 Unesco created the Intergovernmental Oceanographic

Commission to coordinate national oceanographic programs and to conduct the day-to-day activities that SCOR was not equipped (and not particularly willing) to do. In his form letter to oceanographers throughout the world, IOC's first secretary Warren Wooster described the focus of the IOC as being "in the implementation of international programs . . . and in the coordination of national programs relating to the exchange of data and information and the standardization and intercomparison of methods and equipment."[85] Over the next several years, the IOC became the center of international oceanographic activity, with a healthy blend of genuine advisory expertise, political posturing, and discontent among scientists.

Oceanographers' attention to the practical needs of the world only increased after the creation of the IOC. This organization coordinated not only the IIOE but also other cooperative enterprises of the 1960s such as the International Cooperative Investigations of the Tropical Atlantic (ICITA) and the Cooperative Study of the Kuroshio and Adjacent Regions (CSK), involving several nations and dozens of research vessels. The purpose of the IOC, as stipulated in 1960, was to "promote scientific investigation with a view to learning more about the nature and resources of the oceans through the concerted action of its members."[86] The IOC prided itself on coordinating the IIOE, the magnitude of which, in terms of people, ships, money, and countries, made it the greatest oceanographic endeavor to date. Beyond the actual science, the IOC counted among its most important successes the number of educational opportunities that the IIOE provided to scientists in developing countries. It also felt that it had stimulated not only continued interest in marine science in those countries but also a greater awareness of oceanographic problems in many countries of the world.

The IOC created a vision of cooperation that incorporated many of the humanitarian, service-oriented tenets of the United Nations as a whole. Many scientists welcomed this vision. Prior to its first session, the IOC Secretariat received a joint declaration from Argentina, Brazil, Cuba, Ecuador, Mexico, Uruguay, and the Dominican Republic announcing that the IOC should provide more support for Latin American countries in the field of oceanography. Their declaration expressed an awareness of the progress already achieved in industrialized countries but acknowledged that "nations in the process of development, despite their efforts to intensify their activities in this field, are increasingly failing to keep abreast of this process because of their inadequate human and economic resources." The declaration pointed specifically to Latin American countries where interest in oceanography was high, but the means to pursue it were very low. An organization such as the IOC could rectify that situation by directing the financial resources of Unesco toward them.[87]

Despite helping to make international cooperation a reality amid geopolitical tensions, the IOC abandoned some of the earlier ideals of cooperation. During the early 1950s, one of the goals for international science set forth in the first issue of *Deep-Sea Research* had been to promote cooperation at a personal level. One example of this would be the inclusion of foreign scientists on national vessels, which the Americans did with the Japanese during the early 1950s and the Soviets did with the Indians prior to the IIOE. Many scientists hoped that the ultimate result would be an international vessel operated by scientists of several different countries—a microcosm of the United Nations symbolizing the ideal of peaceful cooperation. But upon taking control of international oceanography, the IOC rejected the concept almost immediately. It reasoned that one ship, even an international one, could not expect to contribute much to the study of the oceans. Despite the potential political benefits of a research ship flying the flag of the United Nations, there were more scientific gains to be had in doing things rather differently. International *coordination* promised to divide the work of the world's oceans, almost three-fourths of the earth's surface, between nations that did their work autonomously. By 1966 the IOC had coordinated about 250 cruises over a wide area, far more than could have been accomplished by one international research vessel. The IOC defended itself thus:

> No doubt one may attach a certain moral value to an international oceanographic vessel which, by visiting the ports of various developing countries would bring to them a positive example of international co-operation. But the activities of the Intergovernmental Oceanographic Commission have provided more such examples.[88]

These other "examples" consisted primarily of the establishment of general procedures, namely, the coordination of research and the reporting of results to data centers, modeled upon those of the IGY, throughout the world.

Although coordination and subsequent reporting to data centers appeared to be a reasonable and more productive system of cooperation than scientists' working directly with one another, some scientists felt that this practice did in fact hinder cooperative work at a personal level. Although it was relatively easy for Americans to include Japanese scientists, or any scientists of allied or nonaligned countries, aboard their vessels, more ambitious overtures of cooperation, namely, those between Americans and their Soviet colleagues, were rather difficult. Scripps marine geologist H. W. Menard, for example, tried to bring his Soviet colleague Gleb Udintsev aboard an American ship during the Monsoon expedition (a component of the American IIOE program). Menard reasoned that the American government would offer no resistance, since a scientist visiting a ship at sea would not entail the same bureaucratic headaches (i.e., obtaining the appropriate visa) as one visiting the United States. Menard claimed that because he planned to include Udintsev on a leg of the expedition that did not enter an American port, he needed to make no official arrangements. But he was wrong; both the U.S. Navy and the National Academy of Sciences moved to block his action, the former because Udintsev posed a potential threat to secrecy (the Navy claimed that the ship used some classified equipment) and the latter because the Soviets had offered no reciprocity to the action.[89] The IOC's preference for coordination over personal cooperation provided no international rationale to overcome such obstacles, making it more difficult for scientists to justify personal cooperation. The emphasis upon data coordination and not personal cooperation made scientific bridges between political enemies unnecessary. The IIOE, as Roger Revelle later recalled, "wasn't a well-planned, completely integrated operation. It was a lot of ships from a lot of countries, everybody doing his own thing, telling each other what they were doing but not necessarily working together."[90] Even the guidelines for international cooperation, as shaped by SCOR and by IOC, tended to de-emphasize working together. They consisted of separate national programs, loosely coordinated in an international forum comprising national representatives, whose sole cooperative responsibilities were to report findings to data centers, where others would compile and disseminate findings for anyone interested in them. Such activities did not require that scientists work together for a common purpose but instead established an international arrangement in which they did not have to work together at all.

The IOC soon acquired a bureaucratic character that would cause the enthusiasm of many oceanographers to turn sour. Its very nature as an intergovernmental commission meant that the IOC was made up not of individual scientists but rather of member states. The commission made an effort to have close working relationships with several other global or regional scientific organizations, and naturally with SCOR. But its main source of influence was its direct connection to administrative arms of various governments, making IOC a more powerful organization than SCOR. Because of these government connections, the organization was intrinsically one that favored cooperation for the good of all nations. As the IOC expanded, however, it got bogged down in negotiations, formalities, and above all politics. After about a decade, scientists would find themselves losing the initiative; they could not direct cooperation with the same influence that they enjoyed during the 1950s and early 1960s. Revelle, who presided over SCOR in its infant stages and watched the IOC implement the IIOE and other projects, later lamented:

> When we organized the Intergovernmental Oceanographic Commission its rules were intended to be exclusive, that is, its charter says that membership in the commission will be open to those countries that wish to cooperate in international oceanographic research. To us, that meant having ships and doing oceanographic work in the high seas, big oceanography, big science. It was intended to be cooperation with the Soviets, with the Japanese, with the French, with the Germans, with the Canadians, hopefully with the Indians and the Australians and the South Africans. Now the IOC has 120 members, something like that. Most of them are developing countries that don't even know what oceanography is, or know very little about what oceanography is.[91]

The IOC would, over the years, take the control of international science out of the hands of industrialized countries and place more of it in the hands of developing countries. By the early 1960s, this had not yet occurred. But international cooperation in oceanography nevertheless had found a permanent home within the offices of Unesco, thus appropriating rather overtly an agenda that reached out to the developing world.

The early 1960s was a period of intense public appreciation for the role of science in international relations. Oceanographers, however, lamented that they did not share the limelight, in terms of prestige and funding, with scientists engaged in the space race that followed the launch of *Sputnik*. Evidently oceanography had not proven itself of comparable worth in the public eye. Even the IOC commented that, although its efforts had far surpassed those of the International Geophysical Year, it had yet to enjoy its "due share of public interest and attention."[92] Certainly one of the reasons for this was that both the IGY and the subsequent space race engendered a great degree of national competition, not cooperation. The next chapter addresses how marine science was influenced by the intense scientific and technological competitiveness between the United States and the Soviet Union.

5 COMPETITION AND COOPERATION IN THE 1960S

Programs such as the IIOE attested to the desire of scientists, Americans and non-Americans alike, to resist the imperative of scientific and technological competition with the Soviet Union and to press on with cooperative ventures. These efforts, however, were not universally pursued. Other scientists looked inward at the United States, frustrated by the well-publicized technological successes of their Cold War enemies, the Soviets. American science, and its leadership in the international community, had been challenged by communist successes. Like their colleagues in other scientific fields, oceanographers in the United States craved a renewed focus on national strength in the face of the Soviet challenge. Especially after 1957, the fate of science was tied closely to the affairs of the American government and its relation to the rest of the world. For marine science, the two most significant events of this period were the launch of *Sputnik* in 1957 and the exit of Dwight Eisenhower from the White House in 1961. Each marked a point of departure for marine science policy. After *Sputnik*'s launch, the National Academy of Sciences engaged in a shameless campaign for oceanography that made even other scientists uneasy. Oceanographers gathered support from influential members of Congress who did not wish to see American oceanography fall behind its Soviet counterpart. Initially, their toughest opponent was the president himself, whose closest scientific advisors resented the oceanographers' attempt to capitalize on the nation's panic. But after Eisenhower left office, pro-oceanography men such as President John Kennedy and his new science advisor Jerome Wiesner helped give oceanographers the financial and official moral support they wanted.

An important and paradoxical consequence of the ascendancy of national support for oceanography, from 1958 to 1965, was that it occasionally weakened America's position in international science. The trigger for the escalated support was the publication of a report by the National Academy of

Sciences Committee on Oceanography (NASCO), detailing the status of American oceanography and the need to stay ahead of the Soviet Union. Although international cooperation was mentioned briefly, the report and the resulting organizational structure emphasized competition—a race with the Soviets. In programs that might have been successors to the IGY in terms of international cooperation, such as Project Mohole, American scientists spurned cooperation in favor of achieving a "first." However, this competition-based method of achieving international leadership was self-defeating. By retreating into national programs, the United States ran the risk of abandoning the initiative in proposing and leading truly international projects.

OUTLINES FOR RENEWED AMERICAN EFFORT

The cooperative relationship between the Navy and oceanographers, growing stronger since the beginning of the decade and strengthened during the IGY, helped to improve the support structure for national science. The launch of *Sputnik* resulted in sharp funding increases for many scientific disciplines, in an effort to keep America "ahead," a term rarely defined specifically, of the Soviet Union. Oceanography was no exception. The Navy already had begun to plan a deeper collaboration between itself and civilian scientists, to help address its strategic goals. In 1956 scientists under Project Nobska had made detailed recommendations about the development of submarine-launched ballistic missiles, which spurred another era of close interaction between science and the Navy in research and development for the first of this new breed of missile: *Polaris*. The Navy codified its recommendations in a report outlining a long-range program for oceanographic research. This report, dubbed the TENOC (Ten Years in Oceanography) report, envisaged a massive expansion in scientific research to serve the Navy's purposes. The report observed that the growth of oceanography in the United States, and its now recognized importance for national security, stemmed essentially from the Navy's interest in the oceans. "With no exceptions," the report confirmed, "the Department of the Navy is the largest supporter and user of the oceanographic research."[1] For 1959, the research and development budget for oceanography had been $7.6 million. The TENOC report recommended an annual increase of approximately $2 million, resulting in a budget of roughly $27.8 million by 1969. In addition to these expenditures, the TENOC report recommended that $11.8 million be spent on buildings and an unprecedented $51.6 million on new research vessels. These plans were endorsed by the chief of Naval Operations, Arleigh Burke, on the premise that they ought to have priority because of the direct bearing they had upon antisubmarine warfare.[2]

The TENOC report was authored by scientists at ONR (Gordon Lill, Arthur Maxwell, and F. D. Jennings) who supported the Navy's patronage of oceanography and sought to expand it further. It confined its purview to the civilian laboratories such as Scripps, Woods Hole, Hudson Laboratory (which already drew 100 percent of its funding from the Navy), Lamont, several other smaller laboratories, and new laboratories envisioned for the future, one on each coast to be dedicated explicitly to the Navy's specific needs.[3] The Navy already planned to spend $2 million each on Scripps, Woods Hole, and Hudson, but the authors felt that the Navy ought to help these institutions expand, particularly by providing funds to hire people and build new facilities. The report, the authors claimed, had been spurred by the authors' "long felt need for steadily increased emphasis on oceanography by the Navy." Indeed, they claimed that 80–90 percent of the scientific programs in the United States were supported directly by the Navy through its different bureaus and through ONR, and the Navy ought to help these civilian institutions better address the Navy's needs. The report outlined the current research programs of each institution, including the number of personnel, ships, and other facilities, then went on to describe how this work could be expanded to meet the requirements of antisubmarine warfare research in the 1960s. Of the $27.8 million budget it envisioned for 1969, about half would be taken up by Scripps and Woods Hole, with Hudson Laboratory ranking a strong third with about $5 million.[4]

The TENOC report echoed some of the concerns voiced by Chief of Naval Operations Arleigh Burke some years earlier, namely, that the U.S. Navy was poorly prepared to operate in Arctic areas. The pressing need to conduct oceanographic research in this area had led to Project Ice Pick, a clandestine research operation for military purposes (see chapter 2). The Navy learned then how valuable international cooperation could be for gathering much-needed data. This lesson was not forgotten by the TENOC authors, who stated that there existed "an urgent need for cooperative research in oceanography with England, Canada and Norway in strategic areas where POLARIS . . . submarines may soon be operating." Research programs in all strategic areas could be accomplished much faster, and at smaller cost, with the cooperation of scientists in Europe, Canada, and Japan. Because "the time element is considered to be of overriding importance," the TENOC authors felt that information exchanges with these countries should be encouraged.[5]

Actions to strengthen American oceanography vis-à-vis the Soviet Union were not confined to the Navy itself. A similar projection of the nation's needs came out of the National Academy of Sciences. Its committee on oceanography (NASCO) had been created to coordinate national activity in oceanography during the IGY. The national support structure it helped to create would not, however, be directed toward international cooperation, but toward strengthening national efforts in the Navy and elsewhere to ensure leadership against Soviet challenges. This is not to deny that many of NASCO's members earnestly felt that scientific cooperation was a crucial aspect of scientific activity. But even these internationally minded scientists placed a higher priority on American strength. Athelstan Spilhaus, NASCO's first vice-chairman, had been an advocate of putting up a satellite during the IGY because "America needs a first," to improve its morale in the face of the growing strength of communism during the late 1950s. When the United States did not put up its satellite first, NASCO decided to promote oceanography in terms of the need to maintain American scientific leadership in marine science.[6]

When NASCO developed its recommendations for oceanography, it envisioned a major national oceanographic program to ensure the country's ability to lead the world in naval technology and in exploitation of the sea's resources. To chair the committee, National Academy of Sciences president Detlev Bronk felt he needed a man who could help put its recommendations into action, and he appointed geochemist Harrison Brown. Because he was not an oceanographer, the choice of Brown may have seemed an odd one; however, it was his very lack of connection to the oceanographic community that made him attractive. Leaders of the major oceanographic institutions felt that none of the country's better-known oceanographers should be appointed to the position, because the rivalry between institutions was too great.[7] In addition, it would provide the committee with greater credibility when it promoted findings that begged for more money for oceanographic institutions. Brown himself doubted that NASCO's recommendations would do more than gather dust on a desk somewhere in Washington. To prevent this, he asked Bronk if he could propagandize the committee's findings. A little publicity, he reasoned, might help give oceanography a boost, much needed in the face of public clamoring for supporting space science. Aware of the disproportionate attention to rockets and satellites, Bronk conceded that a little marketing was appropriate under the circumstances, if oceanography hoped to garner its fair share of the post-*Sputnik* funding for science.

The first installment of NASCO's report, "Oceanography, 1960–1970," released in 1959, outlined in detail the requirements of the nation amid growing conflict with the Soviet Union.[8] International cooperation took a backseat to the national programs designed to improve America's standing as the premier oceanographic nation. A significant portion of its recommendations on international cooperation was devoted to discussing international politics, not international science. It envisioned an era in which knowledge of the sea would become more important in establishing international agreements and norms for fisheries, mineral exploitation, and monitoring pollution by radioactive waste disposal and undersea nuclear weapons testing.[9] The NASCO report also acknowledged the coming plight of all peoples, which it identified as the inability of food resources to keep up with the world's growing population (particularly in Asia, whose population NASCO projected could increase ten times more than anywhere else).[10] Using science to feed starving people would shortly become one of the rhetorical focal points of the International Indian Ocean Expedition. However, in 1959, NASCO's agenda was not to help starving populations abroad, but rather to make the United States less dependent on foreign sources for food. The solution was to expand America's knowledge of fisheries throughout the world's oceans. "Our committee's problem," the report explained, "begins at home with needs that will become increasingly evident over the course of the next 20 years."[11] As for international scientific activities, the report mirrored the priorities set by SCOR, and it urged the National Science Foundation to contribute a reasonable sum (approximately twenty thousand dollars per year) to that body. It envisioned a declining role for Americans in spearheading international studies, anticipating that the responsibility should fall to a "World Oceanographic Organization" modeled on the World Meteorological Organization. This idea shortly came to fruition with the formation of the Intergovernmental Oceanographic Commission, but at the time of the NASCO report, planners knew for certain only that the American oceanographic community should have the same relationship to oceanographic cooperation as the United States had to the United Nations. In other words, international cooperation in the NASCO report was a matter of honoring commitments.[12]

The report placed most of its emphasis on the importance of marine science for national security. It recommended theoretical work to help develop new technology, and data collection to provide for accurate oceanic and atmospheric forecasting. "Practically every weapon system now used or being used by the Navy," the report asserted, "could be markedly improved in effectiveness through more reliable environmental forecasts." In the area of undersea warfare, the need for forecasting data was especially pressing. The NASCO report estimated that the effectiveness of present forces could be doubled if the Navy could take the environmental factors of weapons and detection systems into greater consideration.[13] Some of these claims were made only in the longer versions of the report, released after the publication of the first installment in early 1959. But even the initial report minced no words in its plea for America to set its priorities straight in terms of military spending:

> From the point of view of military operations there is no comparison between the urgencies of the problems of the oceans and those of outer space. The submarine armed with long range missiles is probably the most potent weapon system threatening our security today. It seems clear that the pressures of establishing effective bases, and of protecting ourselves from attack, are relentlessly driving us into the oceans.[14]

The report argued that Americans should place greater national emphasis on research in physical oceanography, geology, and geophysics, to make the oceans transparent to American eyes.

The pleas of NASCO met very different receptions in the executive and legislative branches of the United States government. The president made no particular show of support, whereas Congress enthusiastically supported taking major steps to improve and expand oceanography in the United States. One reason for the president's disinterest was that the three scientific arms of government designed to advise him (the Office of the Special Assistant, the President's Science Advisory Committee, and the Federal Council for Science and Technology) were all headed by the same man, chemist George Kistiakowsky. According to Edward Wenk, Jr., who later would become the first secretary of the Marine Sciences Council, Kistiakowsky was particularly sour on oceanographers. He resented NASCO's entrepreneurial attitude toward science, feeling that the attention it gained through the 1959 report had stretched the traditions of scientific ethics. He disapproved of NASCO's unabashed marketing of its findings to Congress and was decidedly unfriendly to the aspirations of the oceanographic establishment. As long as Kistiakowsky was the president's top advisor, NASCO had to find its advocates elsewhere.[15]

Oceanography was fast becoming a vehicle to debate larger and more politically charged issues about American national security. Leading Democrats used the NASCO recommendations to attack the policies of the Eisenhower administration, which had tended to devalue the importance of conventional forces, particularly the Navy, throughout the 1950s. Senator Hubert H. Humphrey of Minnesota long had argued that the policy of "massive retaliation" against the Soviet Union with nuclear weapons severely limited American military and political power. Recent crises in Lebanon and in the Formosa Strait had proven that conventional forces were "fast rebounding to center stage." Shortly after the release of the NASCO report, Humphrey gave a speech to the Senate, claiming that "if we insist on concentrating all our thinking and planning in one direction, we are going to go the way of France in 1940." The Soviet Union, he claimed, was investing not only in missiles but also in oceanography, as part of an intensive program on submarine development and construction. Moscow, Humphrey insisted, "seeks to make the ocean its great base of operations—a base which covers three-fifths of the earth's surface." Research in oceanography was a critical means to ameliorate what he perceived as the two great gaps in "our national armor," namely, the lack of sufficient conventional forces to deter and respond to communist-inspired attacks and the lack of a true deterrent to a massive strategic attack. He argued:

> This is the gap caused by our ignorance of the secrets of the oceans, the marked failure of the antisubmarine science to keep up with the progress being made by the world's submariners, and the all-out construction program of the Soviet undersea force.[16]

Research in oceanography held the key to both, through developing a stronger Navy and an undersea nuclear force. Ever since the launch of *Sputnik*, Humphrey claimed, Americans had been disabused of their notions that strategic air forces alone could possibly act as a true deterrent. He reminded the Senate that if the "Atlantic Community" did not begin to pay closer attention to the sea, they risked being left "in the wake of another surge of Soviet scientific advance."[17]

Humphrey was not alone in using the NASCO findings to attack the Eisenhower administration's neglect of the sea. Contrasting sharply to the White House's cold reception, Congress was immediately galvanized by the report. Indeed, some congressmen, informed of its contents beforehand as part of NASCO's promotion strategy, had been waiting impatiently for the report's release so that they could act upon it. As early as March 1959, Congress was sufficiently moved "under the immediate inspiration of the provocative report recently released" to initiate hearings to see how best to implement NASCO's recommendations. The Committee on Merchant Marine and

Fisheries listened to further evidence of the perils facing America. Richard C. Vetter, NASCO executive secretary, explained that, if one were to construct a chart of all oceanic observations and color them according to the nationality of scientists, it would reveal a rough parity among the United States, Japan, the United Kingdom, and the Soviet Union. However, that parity existed at the expense of the United States' clear leadership just a few years before. The IGY had stimulated interest worldwide. Since then, all three other nations had stepped up their oceanographic programs considerably, dwarfing the rate of growth in American oceanography. Although the United States was still at the forefront of research, Vetter warned Congress, "we are not advancing as rapidly as Russia." He instructed the committee to consider the small amount being done already and "to determine to your own satisfaction whether this amount of activity in these fields is adequate." In NASCO's estimation, it most certainly was not.[18]

Through Vetter and others, NASCO testified a number of times before Congress to make its recommendations heard. West Virginia congressman Ken Hechler asked Vetter whether knowledge of the oceans, "which we do not now have," could truly be critical to the future national defense of the United States. Vetter estimated that obtaining more information about the ocean environment would increase the efficiency both of the Navy and of fisheries groups by 100–200 percent. Not only that, but oceanographic data were not static; they required constant refreshing and updating. Vetter reminded the congressmen that "the Navy is entirely dependent upon a vast amount of oceanographic information for the effectiveness of its operations." Such data needed to be gathered on a continual basis. Vetter tried to make a comparison:

> Perhaps the homeliest analogy you can draw is to compare the success that the Indians had in resisting, and in fighting with, the settlers. They were able with inferior weapons and with inferior communications to quite often very seriously inflict damage upon more highly organized and well-equipped troops simply because they knew their way around the woods much better and could find hiding places and mechanisms of attacks that were unknown and unappreciated by the paleface from the city. It is the same in the oceans.[19]

Vetter's emphasis was not upon new weapons systems, but upon basic knowledge of the sea. The nation that could learn the most about the oceanic environment, Vetter explained, would naturally have a decisive advantage over its enemy. This would require immediate and sustained financial support for American oceanographers collecting data throughout the world.

Congress began initiatives to expand the oceanographic and geologic responsibilities of several government agencies, including the Coast Guard, the Geological Survey, and the Coast and Geodetic Survey.[20] Oceanography was such a hot issue that two committees in the House of Representatives vied for jurisdiction over it, each hoping to be the first to introduce legislation in its favor. In 1960, the Science and Astronautics Committee under Louisiana congressman Overton Brooks issued a report insisting that the Soviet Union had begun a major program to surpass the oceanographic efforts of the United States and that the world should expect major propaganda initiatives—akin to the *Sputnik* launch—in oceanography in the near future. The report, titled *Ocean Sciences and National Security,* went even further than the NASCO report in its recommendations to strengthen oceanography in order to stay ahead of the Soviet Union.[21] In addition to the plenitude of supporters in the House of Representatives, the most significant advocate for oceanography was Washington State senator Warren G. Magnuson, the chairman of the Senate Commerce Committee. Described by science policy advisor Edward Wenk as "Mr. Oceanography," Magnuson was a font of oceanographic legislation. He introduced a resolution (which subsequently passed) to increase spending in oceanic research as a whole, because American oceanography was lagging, "whereas several other nations, particularly the USSR, are presently conducting oceanic studies of unprecedented magnitude on a worldwide basis."[22] The resolution also reiterated the need to engage in international cooperative efforts but noted that such arrangements should be more carefully supervised to ensure reciprocity on the part of the Soviet Union. This was only the beginning of several years of advocacy in Congress, not always successful, during which Magnuson and others sought to establish long-lasting legislation to solidify not only the relation between science and the Navy but also the role of marine science in national security.

The increased influence of marine scientists in Washington, the reorganization of the science's national infrastructure, and the ever-present comparison to the "state" of the science in the Soviet Union, all had profound ramifications for oceanographers in the United States. For one, all three seemed to define scientists' role increasingly as one of partnership with the Navy. After reading the Navy's TENOC report, Woods Hole director Paul Fye wrote to Gordon Lill at ONR to congratulate him on the impressive document. Finally, he felt, oceanography was going to receive the support it needed. "I only hope that we here at Woods Hole," he wrote, "can do our part in the next decade to fulfill the requirements visualized."[23] Fye himself was a strong advocate for those wanting to cement oceanographers' partnership with the Navy. Although it may be tempting to say that scientists merely played up the defense aspects of their research in order to receive adequate funding, this was certainly not the case for Fye, who harbored genuine views about strengthening the partnership between scientists and the Navy. In a December 1958 letter to the Office of Naval Research's research director, he outlined the main areas in which this partnership could be cultivated to help both scientists and the Navy. First of all, oceanographers had "an obligation to assist in training the submariners so that they can understand the environment in which they must live." Second, Woods Hole scientists should devote considerable research effort toward bottom topography, or mapping the seafloor. Third, research on water movements would be critical for identifying the conditions inhibiting sound propagation. This would be of the greatest use to the submariner, who "must be able to find the safest place in which he can operate and hide." Fourth, Woods Hole scientists should develop new equipment and techniques to enable submariners to conduct their own measurements at sea, without the direct assistance of scientists. Fifth, and last, research on weather forecasting would, for obvious reasons, be useful for the Navy.[24]

Others were not so comfortable with the potential ramifications of such support and feared it might spur an unbridled expansion of oceanography for military and economic purposes. Scientists might be less able to focus on academic problems and instead would be confined to large-scale surveys, providing data for programs designed by others. The promise of increased patronage over the next decade worried some scientists, who knew that much of the lobbying in Washington had been done on the premise that oceanography served the "national interest." One of Woods Hole's oceanographers, William von Arx, wrote to the director and members of a committee on science policy that the institution should be wary of making sweeping changes in favor of Navy work. He suggested that Woods Hole, in the face of growing political support (and consequent pressure), should "rededicate our efforts to scientific innovation rather than to allow ourselves to become overblown with people and to amass equipment that inevitably will steer the course of our efforts toward large-scale undertakings of a survey or developmental nature." Von Arx was worried that Woods Hole would undertake policies that stripped scientists of their freedom of inquiry, policies that might "divert or dilute our intellectual resources or reconstitute the scientific atmosphere of this Institution."[25]

Fye, however, strongly supported not only the increased patronage but also the purpose it seemed poised to serve. He had been among those arguing for it, and now he began a policy review to convince his colleagues to rededicate themselves to oceanography in service of the Navy. He sent a memorandum to Woods Hole's Executive Committee and laid out his own philosophy of research and his plans for the institution. In his review of the issues at stake, Fye made his own agenda transparent. He envisaged a close collaboration with the Navy and a sustained expansion of the institution along lines compatible with defense projects. He pointed to the TENOC report, which had the strong support of the chief of Naval Operations, and the Navy's need for oceanographers to provide the research to improve the operational capabilities of nuclear-powered submarines below the surface of the sea. "Already the undersea ships and armament have outpaced our understanding of the medium in which they function," Fye argued. What he called the "need-to-know" about the oceans had increased by an order of magnitude in just the past year or so. Granted, there were many levels of "need-to-know," from food and power resources to waste disposal, but he emphasized the military requirements "because of their established urgency, because oceanography has so far made its most outstanding contributions in this field and because some within the Institution are reluctant to recognize the important contribution to national and world security that oceanography must make."[26]

Fye was confident that the problems many perceived with a rapid expansion would be trivial compared to the benefits. Some of his colleagues had argued that such an expansion would translate into an emphasis of quantity over quality. But their solution, to restrict Woods Hole's growth carefully, was anathema to Fye. He chided them for wanting to condemn the institution to obscurity during what he considered to be a period of revolution in oceanography. "To some, the attractions of the pastoral life are irresistible," he wrote. But if they did not assume a role of leadership during the times of change, that leadership surely would pass to others. In ten years, he argued, Woods Hole no longer would be one of two or three major research institutions in the United States. That monopoly would end, and each existing institution would have to make the most significant choice in its history. "I believe" he wrote, "those who fail to understand what is happening to oceanography—or seeing it, try to shelter from its impact—will become second-rate, pedestrian institutions within five years."[27]

The director's vision of the future met a cautious reception among other scientists at Woods Hole. Von Arx agreed that *Polaris* was one case in which the technology was driving the science, but that there were many scientists

Competition and Cooperation in the 1960s

who were exploring aspects of nature "for which we don't even realize there is a 'need to know,'" and they should be encouraged.[28] Allyn C. Vine agreed that fast growth for the Navy's sake was probably the best way to attract interest from the government, until the United States finally began to appreciate the ocean for more positive motivations. "The center of gravity of oceanography," he wrote to Fye, "is clearly moving toward Washington." At the same time, he feared that making the purpose too explicit might scare off some good scientists. Vine felt that latching onto the *Polaris* system would benefit them probably until the late 1960s, but he also felt that Woods Hole should make a strong effort to encourage talented individuals to follow their own intellectual interests.[29] E. Bright Wilson, Jr., agreed that the institution should not abandon its image as a place where science flourishes for its own sake. "We earn our bread and butter this way," Wilson admitted of Navy work, "and have to keep doing it for money and for patriotic reasons, but I would hope we do not lose sight of the fact that we are fundamentally a scientific and not an engineering laboratory."[30]

Henry Stommel tried to put the matter in the context of the international community. He particularly felt that military objectives ought not to be so prioritized and that the general development of the science was more important. He reminded Fye not to forget the international dimensions of science and suggested that, rather than close ranks, Woods Hole ought to try to stimulate more international efforts, not only with the National Institute of Oceanography in Britain, but with the Soviets and the Indians as well. Young scientists, Stommel argued, tended to shy away from areas of research that are explicitly defense oriented. His own students, he wrote, were suspicious of the sources of money at Woods Hole, and they were more eager to establish an individual reputation that was recognized internationally, rather than be allotted to teams of researchers who tackled problems of particular importance to national security. He wrote:

> We are not all hurrying together on a great well-regulated four-lane highway into a glorious future. I think there is still jungle and undergrowth and unexplored territory where we need individual pathfinders on foot.[31]

If indeed the spirit of Fye's memorandum was well founded, Stommel wrote, then at least there ought to be two major subdivisions of the institution, one for the practical sciences and one in which academic individuals would feel most comfortable. He suggested that Fye "separate them physically, morally, financially."[32]

Fye's memorandum had struck a particularly sensitive cord for Stommel, who during sixteen years at Woods Hole had felt that most of his problems there had stemmed from the attitude exemplified in Fye's words. In Stommel's view, scientists needed freedom, and the "pastoral" attitude so condemned by Fye was precisely what led institutions toward leadership. He went on to cite past examples of scientists whose primary difficulty was the lack of autonomy. In 1940, for instance, the institution had lost Carl-Gustaf Rossby, who left to start the Institute of Meteorology in Chicago. In 1950, it had lost W. Maurice Ewing, who left to become the director of the Lamont Geological Observatory. Such natural leaders craved personal autonomy, Stommel wrote, and they ought to be granted it because they will be the path breakers of oceanography, and they will not only inspire their colleagues but also establish a favorable image of the field in the public's view. "One cannot preach to such proud men," Stommel wrote, "that their first duty is national defense." Rather, they should be encouraged to wrestle with the problems of the science in their own way. He urged Fye to reassure his solitary "pastoral" scientists and to provide them the freedom they needed.[33] Stommel's plea tied a scientist's reputation to the approbation of the international community, not to the patriotic needs of one country. The problem, of course, was that the needs of the Navy were the reason that oceanography was experiencing such growth, and these connections could not be ignored.

THE INTERAGENCY COMMITTEE ON OCEANOGRAPHY

Along with the ever-present role of the Navy, other patrons took an escalated interest in oceanography, so much that it became difficult to coordinate all of them. Oceanographers wanted to give everyone a stake in the oceans. When ONR's Gordon Lill became a corporate research advisor for Lockheed Aircraft Corporation, for example, he tried to make the organization operate with a background solidly based upon science. Even industry, he wrote to the president's science advisor, Jerome Wiesner, "must have an oceanographic esprit de corps if it is to be effective in ASW [antisubmarine warfare], mining, ocean systems, exploration and sea launched rocketry." Lill claimed that he and others were trying to convince scientists within industry that they belonged to the oceanographic scientific community and that they too should have a role in the development of the oceans.[34] With the help of the TENOC and NASCO reports and the enthusiasm of Congress, oceanographers succeeded in creating new disciples of marine science throughout the government. In fact, the biggest problem facing oceanography was not funding, but how to organize a coordinated "national program." Ultimately this led to the establishment of the Interagency Committee on Oceanography. Although virtually all scientists felt that more should be done in oceanography, and that it was essential to devise a national program, they resisted efforts to centralize control or to establish a new agency. The various government agencies hoped they could have a hand in an organization that brought them all together. Rear Admiral H. Arnold Karo, director of the Coast and Geodetic Survey, spoke at length before Congress on the avenues of research supported by his agency, all of which needed to be taken into account in any comprehensive oceanographic program. One committee chairman, North Carolina congressman Herbert C. Bonner, thought that Karo might have been trying to persuade the committee to put all new oceanographic programs under his control. But nothing could have been further from Karo's intentions. He was no scientist, but he led an agency that depended upon knowledge of the sea; like the leaders of many other government agencies related to oceanography, he wanted better coordination.[35] Scientists likewise felt that it would be senseless to try to fix the problem confronting America by putting together a new oceanographic institution. As Lamont director W. Maurice Ewing pointed out to Senator Magnuson, "a research institution cannot simply be assembled by effective, adequately financed management. No matter how excellent the management, how good the intentions, how great the need, or how adequate the budget, the fact is that for achievement in any branch of science, the basic need is for creative, scholarly individuals." Ewing and others wanted continuity of support for existing institutions and better coordination with those who needed what those institutions produced.[36]

The prospect of having no new agency and no new scientific institution was puzzling to Congress. It envisioned a crash program in oceanography, much like it had initiated for the space program. It also envisioned a new agency or institution in which it could focus its efforts. Some members of Congress expected a NASA for the oceans. But few of the principals of oceanographic research, either the producers or the consumers, seemed to want this. California congressman George P. Miller confessed to Navy Rear Admiral John T. Hayward "that this committee at present is groping. . . . We realize that, whereas [a national oceanographic program] is of maximum interest to Government, no specific agency of Government is charged with all of its ramifications."[37] But these ramifications were so diverse that it would be impossible—and counterproductive—to centralize authority. The Navy needed oceanographic research to improve the effectiveness of its surface ships and submarines, as well as improving antisubmarine, mine, and amphibious operations. But the Navy was not alone in needing oceanographic data. The various agencies of the government—including the Bureau of Commercial Fisheries, the Atomic Energy Commission, and even the State Department—all had specific responsibilities and needs.

Eventually the scientists and various government agencies found common ground in advocating a new coordinating committee made up of representatives from various agencies. Richard Vetter, NASCO executive secretary, testified that diverse interests in oceanographic research should be encouraged to proliferate and should not be reined in. In oceanography, he claimed, "there needs to be a close association between the groups that need the information and the groups that are producing the information."[38] This association would engage universities, oceanographic institutions, government agencies, and military organizations in an ongoing conversation about the requirements of science and its potential applications. Harrison Brown testified before Congress, for example, that there were no mechanisms for governmental groups to provide scholarships in a scientific area in which the United States was perceived to be lagging behind the Soviet Union. The United States needed a body to assess these needs. Congressman Miller interrupted to say, "That is what the Russians would do," and Brown concurred, yes, "that is exactly what they are doing." When the Soviets wished to build up a scientific field, he explained, they simply put a lot of money and effort in that field, to the exclusion of many others, easily persuading scientists to enter the field. Many Americans, particularly in Congress, also wanted a crash program. But instead, NASCO appealed for a sustained expansion of research, to be guided by scientists and government officials alike over a long period.[39]

Although Congress would continue to grope with this issue until the creation of the Marine Sciences Council in 1966 (in fact, even after that), coordination between scientists and government agencies was already in the works. It began as an oceanographic subcommittee of the Federal Council for Science and Technology, and by 1960, it acquired full status as the Interagency Committee on Oceanography (ICO).[40] The interagency committee was not merely a way to coordinate; it was also a fruitful alliance of oceanographers, the military, and government agencies. The coordinated voice provided by the ICO found a receptive ear in the White House after Eisenhower left. Jerome Wiesner, who replaced Kistiakowsky as the chief scientific advisor to the president, was more sympathetic to the needs of oceanography, as was the new president. In his address to Congress on February 23, 1961,

President Kennedy pointed out that the sea was one of America's most important resources. A month later, he addressed the subject of oceanography in more detail in an executive communication to the Speaker of the House. He called for more ships, for more shore facilities and data centers, and for a renewed commitment to basic and applied research in oceanography. "Knowledge of the ocean," the president asserted, "is more than a matter of curiosity. Our very survival may hinge upon it."[41] Between fiscal years 1961 and 1962, the oceanographic budget estimated for the Department of Defense nearly doubled, and for the Department of Commerce it almost tripled. In general there were great increases earmarked under the ICO's "National Program in Oceanography" for the Department of Interior, National Science Foundation, Atomic Energy Commission, and the Department of Health, Education, and Welfare.[42] The establishment of the ICO was part of NASCO's strategy of enrolling several disparate segments of the American economy and government into the cause of oceanography. By the end of 1962, Wiesner was admiring how the nation's efforts in oceanography had grown rapidly over the past few years. He credited that growth to the oceanographic community's government-wide approach to planning and implementation of programs, particularly the ICO, which was "a model for interagency coordination in other fields of science."[43]

As "national" programs in oceanography gathered momentum, American oceanographers during the early 1960s were Janus-faced about international cooperation. They pursued competitive national ventures while still promoting international cooperation. Granted, the American oceanographic community was not monolithic in its attitude toward cooperation. Its leaders never had been known for congruence of opinion. As Athelstan Spilhaus later said, "they were like a bunch of artists who get together. You can't have a union of artists; everybody hates each other's work and they're all as competitive as hell."[44] The same was true about their attitudes toward the role of international cooperation in U.S. science. The idea that it was a central component of research was not championed by all of them. But one of the most influential of them, Roger Revelle, consistently advocated for it. Spilhaus later recalled that "Roger would believe in any effort, like the International Geophysical Year, whatever it was doing. Roger was that kind of guy; he just believes in international things."[45] Indeed, even as the nation was responding to the Cold War anxiety stimulated by the NASCO report, Revelle retained his outspoken support for international cooperation. When testifying before Magnuson's committee, Revelle could not help himself from urging Congress to support his view that the United States should exchange information with the Soviets. As he had argued many times during the 1950s, Revelle still held fast to the notion that such exchange would do more good for the United States than it would for the Soviet Union.[46] Revelle also exercised considerable influence over the oceanographic community. Not only had Revelle been director of Scripps and a member of NASCO, he was the leading spokesman for American oceanography in international organizations. In addition, he and National Academy of Sciences president Detlev Bronk were great friends; as Walter Munk once recalled, they were "one of a kind."[47] Leaders of the scientific community, both at the government and at the university levels, supported international cooperation as an integral part of scientific work.

Despite continued attention to international cooperation, national programs emphasizing American strength and leadership were ascendant in the early 1960s. This was due largely to the renewed interest in oceanography by the American president. Senator Magnuson could assure Roger Revelle in 1961 that "there is a greater interest in this field at present than there was 2, 4, or 6 years ago." Revelle agreed that there was a greater realization of the needs of oceanography, particularly because of national security concerns, in the Kennedy administration.[48] But oceanography became a key area of international scientific competition, not necessarily cooperation. Americans constantly tried to determine accurately the "state" of oceanography in the Soviet Union. Scientists did this reconnaissance often at international meetings, as during the early 1950s while preparing for the IGY. Scientists on both sides of the iron curtain engaged in it. The 1959 congress in New York, for instance, was a time for great numbers of American and Soviet scientists to intermingle and to learn of the latest techniques and equipment. Scripps oceanographer Walter Munk later recalled that one of the Americans' principal purposes in cooperation was to examine their enemies— simply to find out what they were doing. Although Munk was not a major participant in the IGY, he felt that the IGY meetings, and East-West scientific meetings in general afterward, were not particularly science oriented. Learning actual oceanography from the Soviets was only a secondary purpose compared to observing what the Cold War enemy was doing.[49] It was a very exciting time for American science, Munk recalled, having an enemy with high technical capabilities. He enjoyed being associated with things that were kept from the Soviets, and he felt that the competition helped to define the problems for scientists to address.[50]

Did American oceanographers truly believe that their Soviet colleagues were about to usurp their leadership in the field of oceanography? Even prior

Roger Revelle (left) and Senator Warren Magnuson in 1961. Revelle was giving testimony before the Senate Committee on Interstate and Foreign Commerce. Courtesy Scripps Institution of Oceanography Archives

to *Sputnik*'s launch, many warned against the West's complacent attitude that science and totalitarianism were by definition incompatible.[51] Exploring the period after *Sputnik*'s launch, it is particularly difficult for historians to assess the difference between rhetorical exaggeration and earnest concern. Oceanography had its own equivalents to the illusory "missile gap." Many scientists felt, for example, that the 1959 NASCO report went too far in emphasizing the national security imperatives of supporting oceanography. But concern about Soviet activities cannot be denied. Reports of Soviet oceanographic cruises, for example, were typically tracked in several sources; these were not only government agencies but also commercial publications. For example, for a number of years the journal *Deep-Sea Research* published translations of articles from the Soviet journal *Okeanologiya*. The editors of the journal did this with the expectation "that the publication of these will prove of value to scientists engaged on similar work in other parts of the world."[52] But the articles were also useful for tracking the activities and priorities of Soviet research vessels.

Americans watched with anxiety following the IGY, as Soviet oceanographers moved quickly into regions that other countries would enter only at a much slower pace. The nonmagnetic schooner *Zarya*, having achieved some notoriety during the IGY as the only oceanic vessel capable of measuring magnetic data accurately at sea, continued its voyages in 1959 and 1960. Although the *Zarya* cruises provided the opportunity to acquire a larger picture of the earth's magnetic field over the oceans, the principal scientific value of these voyages was cartographic—the data allowed magnetic anomalies to be mapped. The *Zarya* concentrated its efforts in the Atlantic and Indian oceans during the IGY, but then it made an extensive voyage in the Pacific Ocean.[53] The *Vityaz,* perhaps the Soviet Union's best-known research vessel, cruised into the Indian Ocean during the same period. Its scientists visited colleagues in a number of countries, including Indonesia, Australia, Ceylon, India, Madagascar, and Zanzibar. The Soviets were the most impressed with India, whose oceanographic work was conducted for fisheries, naval, and university purposes, much like that of the United States and Soviet Union. Like the Americans, the Soviets would attempt to cultivate relationships with Indian scientists over the next few years, particularly during the International Indian Ocean Expedition.[54]

Despite their moves to coordinate national efforts, American oceanographers came under fire for allowing the Soviets to usurp America's power at sea. Particularly, the ICO was criticized. *Ocean Science News* concluded that, although "ICO has done a great deal of yeoman work to advance the cause of oceanic sciences and research, it is nevertheless a boy trying to do a man's job."[55] But staying ahead in oceanography because of national security concerns was not the only issue. The Soviets increasingly felt that scientific competition was healthy, and like the Americans they attached a great deal of importance to scientific prestige. An American visitor to the First Geneva Conference on Peaceful Uses of Atomic Energy wrote that a Soviet minister had chastised him for using the word *konkurentsia* for competition:

> The Soviet Minister replied, 'Do not use that word! It has a bad connotation in the Soviet Union. It means getting ahead by crushing and destroying the opponent. We have a new approach to competition and a new word to designate it. The word *sorevnovaniye* means getting ahead by climbing on the opponent's shoulders.[56]

Americans were fast gaining an appreciation for this conception of competition, particularly in areas such as science, which were symbolic of a nation's power. The Soviets were benefiting from American science, building upon it, and then appearing to take the lead. American oceanography became a matter of national prestige, just as space science had. Athelstan Spilhaus recalled making the comparison in congressional testimony, drumming on his table before the Senate, saying, "Gentlemen, isn't the ocean's bottom at least as interesting as the back side of the moon?" In Spilhaus's view, that testimony was "the shot that was heard around the world" for oceanography.[57] Along with the impact of the NASCO report, Spilhaus's testimony provided an impetus not only for more research money but also for "prestige" efforts in oceanography.

MOHOLE, SOVMOHOLE, NO HOLE, AND THE UPPER MANTLE PROJECT

Project Mohole was the most significant of such prestige programs for oceanic science, and for earth science in general. It is now appreciated, justifiably, as an ambitious science and engineering project gone awry in the face of poor scientific management, corporate inexperience, and government financial constraints. Conceived during the IGY, it immediately took on the tone of Cold War competition. At one of the IGY meetings, a Soviet geophysicist claimed that his country already had the capability to drill a hole ten miles into the ground. This introduced the question of whether the Soviets were beating the Americans as much in drilling technology as they apparently were in rocket technology. When considering a new large-scale project that could be a worthy successor to the IGY, Scripps's Walter Munk and Princeton's Harry Hess discussed the possibility of drilling a hole to sample the mantle, past the "Mohorovičić discontinuity" that served as a border between the mantle and the crust. The idea was sponsored by an informal group of scientists with diverse interests who called themselves the American Miscellaneous Society (AMSOC). Eventually the idea took on huge proportions and became known as Project Mohole. It achieved considerable notoriety in its short duration, largely through the efforts of Willard Bascom, who wrote a book to promote it. He invited John Steinbeck to join an early phase of the project, and the novelist glamorized the project in the pages of *Life* magazine. The whole affair was also covered by science journalist Daniel Greenberg on the pages of *Science,* and he devoted a whole chapter to the failed project in his influential 1967 book *The Politics of Pure Science.*

Despite his dislike for ocean scientists' proselytizing, George Kistiakowsky, chairman of the President's Science Advisory Committee, gave Project Mohole a fair hearing. He did so largely because it had the backing of a number of scientists not known principally as oceanographers, particularly Princeton geologist Harry Hess. Hess claimed that it was foolish to plan trips to the moon, Mars, and Venus when scientists did not even know what made up the interior of the earth. He wrote:

> ... there is no reason why the USSR should not be able to drill such a hole if we don't. While it is distasteful to me to suggest that we do it for a propaganda motive, it would be somewhat more distasteful to me to listen to the Russian propaganda if they do it first.[58]

Project Mohole took on a character rather similar to the more widely appreciated space race. In 1960, Soviet scientists had begun to clamor for a Soviet project to compete with Project Mohole. By mid-1961 they were rumored to have devised plans for major drilling projects. After American scientists succeeded in taking samples of the basalt underneath the sediment on the seafloor, geophysicist Vladimir Beloussov announced—while praising the American achievement—a plan by his own countrymen to sample the basalt as well.[59] As the executive secretary to the AMSOC committee for Project Mohole, Hess and others wished to hire William L. Petrie, whose résumé had him working for the "Cuban Invasion Agency" (presumably the Central Intelligence Agency, or CIA) since 1954. At the CIA, Petrie had followed Mohole's Soviet counterpart, "Project SOVMOHOLE," with great interest.[60] Although Petrie remained connected to the project, it was the marine engineer Willard Bascom who eventually took the job. Bascom popularized the name "Mohole," a catchy way of describing the hole-drilling project while avoiding the need to pronounce correctly the Mohorovičić discontinuity. Because the crust on land is very thick, compared to that at sea, the AMSOC committee decided to drill the hole at sea. As this had never been accomplished before, Project Mohole became an ocean engineering task of unprecedented proportions.

On September 10, 1961, the *New York Times* ran a story announcing a Soviet plan to drill five holes (on land, not at sea) at depths of six to nine miles, challenging the American project.[61] Hess wrote to a colleague at the Navy's Hydrographic Office that the recent announcement of a deep drilling program by the Soviet Union made it essential that the American project "get going without delay or we may appear to be second again in a competing scientific program."[62] Bascom wrote to the newspaper's editor that the drilling projects served as a reminder that the United States was constantly "engaged in a scientific olympics with the USSR." Although healthy competition could be good for science, the friendly competition was "deadly serious just the same, for it is run in the shadow of a barely stable political situation." The Americans seemed safe: on land, they had already drilled a hole more than 25,000 feet deep, while the Soviets had achieved as yet only about 17,000 feet. At sea, where the Americans were planning to drill for Project Mohole, the American record of 11,700 feet was almost a hundred times the Soviet record. But Bascom shunned complacency:

> Nevertheless, I am concerned that another hare and tortoise race will develop. Clearly our opponents are dedicated persevering people who play games by their own rules. Although they have a long way to go, they have accepted the conditions of the race. Doubtless they would enjoy beating us at our own game and adding this laurel to their space firsts. We must also be wary that even in areas where they are not superior, propaganda may make them seem so. We must run, and win, and tell the world we have won.[63]

The leaders of Project Mohole were convinced that deep-sea drilling had become an arena in which to prove American scientific superiority, and they dreaded the possibility of a cosmetic victory that the Soviets could turn into a major propaganda coup.

Mohole's project leaders insisted that it be an American effort alone. Hess was adamant that the Americans should "avoid the hot air and nonsense that went along with so much of the IGY program."[64] Some criticized this move, saying that the Americans should pay more attention to international bodies such as the International Union of Geodesy and Geophysics, which had proposed an international study of the earth's crust and mantle. Particularly in the wake of the IGY, it seemed appropriate to coordinate research with other nations. Office of Naval Research scientist Gordon Lill tried to defuse such criticism, claiming that Project Mohole ought to be construed as the contribution of the United States to such an international program. He added that deep-sea drilling was too complex an endeavor to attempt management by an international body. Once the major drilling began, Lill promised, the Americans would invite observers from other countries.[65] When NASCO drafted a report mentioning that drilling to the mantle presented many technical difficulties that might benefit from international cooperation, Bascom wrote to Richard Vetter expressing his severe disapproval. He felt that the United States needed no such help at all. It would be naïve to believe, Bascom urged, that even an international conference could have any benefit for the project. He disagreed with the notion that international cooperation was imperative for success, and he also disagreed that it was wrong to use science to promote national competition.[66]

What irked the Mohole organizers most about suggestions to "internationalize" the project was that they were making real strides in ocean drilling, whereas the Soviets had made none. What would cooperation gain them? They were genuinely afraid that the Soviets would exploit any cooperative venture and somehow achieve another "first." Certainly the Soviets had begun their own plans to drill deep into the earth's crust, but all their holes (in 1961 they announced plans for five) were to be drilled on land, and only one was slated to penetrate the Mohorovičić discontinuity and probe the mantle itself. That hole, on one of the Kurile Islands, was to be drilled last. Americans were skeptical of this plan, because it would require a hole almost 33,000 feet deep, and the deepest Soviet hole, drilled by the state-operated oil industry, thus far had been only 17,000 feet deep. However, AMSOC warned that given its ability to concentrate its efforts, as revealed by the space "firsts," the Soviets could still "make tremendous progress in that direction albeit at the expense of consumer items and living standards."[67] Despite these fears, the Soviets actually had no plans to drill at sea; indeed they were skeptical that it was even possible, because drilling from a floating vessel would be too unstable. The Soviets actually seemed to be toying with the idea of a geological probe modeled after spacecraft, designed to radio back information about the earth's interior. Beyond this fantasy, however, they were working with land-based drilling equipment that was initially designed for the oil industry.[68] Nevertheless, Bascom and others felt that international cooperation would pull the carpet from beneath them when they already were on a clear path to a scientific and technological "first" in the race with the Soviets. To cooperate with them, even to ease international tensions, seemed foolish. "If the USSR has the slightest intention of becoming more friendly," Bascom wrote, "there are plenty of places for it to begin—and these need not include the area of drilling technology where the US has a substantial technical superiority."[69]

In his 1961 book on Project Mohole, *A Hole in the Bottom of the Sea*, Bascom affirmed the project's place as a national effort comparable to the space race. The United States was competing, he wrote, for scientific and technical supremacy with the Soviet Union, and the primary battleground thus far had been outer space. However, he wrote, "if one disregards the romance and fun of landing on the moon or Mars and honestly answers the question: What is the best way to search for new evidence about the solar system? drilling down will easily win over rocketing out."[70] It was proper for two great nations to compete in this way, he wrote; besides, the United States was well in the forefront of drilling in the oceans. Harry Hess also expressed his consternation at the prospect of bringing the Soviets in on Project Mohole, and he urged the National Academy of Sciences to block any such proposals. Its president, Detlev Bronk, agreed with Hess and Bascom that there were many reasons not to include the Soviets, "not the least of which would be their desire to ride on our success" in drilling to the mantle.[71] On October 7, 1961, AMSOC passed a resolution affirming that international cooperation on Project Mohole was unnecessary, but that Mohole should be considered simply the American contribution to a larger (but not coordinated) international effort.[72]

Naturally the visceral reaction against international cooperation, and the belief that Americans did not need outside help, applied only to the Soviet Union and its allies. Many scientists from other nations, such as France, the United Kingdom, and South Africa, participated in the project on a personal basis (not as representatives of national bodies). After AMSOC determined that one of the best sites for drilling was on islands owned by Brazil, its attitude toward cooperation brightened. William Petrie wrote to a Brazilian colleague that "since this is a scientific project of international significance we would be most happy to have the cooperation and participation of our scientific colleagues in Brazil."[73] Opposition to international cooperation had nothing to do with a belief that it was unnecessary, but rather was a way to keep the Soviets out of the project.

Despite this early confidence, America's plans began to falter by the mid-1960s. Granted, President Johnson had reaffirmed his predecessor's support for oceanographic research. His proposed budget for the national program in 1965—$138 million—was 11 percent more than the appropriations for 1964. He called this the "absolute minimum necessary" to carry out America's objectives in investigating the sea.[74] Still, it was also during the Johnson administration that Project Mohole was canceled, after years of effort by many influential figures in the American earth sciences community. Although the funding and administration of Project Mohole, not to mention the decision even to pursue it, was smitten with controversy throughout its short history, ultimately one can blame the fiscal constraints caused by the Vietnam War for its demise. The details of the Mohole affair, along with scientists' general disgust at how the National Science Foundation handled it, are chronicled elsewhere.[75] Suffice to say that it was a major example of a large-scale project that might have been an international one, but was not, because of American anxieties over leadership. Moreover, in this case, the Americans who guarded their project so fervently against international status failed to achieve the all-important "first," making Mohole that much more of an occasion for bitterness. The Soviets, for their part, never came close.

As the United States focused its efforts on Mohole, many prominent non-American scientists made the earth's mantle an international priority. The International Upper Mantle Project was one that the Soviets did dominate, at least in the beginning. Describing the project, Canadian geophysicist J. Tuzo Wilson likened the interior of the earth to a soft-boiled egg. Like the yolk, its core is fluid. Like the egg white, the mantle is solid. Like the eggshell, the crust is brittle. He and others believed that the crust itself, with all the world's natural resources, was formed from the upper part of the mantle. That region was a subject deserving wide study on a large scale. While leading American scientists busied themselves with Project Mohole, others with a less technological focus took up essentially the same idea and sought to coordinate study at an international level. At the 1960 General Assembly of the International Union of Geodesy and Geophysics (IUGG) in Helsinki, the Soviet Union proposed a broad program of research to be pursued at an international level, directed toward studying the upper mantle and its interactions with the earth's crust.[76] Earth scientists worldwide rallied behind the plan to study the upper mantle, and the International Council of Scientific Unions accepted it as one of its programs at its executive board meeting in Prague in 1962.

In 1963, Soviet geophysicist Vladimir Beloussov took the reins of the project, during the general assembly of the IUGG in Berkeley, by organizing the International Upper Mantle Committee with himself as chair and American scientist Leon Knopoff as general secretary. The committee perceived itself as the organization poised to lead in the investigation of fundamental geophysical and geologic problems that "transcend international boundaries," specifically large-scale studies of the earth's tectonic features such as rift zones, oceanic ridges, and areas of sustained seismic activity.[77] In May 1964, the committee met in Moscow to formulate a scientific plan of action. It decided to focus on problems that could be explored internationally, namely, studies of continental margins and island arcs, the worldwide rift system recently discovered on the floors of the oceans, and tide studies to reveal the upper mantle's mechanical behavior. The National Academy of Sciences contributed a large portion of financial support for the committee's secretariat, but support for the actual scientific work was provided initially by Unesco with a sum of about thirty thousand dollars. These funds were to be spent primarily on a cooperative study of the East African Rift zone in the Red Sea area—a program proposed by Beloussov, the Soviet scientific spokesperson.[78]

The Americans tried to wrest leadership of this international project from the Soviets by suggesting alternatives. But the American project ideas were self-serving, designed to fit preexisting efforts already instigated by Americans and requiring only a minimum commitment to the Upper Mantle Project itself. For example, Americans dubbed the Soviets' East African Rift zone proposal as "too provincial" in scope and called for a comprehensive study of the entire rift system. This was really just a convenient ploy by the Americans, because they already were working on the San Andreas Rift system with Mexico as a partner, and they could call this project their contribution to the international effort. The Americans were also keen to promote an international investigation of the Aleutian-Kurile arcs and the Japanese islands, with particular regard for the deep-sea trenches in these areas. Such a proposal could be tied handily to preexisting cooperation between the United States and Japan.[79] But the Upper Mantle Committee pressed on with its specific program in the East African Rift zone, despite the weak objections of some American scientists. Beloussov's chairmanship of the committee and influence in the IUGG made him a credible leader. Because of the location of the zone, the IUGG planned to include local scientists as much as possible, including the development of training programs for scientists from Ethiopia, Ghana, Kenya, Nigeria, and Sierra Leone. Madagascar, Nigeria, Sierra Leone, and Southern Rhodesia were granted provisional admission to the union (Ghana was already a member).[80] Because it was a Soviet plan and because of Beloussov's role in spearheading it, these developments placed the Soviet Union firmly as the leader of the most significant international endeavor in earth science at that time. Not only did the project seek to widen the scientific community by including African scientists, but also it addressed the most fundamental problems of the earth, namely, the mechanics of the mantle and crust. Beloussov and other Soviet scientists had assumed key leadership positions in international science.

The United States initially took a passive role in the Upper Mantle Project, continuing to call Project Mohole the American contribution to the international effort. But as their exclusionary attitudes during Mohole attested, actual international cooperation was limited and not extended to the Soviet Union. In utilizing international venues (such as Unesco meetings) AMSOC sought not to forge cooperative activities but, rather, to conduct intelligence on "SOVMOHOLE," even writing to all the American attendees to make comments (with the option of anonymity) for a comprehensive report on the Soviet program.[81] Soon, however, many American science administrators saw the wisdom of becoming more actively involved in the Upper Mantle Project. As Mohole was grinding to a halt, they did not wish to be left outside of the international effort. Indeed the Department of State was pleased to find out that the National Academy of Sciences was seeking government endorsement to participate in the Upper Mantle Project like it had in the IGY. "Such occasions," wrote director of International Scientific Affairs R. Rollefson, "have provided effective 'bridges' for communication in the fields of science between United States and Soviet scientists."[82] Science administrators such as Merle Tuve (director, Department of Terrestrial Magnetism, Carnegie Institution of Washington) and Alan T. Waterman (director, National Science Foundation) felt that all of America's efforts (not just Project Mohole) to study the mantle would be more successful if tied to an international effort. Government endorsement of the Upper Mantle Project would help to gain domestic support for the program, and tying it to an international effort might make it more difficult to cancel. It also would allow the Americans, as Tuve noted, "to speak more authoritatively about our contributions to the total world effort."[83] In other words, if the government could actually acknowledge the Upper Mantle Project, the Americans' claim that their projects were "contributions" to the international effort might seem more credible.

All these efforts to participate came late, and the United States could not claim to have any real leadership role in starting the Upper Mantle Project. After it was proposed by the Soviets in 1960, it officially got under way on January 1, 1962 (before Beloussov's committee was formed). Although the United States claimed a desire to participate, it did not earnestly seek government endorsement of American participation until January 1963, when the National Science Foundation (prompted by the National Academy of Sciences) wrote to the president's science advisor, Jerome Wiesner, asking for the president's support. At that time, the American contributions to the project were "heavily based upon the normal programs conducted by Federal agencies and laboratories . . . and by the university research teams with similar interests." The National Science Foundation urged that these efforts be intensified and that new programs be started with explicit connection to the Upper Mantle Project. They reasoned that the United States should do this not only because of the substantive value of the scientific work but also "in view of the opportunity it affords the United States to promote international cooperation and to exercise international scientific leadership."[84]

The National Academy of Sciences noted that the Americans needed to pay closer attention to the nuances of their own leadership, take a more positive role, and "be not only responsive to international recommendations but must take some initiative in this area. The scientific opinions of the US should be made known to the international community, and on a sufficiently timely basis that international positions . . . (for example, the African Rift Project) do not become established through initiative elsewhere and *default on our part*" (emphasis in original).[85] This referred, of course, to the international acceptance of the Soviet plan to have the project focus on the African Rift area. The committee confirmed in early 1964 that it was "important that we take initiative to make constructive recommendations to the international UMC [Upper Mantle Committee] (and not remain in the position solely of reacting to others' suggestions)." The committee wished to regain the initiative not only in the Upper Mantle Project but in international science as a whole, including the existing programs with Japan, South American countries, Mexico, Canada, and of course the IIOE. Because international recommendations seemed to have long-range effects on the research programs of the world's scientists, it behooved the United States to regain the international initiative that it abandoned during the Mohole affair.[86]

THE PROBLEM OF INTERNATIONAL LEADERSHIP

Certainly the early 1960s were divisive years in terms of international cooperation. Not only had oceanography attained an unprecedented stature in both Congress and the White House, but also many of its practitioners had adapted easily to the competitive nature of Cold War antagonism. Addressing the United Nations on December 17, 1963, President Lyndon Johnson stated the nation's objective, saying, "We know what we want: the United States wants to see the cold war end, we want to see it end once and for all." He then went on to equate international morality with the oceanographic community. For example, he said, all nations regard piracy on the high seas as a crime, and all sailors recognize the importance of mutual assistance at sea regardless of national differences. "Because of this tradition," he said, "it appears that positive actions to bring about a peaceful world would be effective if based on scientific activities related to the world's ocean areas. Such activities are encompassed in the subject of 'oceanography.'"[87] Yet those scientists and engineers most concerned with the oceans seemed intent on developing something akin to the space race, situated on the high seas. They wished to ensure American leadership by beating the Soviets at sea, and at the same time their focus on competition blinded them to the ascendancy of the Soviet Union's leadership in the arena of international scientific cooperation. American and Soviet scientists were proving that oceanography was not inherently international, as they made marine science another arena for their ongoing political bickering.

As American scientists perceived the scientific and technological competition with the Soviet Union, Willard Bascom's allusion to the hare and the tortoise—one trying to win quickly and the other doing so slowly and steadily—was a fitting description. It reflected a prevalent view that, although the Soviets had an annoying tendency to achieve "firsts," American science was made of more substance and less flash than Soviet science. Scientists commonly perceived that the Soviets tended to produce large and unsophisticated pieces of technology, designed to achieve limited but dramatic goals. As one commentator noted, Soviet science was to American science what an alarm clock was to a Swiss watch.[88] Yet, despite Americans' belief in their own scientific superiority, many felt it was necessary to compete with the Soviets for the sake of American prestige. Doing so meant that American science would retreat into a national mode, focusing less upon spearheading international projects. This provoked hard questions about the nature of American leadership, as the desire to stay "ahead" seemed to obscure the need to be perceived as shaping the course of international action.

The Soviets' approach to oceanographic research simply propagated the American view of speed and drama over constancy and quality. The larger vessels housed dozens of scientists who carried out a great number of observations; some vessels even contained printing presses so that results could be published swiftly. Walter Munk later recalled that he felt the Soviets had made a mistake in putting so much emphasis upon the size of the ship, because the science was oriented toward "factory-like" surveying instead of more flexible idea-based measurements. When asked whether prominent Soviet oceanologists such as Vladimir Kort preferred the big ships, Munk guessed that scientists themselves did not make such choices in the Soviet Union:

> I think those were political decisions. There was somebody in the government who thought that big ships would add more prestige. Those guys were good scientists, and would have preferred to operate in the American mode. Yes, we did talk about that.[89]

The American ICO concluded that, although the Soviet vessels were not ideal research ships, they were more than adequate to meet the surveying needs that still dominated Soviet oceanographic research.[90]

The Soviets squeezed a lot of prestige out of a few impressive facts. For example, the Soviets had the world's only nonmagnetic schooner, the *Zarya*. This was a source of prestige, despite the fact that other countries simply had stopped building this obsolete type of vessel and had relied on aircraft to make (slightly less accurate) magnetic observations. As testimony to their ability to put advanced technology to sea, the Soviets also had a research submarine, the *Severyanka*. Some of the largest and most modern ships were assigned to basic research; during the early 1960s, the Soviet Union had seven ships of more than 3,500 tons displacement, making its scientists capable of extensive simultaneous observations in different parts of the world. The ICO noted that "[t]here are now sufficient numbers and types of ships to enable the USSR to be a leading participant in any international cooperative oceanographic studies."[91]

The creation of the Intergovernmental Oceanographic Commission (IOC) necessitated interactions with Soviet officials and created new responsibilities for the United States government, because the body consisted of member states, not merely scientists. The Americans, like anyone else, had a delegation, and it was expected to set forth its position on any given issue, as an "American" position. Americans in this international forum were no longer representing themselves, their institutions, or their disciplines, but instead they were representing American interests in international oceanography. Naturally, this meant that the United States had to have some body that could confidently recommend policies at the international level. The only organization, at the time, that could perform this function was the recently created Interagency Committee on Oceanography (ICO), which contained representatives from the National Academy of Sciences, the Office of Naval Research, the Office of the Chief of Naval Operations, the National Science Foundation, the Department of State, and other agencies with an interest in marine research. These "Potomac oceanographers," as some called them, spent a great deal of time in Washington devoting their energies to coordinating the national effort in oceanography.[92] The national programs served two roles. The first was to coordinate action to direct the national effort, to stay ahead of the Soviets. The second role of the national programs was to constitute the American part in the newly formed IOC. The Interagency Committee on Oceanography had the necessary components to do this: capable scientists working in concert with high-level government officials representing the military, foreign policy, and other key American interests.

The choice of who should represent American oceanographic interests fell upon the interagency committee. In December 1961, the IOC's chairman,

Anton Bruun, had died. The National Academy of Sciences Committee on Oceanography felt Bruun's death left a scientific vacuum in the IOC leadership, which ought to be filled by a strong personality from the scientific ranks, and particularly it hoped to send Roger Revelle to represent the United States. But ICO disagreed and was keen to send someone from within the government. Instead of Revelle, it appointed Rear Admiral H. Arnold Karo, the director of the Coast and Geodetic Survey, to represent the United States at the IOC Consultative Committee and to present a program as the Americans' plan of action for the international community. The decision to appoint Karo was a subtle but important step, as it made the spokesman of American oceanography a government official, not a scientist.

At the IOC's first session in October 1961, the United States took a bold position, calling for a long-range intergovernmental program to study the "world ocean" in its entirety. It outlined nine stages of planning that allowed for determining the state of present knowledge, the problems foreseen, the technological capacity for research, and other logistical questions. These should all lead, the Americans said, to the development and implementation of an "international ocean-wide survey effort at the earliest possible date."[93] After that meeting, the "Potomac oceanographers" decided that they needed to take a more aggressive role on the international front. In December 1961, they met and decided to establish a special panel for international programs. This panel, formally established in January 1962, took on the name Panel on International Programs, Interagency Committee on Oceanography (PIPICO). The purpose of the panel was to ensure that the American commitment to international programs, particularly in relation to the international meetings of the IOC, was pursued aggressively, efficiently, and in keeping with the ICO's goals in other domains. The panel would ensure that all interested agencies of the government were kept abreast of the participation of the United States in international programs.

When the panel met for the first time in January 1962, it resolved to take a leading role in formulating broadly conceived oceanographic programs to attract the world's oceanographers to American-backed projects. Reporting on this meeting to Congress, Arthur Maxwell (ONR scientist and chair of the panel) noted that the participants "felt strongly that the results of their work are urgently needed to maintain the position of leadership in international oceanography that the United States has long held."[94] The panel had discovered, through Warren Wooster, that the Soviets were planning to try to initiate the next big international program. To avoid appearing empty-handed, the panel identified a program of investigations in the tropical Atlantic Ocean to present before the IOC at its next meeting.[95] The "Potomac oceanographers" knew very well that they would be competing with the Soviets for defining the focus of the IOC's first original project, and thus for taking the leading role in international oceanography. Initially conceived as a survey of marine organisms in the equatorial Atlantic Ocean, the American proposal, the International Cooperative Investigations of the Tropical Atlantic (ICITA), eventually included a multidisciplinary series of surveys to take current measurements, record meteorological and bathymetric data, and make various other observations between South America and Africa. Enthusiastic at first, Karo wrote to one of ICITA's planners, "I feel strongly that the United States must have a good program to put before the Intergovernmental Oceanographic Commission," and that ICITA was just the right kind of program.[96]

The clash between the United States and the Soviet Union occurred when IOC members met in April 1962 to discuss plans for international programs. Reporting on the meeting to his country's Ministry of Science, the British delegate, Admiral Archibald Day, wrote, "On the whole it was an amicable meeting and I think a useful one, but there was some inevitable sparring between the two colossi."[97] Representatives from the United States and Soviet Union had come prepared with competing proposals, which the new IOC Chairman, Canadian W. M. Cameron, agreed to hear only after some hesitation. He knew that doing so would do little to inspire the spirit of international cooperation. The Americans came with their plans for the tropical Atlantic. The Soviets, however, had taken to heart the Americans' bold statements about outlining a large-scale program. They came prepared with a detailed plan of some seventeen pages, conceived as an initial step in understanding the "world ocean," aimed at the northern section of the Atlantic and Pacific oceans. The utilization of these regions for navigation, communications, and fishing required "all-round knowledge of the physical, chemical, geological and biological phenomena and processes developing in the water mass of the oceans, the atmosphere above it, in the earth crust beneath it; all these aspects should be viewed in their inter-relations and mutual interdependence."[98] The Americans wanted to investigate the tropical Atlantic; the Soviets wanted to survey the entire Northern Hemisphere.

Karo did his best to decry the Soviet plan. He claimed that the American plan was more realistic, taking into account the availability of ships and the short-term financial limitations of participant countries. Soviet representative Konstantin Ryzhikov, however, argued that the IOC ought to try a more ambitious survey plan like the IIOE, in which twenty-three sections of ocean, each five to seven thousand miles long, would be observed in the North Atlantic and North Pacific, requiring over forty vessels. When asked if the United States might be interested in participating, Karo flatly recommended rejection by all parties. Other Americans, less adamantly opposed, privately criticized Karo's undiplomatic disregard for the efforts made by the Soviet delegation. But, as British representative Admiral Day noted, "Karo stuck to his guns." The American reasoned that everyone should take more time to determine what was known, and what they most needed to learn, and should first "try our wings" on something smaller, to see if such cooperation was even feasible. In addition, oceanographic instrumentation was not yet reliable enough to ensure the success of such a vast program, and the United States already had committed most of its ships to other projects. Some of the Soviets became upset at Karo's objections, claiming that their plan had been inspired by the American ideas at the first IOC meeting. Reporting these events, Admiral Day noted with amusement that the Soviets even "took the opportunity to allay misunderstandings and to say it was really a US/USSR plan!"[99]

In reality, the Americans were embarrassed by the presentation of program proposals at the IOC. Milner Schaefer, director of Investigations of the Inter-American Tropical Tuna Commission, felt that the American proposal had not been considered thoroughly enough by scientists outside of the government, and thus its scope might not have been "as comprehensive or useful as it otherwise might be." For many scientists, the first time they saw the proposal was at the IOC meeting itself, where few could obtain copies "because they were snatched up so fast by the Russian delegation."[100] After the meeting, Karo returned to the United States with an appetite for something bigger and bolder, to outshine the Soviet proposal. "Any other course," he told ICO chairman Admiral James Wakelin, Jr., "could only lead to disaster and loss of prestige and world leadership."[101]

The official response by the United States to the Soviet plan tried to be less violently opposed than Karo had been. It agreed with the Soviet proposal in principle, and in particular agreed that the United States ought to coordinate its efforts along such lines in the future. For the present, however, it stressed that the instrumentation of the United States was not advanced to the stage at which such a large-scale effort could be justified. It would make no sense to take measurements with present instrumentation in many of the areas proposed by the Soviets, because it would simply repeat previous observations without the benefit of improved technology. Further, the Soviet-proposed survey was so great in scope that the United States simply could not commit to it, because it would sap the resources from other programs that it felt were more important.[102] The British came to the same conclusion. Their National Committee on Oceanic Research noted that although the plan had many things to recommend it, "it can be criticised as attempting too many things at once" and was not detailed enough in any particular area to solve any outstanding problem facing marine science. Perhaps it might be an appropriate program five or ten years down the line.[103]

The question of who would lead the international community into the next worldwide oceanographic program still remained. The United States threw its weight behind the relatively small-scale ICITA and gently but firmly rejected the ambitious Soviet plan as premature. Nevertheless, during its first session, the IOC had resolved to attempt to develop a much more comprehensive program to study the "world ocean." Before the second session, the United States submitted to the IOC a position paper that proclaimed its belief that "a broad, long-range program for world ocean study is desirable." It recommended that a small permanent working group be established to assist the IOC in making a continual review of any such program. The United States envisioned the study of the tropical Atlantic as the beginning phase of this program, but it professed no actual plan. It left to the working group the task of placing programs such as the ICITA "in their proper context in the overall study of the world's oceans." These statements served to blunt criticism of the American program's lack of scope. They were defensive steps, and Admiral Karo recognized them as such immediately. He was hoping to return to the IOC with an ambitious American plan to trump the Soviet one, and he was appalled when he realized that his fellow "Potomac oceanographers" were making no real effort to create a truly global program with specific recommendations. Incensed to see that the American position paper had recommended only a study committee, he wrote to Admiral Wakelin:

> Further procrastination and delay. Are we never going to learn? Are we always going to be on the defensive? The best defense is a good offense. The time for action is now. With less than a month to go, it will be a Herculean task to come up with any semblance of a real program for a coordinated world oceanographic investigation, one which will stand up and we can be proud to propose.... It must be done if we are to change from the defensive to the offensive and assume our rightful place of world leadership in the all important field of oceanography![104]

Karo saw the political enthusiasm for oceanic research during the early 1960s as a means to ensure the ascendancy of the United States in all things oceanic, whether it was scientific knowledge, technological superiority, or political prestige.

To resolve the question of whether to pursue the Soviet plan, West German oceanographer and SCOR secretary Günther Böhnecke proposed that Henry Stommel at Woods Hole and John Swallow at Britain's National Institute of Oceanography act as "experts" to evaluate the plan. Swallow felt that objectively assessing the plan seemed "an altogether too formidable thing," and he and Stommel agreed that it would be better to have something a bit more informal. Thus they and other scientists met with their Soviet colleagues to try to evaluate together the real scientific potential of the plan.[105] In the end, the Americans managed to convince their Soviet colleagues that the program was simply too ambitious and that the scientific dividends from such a program would not necessarily be as high as a more focused one. The Soviets understandably were frustrated. Oceanographer Konstantin Fedorov, then deputy secretary of the IOC, wrote that perhaps the procedures ought to be changed. As it stood, national committees were expected to make proposals, and that is what the Soviets had done with their program for the northern Atlantic and Pacific oceans. "Would it not be more efficient," he wrote to SCOR president George Humphrey, "to collect opinions from leading specialists in various disciplines of marine science rather than soliciting such information from National Committees[?]" The discussions between the Soviet and American scientists were far more useful, and it might be better to start from there by agreeing on something rather than ending there by killing a program. But though Humphrey agreed that it would be better to give an international group of scientists a greater role in devising a way to tackle world ocean surveys, one also had to think about attracting national laboratories and eliciting the input of scientists not normally associated with international things. "The national committees are the best, but probably still not a very good way to do this," he admitted.[106]

At the IOC's second session, the Soviet plan received some praise but no support. As one British representative put it, "the 'Western' delegates wished to avoid complete rejection of the proposal." The projects that were adopted at the session, notably the American plan for the tropical Atlantic, were officially envisaged as the preliminary steps in realizing the Soviet goal of more extensive investigations, the specific plan for which was yet to be determined. The Soviets, to everyone's relief, not only begrudgingly accepted this but also agreed to take part in the investigation of the tropical Atlantic. Impressed, the other delegates felt that it was quite a gesture by the Soviets, who at least appeared to be sincere in international cooperation. The IOC Bureau, which managed the session, was wary of being perceived as a tool of the United States, and it commended the Soviet oceanographers for their flexibility. W. M. Cameron, chairman of the bureau, wrote to Soviet Admiral V. A. Tchekourov, "I think I can truly say that the contributions of your Dr. Kort and the willingness on the part of all to compromise, were what lifted this proposal from the status of a United States project to that of a truly international one."[107] The IOC Bureau was aware that international programs had suddenly become a point of competition between the Americans and the Soviets. Others were conscious of it as well. Leading British oceanographers judged that "the IOC was getting out of its depth," allowing itself to be politicized by the superpowers. In their view, the IOC itself was playing little genuine role in the development or coordination of projects. Although they also disapproved of the Soviet plan, they felt that the study of the tropical Atlantic was too clearly a United States plan seeking an international imprimatur. In reality, it was not a truly international endeavor, but instead was being organized and directed "from Washington D.C. by a scientist paid by the US Government."[108]

Cameron was particularly disappointed with the politicization of the IOC. In the subsequent months he was very conscious of the fact that the Soviets had developed a distaste for international cooperation, American-style, and their attendance and participation in meetings began to falter. At the same time, some suspected that the Soviets had proposed such an ambitious plan only for its propaganda effect and that they had wanted to make a flamboyant showing during the plenary sessions, without making real efforts to bring the projects to completion. It was becoming clear to many in the IOC, including Cameron, that the recent ambitious proposal by the Soviet national committee reflected "a difference in philosophy as to the function of the IOC in Plenary Session."[109] He sensed that the IOC was becoming a political forum rather than a coordinating body. In writing to Tchekourov, who led the Soviet delegation, Cameron said he was convinced that the success of the IOC "will not derive from formal debates or solid pledges at its Plenary Sessions. Rather, its success will depend on what is accomplished—the work that is done." He thanked Tchekourov, Kort, and other Soviet scientists for making an effort to participate actively in coordination despite sharp disagreements regarding the focus of projects. At the same time, the Soviets were showing a distinct lack of interest in attending some of the lower-level working group meetings arranged by SCOR and IOC, and they recently had requested that several of these meetings be postponed. The Soviet request, Cameron noted, "has generated an alarming spirit of disillusionment." Using the IOC as a political forum, which both superpowers so clearly were doing, did not bode well for the future of a cooperative body that contained both the United States and the Soviet Union.[110]

6 OCEANOGRAPHY, EAST AND WEST

One cannot stress enough that the history of international cooperation in oceanography has been conditioned by geopolitical considerations. It was largely the U.S. Navy that provided the means of ascent for American oceanographers, and it had clear strategic reasons for doing it. Easing tensions was the rhetorical backdrop of the IGY, and economic development became that of subsequent years. Previous chapters have shown how important competition was to the Americans, at home and in international forums. But there was more to the American-Soviet confrontation than competition for leadership. The Soviets were unlike other scientific partners, and they never fully belonged to the international community of oceanographers, despite having contributed a huge amount of resources and time to oceanic surveys and international meetings. They were outsiders, and for a couple of reasons. One is that many Western countries simply did not trust them, which this chapter should make clear. But the other reason, and perhaps a more important one, was that Soviet science did not seem to be on the cutting edge of research. Soviet oceanographers used old techniques, and they preferred to promote surveys for data collection rather than investigations of scientific problems. Why this should be so was a matter of speculation, and many scientists attributed it to flaws in the Soviet system: it was too bureaucratic, too top-down, too controlled by the Communist Party. The fact that the Soviet proposals for international expeditions always were huge in scope but limited in their specific scientific goals only reinforced beliefs that Soviet oceanography was being manipulated by the Soviet government for propaganda purposes and that scientific innovation was stifled; bigger was always better, form took precedence over substance.

At the same time, the intergovernmental and increasingly bureaucratic character of the IOC alienated many scientists, who felt that their own autonomy in conceptualizing research was being threatened. Some scientists blamed the Soviets for this. In addition, Soviet motivations were consistently opened to question by scientists and government officials in the West. Worse, conceptual differences among leading physical oceanographers and especially among geophysicists led to stark contrasts in the research agendas of East and West, making cooperation appear less and less useful. This went beyond differences in instrumentation, and by the late 1960s Soviet and Western scientists were developing strikingly different theories about the oceans. Combined with embarrassing political and social interactions in international forums, this did little to bring the communities together. The Soviets would remain outsiders, despite their domineering—though not dominant—presence in the IOC.

THE INTERGOVERNMENTAL STIGMA

The IOC Secretariat identified three areas in which international cooperation would be crucial. Data exchange was the first. It reasoned that such exchange, if done promptly and completely, would lead to more efficient effort, prevent duplication, and facilitate planning for future work. All of these were aims routinely enumerated by those who favored international research. The second aim of cooperation was to establish common standards, units, and methods in order to make data exchange worthwhile. The third and most complicated area of importance was in cooperative expeditions themselves, which had gained tremendously in importance during the previous decade. Such projects permitted scientists to cover a large area in a short amount of time, thus moving closer to the ideal "synoptic" study, the same rationale that had motivated Roger Revelle to advocate including Japan (as well as Canada) in the NORPAC expedition in 1955. Still, the amount of coordination differed greatly between projects. Large-scale ones such as those of the IGY and the IIOE were built primarily from national programs and the amount of coordination could be quite loose, with little connection between the scientific programs of individual ships, and instead depending upon data exchange as the only specifically cooperative act. Smaller programs between fewer countries, such as the NORPAC expedition, the Naga expedition, and even the ICITA program, allowed close integration of scientific programs, coordinated planning of ship tracks, and often the results themselves were published jointly.[1] Such categories, however, were far from fixed. The idea that the ICITA and small expeditions might be closely coordinated, even directed, by agreement in the IOC struck many scientists as an infringement upon their intellectual freedom.

The rising tide of national and intergovernmental oceanographic programs, while providing new opportunities and increased financial support, appeared rather bureaucratic, and even as a constraint on scientists' freedom of inquiry. Woods Hole oceanographer William von Arx noted that "national programs" had certainly come into vogue to gain financial support. However, they were "hand-crafted by committees or panels of the ablest minds available but who, in general, will not participate directly in the actual research activities of the Program." The decision to embark on a particular national program was "essentially a strategic one" and not necessarily serving the best interests of science. Von Arx complained:

> Questions of its influence on other fields of science, on the political image of the nation, and on the national economy are considered at some length, but not, it seems, the effect of such a Program on the deployment of manpower and creative talent in the science centrally concerned.

Some scientists, he continued, jumped at the chance provided by the funding of a "national program" and then found themselves, to their dismay, becoming administrators instead of scientists. In an eight-page memo titled "A Science in Bondage," von Arx asked: were all the benefits to adjacent fields, the image of the U.S. abroad, economic development, and technological innovations "really worth the crippling effect of the National Program method of enhancing activity in marine science?" He called for oceanographers at Woods Hole to resist the temptation to point out the political uses of science. He acknowledged that such restraint may seem idealistic, but that it must be done if progress in oceanography was to be defined by more than nutritional and defensive needs.[2]

Faced with indifference and even opposition to micromanaged large-scale projects, the IOC did not set off on an impressive start in coordinating international action between scientists. Few countries bothered to send the IOC detailed information about their actions. For example, the IOC began a series of information papers on the IIOE, and as late as 1963, it had sent out only one issue. Some scientists complained, believing that IOC was wasting time trying to put the information together in an elegant way when it ought to be focusing on promptness over aesthetics. Quite to the contrary, IOC secretary Warren Wooster explained, the delays in publishing were not the result of "unnecessary refinement of manuscript but by the fact that in general we are not receiving much information." Despite their hope to receive preliminary cruise reports, few countries sent them.[3] This complaint, incidentally, came from the British, and it was a symptom of their overall agitation with the bureaucratization of research at the international level. At one meeting of the Royal Society, British scientists complained that the IOC was more concerned about what might look good on paper than about providing the required coordination. The expectation that each country should submit a fixed, comprehensive national plan seemed idiotic to some British scientists, who preferred a bit more flexibility. They often had to compromise between scientists and institutions and develop scientific programs that might contribute to their long-term goals and obligations. "Oceanography must advance," the Royal Society's British National Committee for Oceanic Research agreed, "like any other science, through growing points." It could not afford to support everything, and it had to be very selective about its research program, basing it upon the needs of its scientists and upon what was being done by others. The committee complained:

> Such compromise does not look very orderly or efficient to scientific administration in countries that have abundant facilities or to small countries that have not yet had to face the task of doing large things with inadequate resources, but the UK has to work this way, and so in fact do most other countries.[4]

What British oceanographers intended to do in their national program would depend greatly on what other scientists were doing, thus making coordination extremely important and thus making them annoyed at the IOC's slowness. Of course, they were also very sensitive to the lack of attention given by the United States and the Soviet Union to financial constraints. The British detested the IOC's penchant for ambitious proposals, backed alternatively by the two superpowers, as the IIOE was expensive enough and there was no way to keep participating in projects of ever-increasing scope.[5]

Membership in the IOC required no particular commitment on the purse strings of any country. Certainly its most extensive (and its first) major project, the IIOE, included a great many countries. But afterward, participation was less. During the ICITA, only a few nations participated, including the United States and the Soviet Union, but the British were notably missing. During the Cooperative Study of the Kuroshio and Adjacent Regions (CSK), which would extend into the 1970s, the United States participated but not as a driving force. In other words, although the IOC became a forum for discussing the relative merits of international programs, and for planning coordination of participants, it certainly was not a binding organization. As its administrators were to discover, its muscle depended almost solely upon the strength of the national programs of particular members and their active willingness to coordinate with others. For the ICITA, that country was the United States, and for the CSK, it was Japan.

The emphasis on large-scale expeditions naturally put the locus of decision-making on those who had money. Many oceanographers were caught between the desires of superpowers and the growing voice of the developing world. The director of Britain's National Institute of Oceanography, George Deacon, complained that one of the difficulties in these meetings was the tendency for the American and Soviet delegates to try to arrange things between themselves. Scientists of other nations did not dare to meddle in such negotiations, knowing quite well that nothing could ever be accomplished without the agreement of the superpowers. "Perhaps this sort of thing," Deacon wrote to D. C. Martin at the Royal Society, "has contributed to rather noticeable deterioration of standards in international scientific meetings." In his view, everyone knew that the real obstacles would be practical issues, but the political importance of leadership delayed such discussions and added unnecessary problems. Both superpowers, and particularly (in his view) the Soviet Union, tended to mix policy and prestige with science.

Even worse, the IOC really stretched the meaning of what it meant to be engaged actively in oceanographic research, which sometimes could mean as little as operating a tide gauge or taking an interest in fisheries.[6] What irritated Deacon more than the superpower dominance was the growing voice of developing countries, which seemed to enjoy more influence than they deserved. Countries such as Britain found themselves caught between the countries that dominated oceanography and those that were barely making a start. The IOC should not listen only to these countries, he felt, but should follow a course of action recommended by the most active scientists from all countries. Such scientists, Deacon told Admiral Day, "are not necessarily the people in the big USA and USSR delegations or the host of small countries that like to call the tune without paying the piper." To Deacon, these active scientists typically meant those in Britain and in the United States who were trying to keep the focus on specific problems and to "get the subject into a sound state," rather than trying to establish comprehensive programs to look good for international relations. "The scientists," he complained, "begin to feel they are being dictated to." The IOC was fast becoming something more like the United Nations and less like a scientific body.[7]

These irritations deepened in 1963 when a British scientific attaché in Moscow reported a conversation with Vladimir Kort, who suggested that the necessity for a stronger intergovernmental basis for oceanography was being demonstrated by the IOC's experiences. Kort, who directed the Soviet Union's premier oceanographic center, the Institute of Oceanology in Moscow, applauded the investigation of the tropical Atlantic not as an American-inspired plan but as an international effort that was going to succeed because of close coordination. He lamented the fact that organization for the International Indian Ocean Expedition had been loose, and he insisted that they could operate more effectively through closer coordination. Like the British, Soviets such as Kort deplored the lack of frequent distribution of updated reports by the IOC. But although the British wanted to use such reports to coordinate their research and avoid duplication, Kort took this one step further by claiming that the research ought to be more thoroughly planned out from above and carried out by scientists. Although SCOR had been an excellent scientific body, and was to be lauded for developing the initial projects, it had lacked the power and means to compel the ships of any country to work according to an overall plan. This was, in Kort's view, the fundamental weakness of the IIOE. At the same time, this was the great promise of the IOC, because it was poised to provide the intergovernmental teeth that SCOR lacked.[8]

George Deacon, having heard about this conversation, wrote at great length to his country's Ministry of Science to prevent any of Kort's views from taking hold. Already the Royal Society had expressed its concern over the direction of international science, particularly now that SCOR no longer was playing much of a role in coordination. It raised serious policy questions for Britain, whose scientists (and politicians) wondered whether independent national programs might not be of greater interest to the advancement of oceanography within the country. Deacon confirmed that without improved techniques, any survey-type studies such as the IIOE, in areas that had already been studied, would provide few results for the expense. It was still possible to concentrate on smaller areas, as long as the level of accuracy was high and the studies were aimed at developing and testing theories. Such studies, Deacon felt, were being pursued by scientists in Britain and the United States, and they should be encouraged. Their work, Deacon argued, "is being made more difficult rather than easier by what amounts to a rather large noise to signal ratio generated by international oceanographic meetings of the U.N. type." Deacon was convinced that the best contribution Britain could make to the science of oceanography would be a very specific study, for example, of the propagation and decay of vortices, "but we could not hope to sell such a project to 47 nations meeting in Paris." He accused the Soviets in particular of having an affinity for surveys and "rather uncritical observations" and asserted that Britain had better devote its energies to producing work of a more qualitative importance. Without this, he felt, it would not be able to maintain its place with the United States at the forefront of oceanography, or to retain the respect of other nations, or to have its students accepted in the top laboratories.[9]

Deacon went on to disavow Kort's assessment of the IOC as the organization most capable, administratively and scientifically, to arrange the coordination of oceanography. To put such things into its hands would be to stifle ideas and slow down research. "It is far too large and too woolly," Deacon wrote of the IOC. It would be terrible if scientists were compelled to conform to a plan that was handed down by a higher body, without their input and possibly without their agreement, as Kort appeared to suggest. The only reason that the IIOE had been so attractive was because the ships would meet and exchange ideas and methods in oceanic conditions that they had never experienced (and which, because of the monsoons, seemed unique). He recalled the *Indian Ocean Bubble,* the informal American newsletter, in which the idea of compulsory methods and programs had been condemned. The success of the expedition was not the result of any control from above:

> After down to earth, democratic, discussion, in laboratories as well as round tables, where all views were expressed, compromises were achieved which seem to have pleased everyone except the IOC. . . . Perhaps the final plan did not look so good on paper, and it is certainly rather complex for 47 nations to appreciate, but it will tell us more about the oceans.[10]

Deacon argued that surveys should be designed by the scientists who expected to take part in them. The IOC could only do damage to the scientific integrity of these plans by interfering.[11]

Deacon went beyond criticizing the IOC and laid blame for this destructive approach at the feet of the Soviets themselves. Although Kort and Deacon were "quite good friends," he had little respect for Soviet science or the manner in which it proceeded. "Kort may be right in insisting that [the IOC] is a young organisation still finding its feet," he wrote, "but it clearly has a tough assignment if it has to change Soviet leaders to a more western philosophy. I think it is quite hopeless." He added his hope that Britain would insist upon freedom of action and approach by scientists and not commit to any international plan unless the scientists themselves felt that they were "in keeping with modern ideas." It would be wrong to hold the charitable notion, he warned, that Soviet oceanographers "can tell us much about the oceans." The problem was not the Soviet scientists, but the fact that none of the good ones had much input in their overall plans. If the British were to send some outstanding scientists over to the Soviet Union in an exchange program, "Kort and the Soviet expedition leaders will not understand what they are talking about." Deacon was adamant that his country should not be swayed by the Soviet approach, which lately seemed to entail latching onto the IOC as a vehicle for its expensive and unimaginative projects that promised only to sap the meager resources of those nations foolish enough to agree to take part, without much scientific return on their investment. He added, "It would not be wise, I think, to let my frank assessment of Soviet oceanography get to Russia or Unesco."[12]

To the IOC, Deacon was less blunt but equally firm. He wrote to Wooster claiming that he was sympathetic to the need to make something look good on paper. In retrospect, he added, he was not sure what scientists had expected from the IOC to begin with. He admitted that much of their support had to do with thinking it would make the job of securing funds at home much easier. But in reality, his own field of physical oceanography was having quite a difficult time trying to focus on ocean dynamics, whereas the IOC gave "every encouragement to geographical exploration," the kind of bland survey work he detested.[13] This was, incidentally, the exact reason that Deacon at this time fervently promoted the North Atlantic Treaty Organization (NATO) Science Committee (discussed in greater detail in the next chapter), because he saw it as the only international organization that would let physical oceanographers do what they did best. But writing to Wooster, Deacon simply expressed his grave concerns that some scientists wished to use the IOC as a way to compel scientists of other countries to conform to investigations planned from above, and, although he did not mention Kort by name, he noted that he had evidence that some members had become convinced of an "intergovernmental" rather than "international" approach, believing that this was better than letting scientists follow their own ideas.[14]

Kort set forth his views strongly at the IOC Bureau meeting in Moscow in 1963. He urged closer coordination of studies for the IIOE, which the British delegate, Admiral Day, called "a typical Kort contribution." Kort repeatedly argued, against the objections of the United States, Britain, and others, for far closer synchronization of observations and closer IOC management during the study of the tropical Atlantic, and he expected something similar from the Cooperative Study of the Kuroshio and Adjacent Regions (CSK).

When Kort suggested that the CSK be widened to include a larger area, the Japanese delegate was adamant that the Soviets stop meddling with the original plan. Upon seeing this, Day observed, "Some political background, I fancy."[15] He was right. The IOC devoted considerable effort toward Asia throughout the 1960s, particularly with the IIOE but also through the CSK, the Japanese scientists' pet project on the Kuroshio current. The CSK was first proposed in Manila in 1962, at the Second Meeting of Marine Science Experts in East and Southeast Asia, when Japanese scientists hoped to interest oceanographers in the Kuroshio current near their home islands. The CSK's focus was rather specific: studying the sources of water exchange between the Pacific Ocean and the South China Sea and the dynamics of the current. Japan took the lead in this project, although both the United States and Soviet Union participated at some level. But discrepancies in vision between Japanese scientists and their Soviet counterparts caused periodic conflict. They argued over the size of the survey area and the number of traverses to be made; the Soviets wanted to broaden the project, while the Japanese wanted to rein in the project to make it more feasible and economical. The CSK's newsletter (published by the Japanese Oceanographic Data Center) characterized it as a conflict between "the rather realistic and practicable plan as proposed from Japan and a more idealistic one presented by the Soviet experts."[16] As the CSK wore on, the Japanese grew impatient with Soviet interference in a plan that they hoped would improve knowledge of a specific area around their home islands, not just add data to an ambitious compendium of the entire Northern Hemisphere. Japanese scientists initially envisioned cooperating with countries in Southeast Asia to conduct a synoptic survey modeled after the NORPAC expedition of 1955. Because of the limited capabilities of the participating countries, they hoped to start with a couple of surveys per year and then expand to a more elaborate program. However, both the Soviet Union and the United States wanted to participate, so the program expanded considerably, perhaps more than the Japanese wanted.[17]

Faced with conflicting visions of cooperation, the IIOE's International Coordination Group met in Paris in January 1964 to discuss its priorities. It conceded that strict coordination had never really been its primary objective and that its most significant efforts had been in identifying the most pressing scientific problems of the Indian Ocean. The group hearkened back to the days when Robert Snider energetically had focused on planning how to integrate programs. Although some had disliked Snider's role at the time, the group now commended him for his role in publicizing national plans, requesting port facilities, and selling the project to create a sense of international purpose, cooperation, and goodwill.[18] Certainly many were sympathetic to Deacon's attitudes, but his stubbornness with regard to international cooperation annoyed others, particularly the American Roger Revelle, who had been instrumental in SCOR's early days and had worked closely with Deacon in international meetings. By 1964, Revelle was trying, in vain, to persuade his friend Edward Bullard to try to become the next president of SCOR. Bullard was a geophysicist, but had a long-standing interest in the oceans and many personal contacts among the Americans. If Bullard could become president of SCOR, Revelle reasoned, it would diffuse Deacon's dominant position as the spokesperson for British oceanography. Revelle wrote to him, "My personal reason for suggesting you was that we must get British oceanography out of Deacon's hands and into those of someone with a broader and more liberal view, and particularly must associate it more closely with the universities."[19] But Bullard demurred, claiming that oceanography soon would not be confined solely to the National Institute of Oceanography and that universities such as his own (Cambridge University) were having more and more success obtaining what they wanted in terms of ships and funding.[20]

For his part, Deacon widened his criticism to include SCOR. He felt that it ought to concentrate on the IIOE, by helping to make its data available and by trying to improve the quality and design of experiments, observations, and theory. "There is a lot of second-rate work being done," he wrote to D. C. Martin at the Royal Society, complaining that a lot of the information in data centers had to be discarded. In his judgment, SCOR "has flopped a bit: it was all right while it held the Indian Ocean work in its hands and has not achieved much since." By this time he had become adamant in his ideas about not being compelled to do anything by the IOC, even to the point of being rather sensitive to language. He wrote to a colleague in the Royal Society, "It is not likely that we can contribute (I resent your 'submit') data during the next few months," particularly if it was clear that no one wanted it for a specific purpose. "With the best will in the world," he argued, "overworked scientists will not summon up as much enthusiasm for 'centres' as for publication and direct exchanges with people they know." As usual, his sensitivity was directly related to the financial difficulties at the National Institute of Oceanography, and with such challenges to institutional survival, many of the IOC activities seemed irrelevant.[21] Deacon took an active interest in changing IOC into a more scientific organization, but by his own judgment rarely succeeded. Too often the organization fell victim to what he felt were the hallmarks of Soviet oceanography: too much planning from above and uninspired survey work that was far from the cutting edge of research. The IIOE, the ICITA, and the CSK were, in Deacon's view, all projects that some scientists wanted, and thus they had a chance of success. But when the IOC tried to do other things, all of them negotiated at large meetings, "they have almost inevitably been rather dull things: not what the really active young men, who in the end we rely on, know to be most profitable and are keen on doing: useful background activity but not really spearhead attacks."[22]

OCEANOGRAPHY, WEST VERSUS EAST

Irritation with the Soviets continued in other ways. Physical oceanographers had increasingly turned to NATO as their forum of choice, particularly those scientists in Britain, Norway, and other smaller countries with strong programs in physical oceanography. Rather than take an interest in ever-larger survey projects of the Soviet type, NATO oceanographers sought to concentrate their interests in developing effective unmanned instrumented buoys. Such buoys could be left unattended for months, gathering data on the basic variables of physical oceanography: temperature, salinity, and the velocity and direction of currents. With increasing technological sophistication, this kind of program had far more promise than sending ships on survey operations, which would be more expensive and ultimately would yield less interesting results. Unlike data from vessels, buoys had the advantage of making synoptic studies possible on a continuing basis. Vessels then could concentrate on more problem-oriented research.[23]

The buoys themselves raised sticky international issues. In 1966, the British Ministry of Defence was seeking out the assistance of the National Institute of Oceanography to "identify the purpose of a device of Soviet origin which was found floating in the sea." Particularly because of their radio transmitters, such devices were inherently suspect. This specific device was in fact a wave recorder, with a deep sensor suspended below the surface to avoid the action of waves and a buoy on the surface to move up and down with the waves, thus measuring the vertical movements at that point. Scientists at the National Institute of Oceanography knew the device quite well, had seen such equipment aboard Soviet vessels, and assured their government that indeed the Soviets were "doing serious research on waves."[24]

Aside from the problem of identifying legitimate scientific devices, there was also the problem of protecting them. Scientists began to realize that there was no way to ensure that these buoys could accomplish their work unhindered. Many of the buoys were lost at sea, and although some may have sunk as the result of natural causes, clearly some had been either stolen or damaged. Initially, most scientists suspected that curious fishermen had wanted souvenirs or perhaps had hoped to pretend that they had saved the buoys from sinking and could thus earn a salvage reward. The loss of such buoys was rather high by the late 1960s, and according to British fisheries scientist A. J. Lee, the losses had caused at least one institution to call off a research project because of the fear of losing its equipment.[25]

Edward Bullard confided to the Ministry of Defence that many believed that Soviet trawlers had been stealing and sabotaging buoys. The losses, to the British alone, amounted to several thousand British pounds per year. The Royal Society's D. C. Martin wrote that he had a "nasty feeling, however, that there is little that can be done to prevent such occurrences short of a policing service which would nullify the advantages of unmanned mooring stations."[26] Sometimes foul play was irrefutable, as in the case when a transponder was stolen and the electric cable obviously had been cut clean with a knife. Such equipment would not be particularly valuable to fishermen, and it would be difficult for the perpetrators to sell. There was no evidence that the Soviets were responsible, but according to A. Potts of the British Ministry of Defence, "all Soviet ships may be considered as potential intelligence collectors and we would not be surprised if this is where they have gone." Potts added that he doubted whether individual buoys had been targeted, but that perhaps the Soviet Union had made a general request for its ships to retrieve any buoys they might find.[27] The problem of unmanned buoys remained an international problem throughout the 1960s and beyond. The IOC tried to solve the problem by creating working groups to study the safety rules and legal status of buoys and other scientific instruments, dubbed Ocean Data Acquisition Systems (ODAS). Some of the safety issues were defined during the early 1960s, but the legal issue remained, and questions of liability in the case of loss or damage continued to be hazy, even after a convention in 1972 came together to discuss the issue.[28] There is no solid evidence that the Soviets were responsible for the problem, but the fact that many suspected them did little to encourage the spirit of cooperation.

Even when the Soviets were engaged in planning useful projects, they were treated with skepticism. One of these projects, which promised definite practical use for scientists, was the General Bathymetric Chart of the Oceans (GEBCO). Such a chart would compile all oceanic soundings (measurements of depth) from international expeditions. This work normally fell under the auspices of the International Hydrographic Bureau (IHB), based in Monaco, but its work lately had slowed considerably because of a lack of funds and personnel. After 1957, some of the work was completed on a regional basis, but there was no worldwide coverage, as the GEBCO project lacked money and presumably sufficient interest. According to a 1962 report of the British Admiralty, "work on GEBCO has at the moment almost come to a standstill."[29] The Soviet Union, not a member of the IHB, proposed to the IOC that Soviet hydrographers, and not the IHB, should take on the responsibility of compiling the chart, and could do so far faster than the IHB could, simply because they had managed to secure the funding for it. When this proposal was made, the American delegation was surprised and intrigued at the offer; however, American scientists were immediately cynical about the Soviets' motivation. Princeton geologist Harry Hess suspected that this was the Soviet Union's way of avoiding sharing the greater part of its own data. If the Soviets were allowed to compile the chart themselves, no one would ever have a chance to see the data that they had collected, except in the final chart. Hess was particularly opposed because he estimated that 80–90 percent of the soundings had been taken by American ships, and it would be a flagrant case of the United States providing a great deal of data to the Soviet Union and receiving a minute portion in return. Granted, the American data were going to be publicly available anyway, but most hoped that the Soviet data would be available too. Hess even suggested that the United States should take the task upon itself, to ensure the quality and usefulness of the chart to scientists.[30]

Most scientists reluctantly admitted that they had better take the Soviets up on their offer. The "Potomac oceanographers" on the Panel on International Programs, ICO (PIPICO) were sympathetic to Hess's objections, but at the same time, they reasoned that only through this chart would anyone have an opportunity to obtain the Soviet data at all. The Soviets, after all, were not members of IHB and thus would not necessarily be held accountable to participate in the project. More important, PIPICO was concerned about the cost of putting together such a chart, and if the Soviets were willing to absorb the cost, then they ought to be encouraged to do so.[31] The British also felt that the Soviets probably would never publish their results fully unless they were attached to some international endeavor. Their IGY data were being published as adequately as any other country's, and recently the Soviets had become deeply involved in the International Commission of Northwest Atlantic Fisheries and seemed committed to the investigation of the tropical Atlantic. On the question of the Soviet motives, most American and British fisheries scientists felt that because Soviet activities were only going to increase, it was wise to include them as much as possible. However, the issue became moot when France, already a member of IHB, offered to assume responsibility for the project. In the end, France took on the project and the Soviets produced a major chart anyway (of the Pacific Ocean basin), presenting it in 1966 at the Second International Oceanographic Congress in Moscow. The chart, created under the direction of marine geologist Gleb Udintsev, would be widely acclaimed by scientists worldwide as making a major contribution as an atlas of the ocean floor.[32]

The prevailing attitude about the Soviets, despite the difficulties working with them and pervasive suspicions about their intentions, was that they were better in than out. Still, the problem of data exchange often undermined enthusiasm for international cooperation, and it was widespread during the early 1960s. American scientists were particularly wary of the Soviets, whose adherence to data exchange procedures typically meant delays, and often Soviet delegates to the IOC had to explain why, for some administrative or logistical reason or another, data had not yet been sent to the World Data Center in Washington. At the IOC Bureau meeting in Moscow in 1963, for example, delegates from the West gently suggested that the Soviets, in charge of World Data Center B, had not been disseminating the data they received. Many felt that World Data Center A sent a great deal of data, but received hardly anything at all. The Soviet delegate, V. A. Tchekourov, confided to his British counterpart, Admiral Day, that he believed that the Americans were not sending everything they had. Day responded that he doubted it. After some serious discussion, Tchekourov conceded that perhaps the Soviets had not been quite as organized as World Data Center A had been and promised that the data would be forthcoming.[33]

The Soviets were not the only culprits in this regard, but they were certainly the most conspicuous because of the locations of data centers A (Washington) and B (Moscow). This was exacerbated by the fact that the Soviets tended to withhold data whenever they were not required by international agreement to share them. For example, a loophole in these agreements allowed the Soviets to withhold data from the International Association of Physical Oceanography (IAPO) on the mean sea level for any station that they did not declare as part of a "national program." Also, they did not want to share any data for dates prior to the international agreements, even for stations included in the national programs. The first such agreement took effect during 1957, for the International Geophysical Year, and it remained in effect during the less intensive International Geophysical Cooperation

(IGC) that extended some IGY studies into 1959. Soviet data on mean sea level, prior to and after the IGY, were inaccessible to researchers in the West. "Indeed," one IAPO member complained of Soviet mean sea-level information, "the supply of data from the USSR has dried up completely following the IGC." Similar problems plagued the coordination of tide gauge stations during the IIOE, not merely because of poor dissemination of data but because not everyone was apprised of all the stations that existed.[34]

It is unlikely that Soviet scientists were especially to blame for these problems, as they were working within a system that, as a matter of policy, generally blocked the publication of data that were not attached officially to an international endeavor. Soviet oceanographers thus had a double incentive to connect their projects to international programs. Not only did they value the exchange, but also they generally were able to publish from their own data without the constraints of internal security classification. If they could not establish an official connection to an international program, they not only halted international data exchange but also crippled their own ability to publish at home.[35]

To be fair to the Soviets, data exchange was just one manifestation of the IOC's endemic problem of coordination. As PIPICO chairman Arthur Maxwell reported of the IOC, many countries, despite their agreement to do so, had not even bothered to let IOC know of their national oceanographic programs, let alone taken the initiative to share data. Coordination often was a nightmare for IOC officials who had to deal with representatives whose priorities were different and whose levels of preparedness varied considerably. This was not confined to the smaller, less scientifically robust states. A 1964 proposal by an American to establish observations near the Gilbert Islands was a perfect example of the chaotic nature of international coordination. According to Maxwell, "Confusion reigned because England did not realize she owned the Islands, the US did not realize that two Americans had planned the project and the USSR thought the proposal was related to an earlier one they had submitted."[36]

Although there were some genuine ways that the Soviets and Americans cooperated, such as data exchanges, the scientific communities themselves remained very isolated from each other. The lines of demarcation fell along Cold War lines, and the issues that divided them even served to reinforce conceptual divisions as well. This was what many scientists during the IGY had said did not exist, claiming instead that science was the common language of mankind and that it transcended national borders. Others felt that the conceptual divide did exist, but that cooperation would help to eliminate it. But the situation in oceanography was not so simple, and despite some cooperation, Cold War barriers persisted in dividing East and West, blurring the politics of scientists with the concepts of science. The security measures that often made the British-American cooperation so close, for example, existed specifically to prevent relevant research in marine geophysics from reaching scientists in the Soviet Union. Specific cooperative efforts, such as the IGY, IIOE, and the Upper Mantle Project, were exceptions. The research results from all American IGY expeditions were immune to military classification and were freely reported to data centers throughout the world. But such rhetoric of openness, and what one historian called "orgies of international research and cooperation," could not forge an inclusive international community that embraced both East and West.[37]

American scientists often sympathized with the plight of their Soviet colleagues, and vice versa. The Soviets understood the need for classification, of course, but they were impatient with the sluggish publication practices of Western scientists. Institute of Oceanology marine geologist Gleb Udintsev, for example, urged Scripps's H. W. Menard simply to publish his soundings in a form acceptable to the Navy; otherwise, he would never see them. Udintsev, wanting to participate in the interpretation of the seafloor provinces Menard found in the Pacific, lamented, "I regret very much that at this time it is not possible for me to avail myself of the material of some of your expeditions." The bathymetric charts of Downwind (an IGY expedition) had helped him and his colleagues a great deal, but they still awaited, in vain, the publication of the data from the TRANSPAC and Northern Holiday expeditions, which were completed nearly a decade before.[38] The situation got at the heart of the problem faced by an international research community connected only by formal publications and international conference presentations. In this case Menard was hampered by the fact that on these expeditions he did not take data that were of acceptable quality for publication; he had technical difficulties with his equipment, and he felt that any publication would lack scientific integrity.[39] Yet it was on Northern Holiday that Menard surveyed the most prominent fault scarp on earth, the Mendocino escarpment, which was the greatest example of his discovery of "fracture zones." His data, even if not publishable, were certainly both available and extremely valuable to scientists who visited Scripps.[40]

Successful research in oceanography depended upon access to data, which meant that scientists had to have security clearance if they wanted certain kinds of data (especially bathymetric data). Menard guessed that in the 1950s a few insiders had approximately five years to digest and interpret classified research results before they ever became public knowledge.[41] The best data on the Pacific region resided at Scripps, collected in the many expeditions it had conducted in the years following its first major postwar cruise in 1950. The Navy used the data to make bathymetric maps, and scientists with security clearance used the data for their own academic purposes.[42] Similarly, Lamont and Woods Hole scientists held proprietorship over large quantities of unpublished seismic data as well as bathymetric maps made of the Atlantic seafloor. As noted in chapter 2, Menard lumped them together and likened them to an "invisible college," defined by the research interests of a select few scientists.[43] That invisible college often included George Deacon's National Institute of Oceanography and Edward Bullard's Department of Geodesy and Geophysics at Cambridge University. As Bullard recalled later, only these five centers were making valuable contributions in marine geology and geophysics as late as the mid-1960s. Most of the British scientists found employment in American laboratories, as there were few posts available at home. In a 1969 oral history interview, Bullard said, "We're practically running at Cambridge a branch of the American educational system." Aside from J. Tuzo Wilson, a Canadian, Bullard claimed that marine geophysics was really an Anglo-American endeavor. "For some reason or another," he said, "the subject sparked in England and America and didn't spark elsewhere."[44]

DIALECTICAL MATERIALISM AND "GEOPOETRY"

Restrictions around the Anglo-American "invisible college" and the scientific community of the Soviet Union caused a growing divide in conceptual points of view about the cutting edge of scientific work and the problems toward which international activity should be bent. Many physical oceanographers preferred to cooperate under the auspices of NATO, despite the risks of offending Soviet colleagues. This was not in order to work on military problems but, rather, to focus on physical oceanography and tackle specific problems chosen by a few physical oceanographers without interference by government bodies or scientists from too many other disciplines. In addition, many Americans remained disinterested in Soviet views because they felt that philosophical biases skewed Soviet science. One of the fundamental principles of Stalinist dialectical materialism was that the accumulation of experimental and observational material should help to eliminate the idealism in scientific theories. Fervent anti-idealism and an aversion to any science aimed at increasing profits had earned Soviet science a poor reputation in the 1930s, and some of that carried over into the postwar era. Most famously, Western scientists noted with astonishment and disdain the ability of agronomist Trofim Lysenko to orchestrate the official demise of genetics in the 1940s.[45] In oceanography, Soviets admired the conceptual simplicity of certain theories, such as Henry Stommel's theory of deepwater circulation. But theories with such simplicity could not represent reality, and the Soviets thought that more intensive observations were needed in order to gain a more true understanding of the earth. Similarly, when Vladimir Beloussov published his plea to avoid jumping to conclusions about the horizontal mobility of the seafloor, he argued that the emerging American-style model of the earth was too formalistic, too schematized.[46] Such reasoning appeared to fit the anxieties of the stereotypical Soviet scientist. The stereotype was no mere caricature of Soviet science; dialectical materialism did continue to inform marine science in the Soviet Union even into the 1980s.[47]

The modern reader can sympathize with Soviet reticence in accepting idealistic theories. For one, the new ideas about the earth were certainly rather schematized, and their early proponents admitted as much. When Harry Hess first proposed the theory that would be called "seafloor spreading," he did not call it his theory, but rather he preferred to "consider this paper an essay in geopoetry . . ." in which he did not wish to "travel any further into the realm of fantasy than is absolutely necessary."[48] He claimed that new crust was created at distinct spreading centers, likely at the mid-ocean ridge, and the crust traveled outward away from the ridge before being destroyed in the deep trenches found in the ocean floor. The 1960s version of continental drift—soon to be called "plate tectonics"—was based upon simplified models, which showed the fit of the continents and the geometric rotation of plates. Two of Beloussov's colleagues, publishing an alternative theory of the earth in 1969, were proud to write that their theory made no attempt to establish a universal logic, unlike the Western theory.[49] To them, it was an asset to divorce their work from any claim of universality; models were to be mistrusted. Yet to Western observers, these Soviet preferences appeared more ideological than scientific. Old criticisms of Soviet science provided Westerners with a precedent to ignore them.

By the end of the decade, Hess's caveat about his theory being overly simplistic mattered little in the West, for the criteria required to legitimate his theory, despite its schematic simplicity, had been met. In 1960, when Hess first conceived of his idea, the more stubborn minds would require not merely a plausible idea but also independent tests to corroborate the theory before converting to it.[50] In 1963, Fred Vine and Drummond Matthews published an idea that was equally schematized. They stated that the magnetic stripes on the ocean floor were simply bands of mantle material magnetized according to the earth's magnetic field and then pushed away from their source at the mid-ocean ridge. If this were true, the stripes on the seafloor provided a historical record of changes in the earth's magnetic field.[51] In 1966, a ship operated by Lamont scientists, the *Eltanin,* recorded these magnetic reversals as it passed over the mid-ocean ridge. The results for one of the legs showed that the pattern of stripes (recording magnetic reversals) flanking one side of the ridge mirrored the stripes on the other. The discovery shocked scientists at Lamont, indicating to them that Hess's "geopoetry" of seafloor spreading might be correct. The clincher for most scientists in the West came when the epochs recorded by the magnetic stripes were correlated with independent methods of dating the earth's magnetic reversals, through paleontological and radioactive evidence.[52]

Most detractors of seafloor spreading felt that the considerations of mathematical physics were not being applied adequately to the oceanic crust; any mathematician, they argued, knew that such mobility simply was physically impossible. One of the theory's serious detractors was British geologist Harold Jeffreys, whose textbook *The Earth* went into numerous editions since its initial publication in 1924. A longtime opponent of continental drift, Jeffreys in 1963 attacked the evidence for the earth's periodic magnetic pole reversals, claiming that there was far too much leeway for interpretation.[53] Nevertheless, scientists had to reckon with the new evidence brought forth from the ocean floor and elsewhere. Most scientists in the West were slowly moving toward some version of continental drift based upon seafloor spreading. Keith Runcorn, one of the paleomagnetologists who was attempting to correlate the magnetic reversals, sent to J. Tuzo Wilson a newspaper clipping that showcased Jeffreys's mathematical objections. Wilson's only response was that "I am not any longer worried about what Sir Harold says."[54]

The new data and the new ideas had come from North American and British scientists; the Soviets played no role. Part of this was due to the fact that the Anglo-American scientific community excluded them. Perhaps more important were constraints within the Soviet Union. Vladimir Beloussov, a vehement opponent of the new ideas, exercised far more influence in the Soviet Union than his equivalents abroad did in their own countries. He led the Geophysical Committee of the Soviet Academy of Sciences and was charged with approving expeditions and their scientific programs. It was easy for him to eliminate certain investigations (on magnetic anomalies, for example) if he wished, or simply not to approve an entire expedition. It was difficult to oppose his ideas, because it might put one's career in danger. Vladimir Kort, the director of the Institute of Oceanology, did not challenge him, nor did his successor, A. S. Monin. It was not until the 1970s that Soviet ideas about the mobility of the seafloor would be widely discussed, studied, and published.[55]

In the meantime, Beloussov and others developed new theories that took the new evidence into account without accepting the new schematized ideas of the West. Believing that the crust's motion was purely vertical, Beloussov had to face the evidence and develop a plausible explanation, to counter seafloor spreading and preserve the portrait of the earth as it already was known. He explained the magnetic stripes by a process he called "basaltification" or "oceanization." Often when vertical blocks shifted, basalt magma poured out between them, as in the case of the mid-ocean ridge. The magma then poured over the crust, covering the ocean floor. The flow decreased over time, creating a shingle effect. Beloussov claimed that the magnetic lines were simply layers of magma that solidified at different times.[56] The theory of oceanization, as it typically was called, was a more sophisticated version of traditional tectonics that tried to incorporate the new evidence from the oceans without radically changing the theoretical foundations of earth science.

Meanwhile scientists in the West were pursuing research based on the horizontal motion revealed by Hess's "geopoetry." They turned their attention back to Menard's fracture zones, which offset not only the magnetic stripes but also the mid-ocean ridge itself. In 1965 Lamont seismologist Lynn Sykes realized that seismic activity along the great fracture zones occurred only in the zone between the ridge crests.[57] J. Tuzo Wilson proposed that these regions were in fact special kinds of fracture zones: transform faults, where different blocks of crust slid past each other, creating enormous seismic activity. This explained why the parts of fracture zones between ridge crests were so active, while the parts away from the ridge crests appeared to be "dead" faults. Wilson's concept of transform faults bridged the idea of seafloor spreading with a new tectonic theory of massive blocks moving relative to each other. For the United States, this refocus by geophysicists on the sea was one of the most exciting scientific trends of the 1960s. For its contribution to the Upper Mantle Project, the United States had begun an extensive geophysical survey across North America, recording geophysical data along a narrow band of latitude, 4° wide, centering on 37° north. This provided a great deal of new data on the nature of the continents, and in its enthusiasm, the United States Upper Mantle Committee in 1964 decided to extend the band five hundred miles into the sea on both the Atlantic and Pacific coasts. The concept of geophysical traverses proved useful for a wide range of scientists, because it incorporated geological, geochemical, and geophysical data. The same year, the International Upper Mantle Committee recommended that such multidisciplinary traverses ought to be encouraged, and two years later it endorsed the extension of these methods to the sea as well. West and East, however, would each pursue the problem differently.[58]

It was through this particular problem of marine geology and geophysics that the communities diverged most thoroughly. Two influential scientists in collision were Beloussov and the Canadian geophysicist J. Tuzo Wilson. Both men had been presidents of the International Union of Geodesy and Geophysics and were leaders in their field. Both men had been instrumental in starting the Upper Mantle Project, and the Soviet ambassador to Canada in 1961 had personally thanked Wilson not only for his contribution to such an international venture but also for his public praise of Beloussov as a scientist of high merit.[59] Wilson was keenly aware of the leadership role in the international scientific community that his Canadian citizenship allowed him to take. In 1959, for example, the American National Academy of Sciences had asked Wilson if, while traveling to communist China, he could try to find out what the Chinese had done during the IGY. Wilson thought it might be easier for him to do so, for "as a Canadian I cannot be accused of being either a British imperialist or an American imperialist!"[60] Similarly, he felt he had considerable sway over scientists in developing countries. Although countries in South America seemed to dislike the United States, for example, "the fact that these people regard Canada warmly and without jealousy opens up a tremendous opportunity for creating goodwill and providing some leadership" in the international scientific community.[61] It should come as no surprise that it was Wilson who, after making a discovery as important as transform faults to support both seafloor spreading and tectonic motion, was the most vocal in calling the international community to support the theory of the crust's mobility. In doing so, he became a spokesperson for Western geophysics.

Occasionally Wilson's attempts to sell his ideas met with the consternation of his peers. For example, he once submitted an article to *Science* that one of his peer reviewers, Harry Hess, said might have made a good public lecture but was not appropriate for a scientific journal. Hess was insulted that Wilson called it the "Wegenerian scientific revolution," after the early twentieth-century proponent of continental drift, particularly because Hess himself felt responsible for the current idea of seafloor spreading. Hess disliked Wegener's being made into a hero. "I regard him," he wrote to Wilson, "as rather a windbag."[62] Hess wrote to *Science* that although Wilson had made some extremely important contributions to seafloor spreading, "these thoughtful and productive papers are interspersed with flamboyant, poorly documented generalizations such as the present manuscript."[63] Flamboyance indeed was integral to Wilson's style of argument, and despite Hess's criticism, it was his great strength in his confrontation with Beloussov. In the magazine *Geotimes,* Wilson in 1968 detailed how recent studies of the seafloor and of the earth's magnetic field had created a "revolution in earth science." In this polemical article, he wrote, "What an exciting challenge this is! What a chance for great discoveries! What an appeal to young men!"[64] Disturbed by what he considered an irresponsible appeal to emotion rather than rational science, Beloussov wrote an "open letter" to Wilson to be published in *Geotimes*. In the letter, Beloussov critiqued the so-called revolution, claiming that the data could be interpreted differently, perhaps in terms of his oceanization theory. He asked that no rash pronouncements be made.[65] That Wilson's article was an appeal to emotion was true enough, but Beloussov likely also was miffed by Wilson's exhortation to scientists, universities, and industry alike to abandon old concepts and to support by all means a new research program assuming a mobile seafloor. Wilson responded to Beloussov in the same issue of *Geotimes,* thus providing the international community with a text not only by the leading advocates of each side in the debate over the crust's mobility but also by representatives of East and West.

Wilson's position, despite his zealous behavior, had wide appeal. Hitoshi Hattori of the Geological Survey of Japan, who translated the three articles into Japanese, later wrote to Wilson that the articles were being used in colloquia and seminars in Japanese universities. Beloussov's works were well known to Japanese geologists. Many of the students, having previously been convinced that oceanization was more sensible than continental drift, "received a terrible shock by the debate." Still, Hattori was surprised to see so much attention given to how the revolution ought to shape universities and industry, particularly Wilson's urgent call for support along a new research trajectory. Hattori wrote to Wilson, "Your strong appeal raises an alarm signal which has never been made before in a complete paper."[66] Others shared Wilson's role of advocacy. English geophysicist Edward

Bullard's well-known diagrams of continental fits were produced on a large scale with expensive color paper, an unusual occurrence at the time, which he later said was probably responsible for their great impact.[67] Many scientists, such as Robert S. Dietz, then at the Naval Electronics Laboratory, and several scientists at Lamont, went on lecture tours that they called "road shows." Scripps's H. W. Menard later justified these activities, writing that "a scientific revolution requires more than a letter to *Nature*. Even Darwin had his Huxley. A new idea must be advertised and sold."[68]

The new findings on the ocean floor coincided with the publication of Thomas Kuhn's seminal philosophical work, *The Structure of Scientific Revolutions*. Kuhn's book created a framework for understanding the shift in outlook between the ancient and modern worlds, and 1960s earth scientists adapted the model to their own era. For Kuhn, the continuous cycle of scientific activity moved from normal science within an existing paradigm or worldview to a collection of anomalies that did not seem to fit the existing paradigm. Then a crisis occurred in which the existing paradigm simply could not account for the number or magnitude of anomalies. A paradigm shift then ensued in which a new outlook replaced the old, beginning a period of normal science existing within the new paradigm.[69] Menard later recalled that he was already doing "normal science" within the framework of the new theory of a mobile seafloor by 1966.[70] Wilson developed a Kuhnian outlook for his own work, too. He thanked industry and military for helping amass data in the 1950s and 1960s. "Was not Tycho Brahe," he asked, recalling Tycho's massive accumulation of astronomical data in the late sixteenth century, "followed by Kepler and Newton?"[71] In view of such arguments, an exasperated Beloussov asked only that they wait for more data, perhaps to be provided by the international Upper Mantle Project.[72] But calling for more data collection, now routinely derided by Western scientists as a typical Soviet solution, seemed to reinforce the notion of Soviet dependence on dialectical materialism. Wilson rejected Beloussov's conservatism, claiming that if two groups of scientists studied whirlpools and one group refused to acknowledge that the water moved, no amount of data would help resolve the problem. "It's not new data, but a change in outlook that marks a scientific revolution," Wilson remarked, "as T. S. Kuhn... has so elegantly pointed out."[73]

Wilson made such assertions in the exchange of views in *Geotimes;* references to Kuhn presented the philosophical justification for the barriers, provided by the Cold War, between scientific communities of the East and the West. Kuhnian philosophy's espousal of incommensurable paradigms was as much to blame for the rift as the Marxist-Leninist philosophy of dialectical materialism and its abhorrence of idealism. Philosophy allowed Western scientists to pursue their paradigm regardless of the objections Soviet scientists might have had. Wilson claimed that it was the most exciting event in geology for a century and that "every effort in research should be bent toward it."[74] When he began to publish in *Scientific American,* a semipopular magazine with a conservative tradition of allowing only reasonably orthodox concepts onto its pages, Wilson gained further legitimacy for his research agenda.[75] In a preface to a collection of articles from *Scientific American* titled *Continents Adrift,* Wilson again cited Kuhn, who "pointed out that, as the quantity of knowledge increases, each branch of science reaches a stage in which theoreticians reinterpret the lore of practical men into new and subtler formulations." The acceptance of continental drift had "transformed the earth sciences from a group of rather unimaginative studies based upon pedestrian interpretations of natural phenomena into a unified science that is exciting and dynamic and that holds out the promise of great practical advances for the future."[76] Continental drift, the recent version of which would become known as plate tectonics, increasingly appeared as the new paradigm. The Soviets, widely perceived as champions of a defective brand of earth science, could not possibly have a more competitive paradigm in mind. This scientific conflict provides a useful backdrop for understanding relations among marine scientists (specifically marine geophysicists and geologists) who interacted across political lines.

In physical oceanography, scientists were equally dismissive of Soviet work. To follow up the investigation of the tropical Atlantic, the Soviet Union came to the IOC's fourth session in 1965 prepared for another large-scale project, this time on the dynamics of the North Atlantic Ocean. This, the Soviets hoped, would find supporters not only among the IOC but also among the International Council for the Exploration of the Sea (ICES) and the International Commission of Northwest Atlantic Fisheries (ICNAF), both of which took an active interest in the ocean dynamics of the region. The project was accepted by the IOC, and at its fifth session it proposed the formation of a group to coordinate the activities of the three organizations. But by the sixth session in 1969, the Soviets were deeply disappointed that very little had come of it. And in 1971, the Soviet delegation complained that "over the past six years ... nothing has in fact been done to secure the coordination ... of all the oceanographic investigations carried out by the Member States of IOC in this extremely interesting part of the ocean." It called attention to the fact that the United States appeared to have conducted some sixty expedi-

Oceanography, East and West

tions, and the Soviet Union about thirty, in 1970 alone. There were nearly a dozen major international programs in the region, without any genuine effort to coordinate them all to achieve an efficient investigation of the region's scientific problems.[77]

The Soviets were right that very few were interested in their plans or their research. In 1967 they had proposed a major study of the Antarctic region, and most scientists felt that it would not be particularly useful. Despite the fact that the United States was in the process of conceptualizing a long-term "decade" of ocean exploration, scientists in the West denounced Soviet plans as too survey oriented. In addition, as British fisheries oceanographer A. J. Lee noted, the Soviet plans were "based very largely on what we are coming to regard as old-fashioned techniques which are producing rapidly diminishing returns." He looked forward to more problem-based studies, to take a respite from the major surveys that had characterized most of the 1960s since IOC's inception. "Oceanographers in general should be allowed a breathing space of a few years" to perfect some new techniques, he wrote to a colleague at the Royal Society, citing the potential of moored instrument arrays, which might greatly improve upon the effectiveness of ship surveys.[78] The perennial Soviet desire for more and more data, and the apparent disinterest in formulating major theories or solving problems, held little sway over American and British oceanographers.

American scientists' impressions of Soviet science during this period contributed to sustained tensions between the two distinct scientific communities. Western scientists were conscious of the fact that their Soviet counterparts were outsiders, despite their nation's geopolitical importance. This feeling certainly existed during the 1950s, because American scientists received reports that tended to point to the deficiencies of Soviet oceanography, not its strengths. The view that American oceanography far exceeded, in terms of quality and quantity, the efforts being undertaken in the Soviet Union, was pervasive among American oceanographers. The notion that the Soviets were gaining fast, routinely called to mind after the publication of the NASCO report to justify expansion in oceanography, was useful as rhetoric, but it was not real. Even the Soviet Union's great research vessels, the *Mikhail Lomonosov* and the renowned *Vityaz*, were not terribly impressive to American oceanographers, who rather felt that they demonstrated the penchant of the Soviets for size and flair over substance. These ideas were reinforced in 1963 when even a Soviet naval journal praised the American ships and decried the huge Soviet ships, which were "built without a sufficient thorough analysis of their economy and sometimes without a clear understanding of the ship's purpose and the effectiveness of its use."[79] In late 1964, a delegation of American oceanographers visited the Soviet Union as part of an exchange agreement, and they returned certain that the Soviet scientific effort was even less noteworthy than previously realized. Reporting to PIPICO, the delegation claimed that the Soviet Union had some large-scale efforts in action, but none with any real imagination. They seemed to be slighting scientific research in favor of technological applications, "living off capital without providing a program in basic research for the future." In fact, it seemed that oceanography had nowhere near the stature in the Soviet Union that it had acquired in the United States; its scientists, while actively engaged in research, relied heavily on input from other countries.[80]

This contributed not only to Western feelings of Soviet inferiority in scientific results but also to suspicion that the Soviets were trying to exploit others' knowledge or to abuse cooperative relationships for their own national purposes. Geophysicists had similar suspicions of large projects that included the Soviets. The Upper Mantle Project, for example, began to cause some concerns among British officials, particularly because the chairman of the International Upper Mantle Committee, Beloussov, had insisted upon the importance of the African Rift Valley as an object of study. At a 1965 meeting in Nairobi, Kenya, to discuss the project, Beloussov brought along with him a man who claimed to be a seismologist, but who, according to Edward Bullard, never talked about technical matters and insisted on taking a lot of photographs. Still, Bullard wrote to the Ministry of Defence that he thought Beloussov's intentions were sincere and that it was all "not a deep laid plot to introduce a lot of Russians into East Africa."[81] The Ministry of Defence, however, was not so sure, and it believed that Soviet military geologists were highly interested in sending teams to the area to explore Kenya's mineral resources.[82] Questions of what exactly Soviets hoped to gain from these cooperative efforts remained, particularly when the scientific results seemed so out of step with those of the West. This skepticism continued throughout the 1960s, especially during the Second International Oceanographic Congress in Moscow.

THE SECOND INTERNATIONAL OCEANOGRAPHIC CONGRESS, MOSCOW, 1966

At the IOC's third session in 1964, delegates discussed the possibility of holding in Moscow the Second International Oceanographic Congress, a follow-up to the highly successful one in 1959 in New York. Some noted with interest

The Second International Oceanographic Congress, Moscow, 1966. Standing is Vladimir Beloussov; to his right is H. W. Menard. Courtesy Churchill College Archives

that the convening body was not listed as Unesco, but simply as IOC. This seemingly simple difference prompted an American delegate to inquire whether he could be assured that Unesco's regulations and practices would still be followed if the meeting was held in the Soviet Union under the IOC's auspices. At this insinuation, the leader of the Soviet delegation, Tchekourov, became irate. The Soviet Union, he maintained, had hosted many conferences in the past without any problems, and in any case, they were not so keen on being the hosts. He disliked the idea that anyone should think of imposing conditions on the Soviet Union for hosting the event, and particularly he resented the implication that other countries were wary of what might happen without such guarantees. He objected strongly to some of the wording in the proposed resolution about the congress, and ultimately he succeeded in having (what were to him) the most offensive statements of the need for guarantees deleted.[83] It was an inauspicious beginning for the second congress.

In 1966, as seafloor spreading gained more and more adherents in the West, the Soviet Union convened the congress. Taking place from May 30 to June 9 in Moscow, this meeting was intended to be even more impressive than the first one, held at the United Nations building in New York City in 1959. In 1959, the participation of the Soviets had seemed to bode well for international cooperation, particularly because of the personal contacts that Western scientists cultivated among their Soviet colleagues. However, the spirit of personal collegiality, so evident in 1959, had declined by 1966. Despite the Soviet scientists' willingness to include colleagues from all countries, the Soviet government barred entry to some. Representatives from Taiwan and South Korea, for example, experienced such delays in acquiring visas that it was impossible for them to attend. The Korean delegate to Unesco demanded an apology from the Soviet Union and requested that no further conferences be held there. The American representative to Unesco wrote that "the Soviet Union has violated both the spirit and the letter of the contract it made with Unesco," and he urged Unesco to consider withdrawing financial and logistical support for the meeting. Unesco's director-general, René Maheu, hedged on the issue, saying that although Unesco provided some funding, the logistics were being left up to the Soviets' national arrangements committee. This was familiar territory for many scientists; at an IOC meeting held in Moscow in 1963, the delegation from West Germany had not been able to attend because of delays in granting visas. In that case, the visas had been authorized two days prior to the opening of the meeting.[84] Another great disappointment was the fact that the American ship *Silas Bent* was refused entry to Leningrad at the last moment, on the grounds that it was a naval vessel, not a purely scientific one. Scientists had hoped that the *Silas Bent*'s presence would contribute to the spirit of cooperation in much the same way that the *Mikhail Lomonosov* had in 1959. The Soviet government's refusal embarrassed many of the congress organizers, who had even had postcards printed in English and Russian commemorating the ship's visit.[85]

After the Congress, NASCO sent questionnaires to the American participants to collect views of their experiences in the Soviet Union. The attention paid to these questionnaires varied; some answered briefly, while others included additional comments ranging in length from one page to twenty-seven pages (as in the case of Bureau of Commercial Fisheries scientist Robert E. Stevenson). The responses offer a portrait, though probably not a balanced one, of American oceanographers' scientific and political impressions of their Soviet colleagues. If scientists expected to be disabused of their notions of Soviet incompetence, the questionnaires indicated that the logistical performance at the congress did little to help. Most attendees reported some form of a frustrating experience in the Soviet Union, rein-

Oceanography, East and West

forcing already-held beliefs about the Soviet system. Although many were charmed by some of the social events, the Americans generally complained about the many logistical shortcomings of their journeys. For example, virtually all the questionnaire respondents felt that, other than the caviar, the food was bad. The Intourist hotel service was frustrating; the participants often had no idea where other participants were staying. One congress participant, Leon Knopoff, spent a day of anxiety as Intourist literally lost his wife, whom they had arranged for him to meet in Leningrad.[86] Geophysicist Hugh Bradner put it thus:

> I feel that a trip to Russia should be required education for all Americans; and I feel that the trip should deliberately include an attempt to make some unscheduled small everyday activity of the Western world, such as changing hotel reservations, changing plane reservations, trying to pick up some nonpolitical literature, etc.[87]

Translations at the sessions were reported to be universally poor, and when one Soviet scientist decided to conduct the session he chaired in English, because it was spoken by most of the attendees, he was reprimanded by congress organizers.[88]

These negative reactions were compounded by what could only be called the "runaround." The extent of personal interactions of the 1959 congress simply never happened at the 1966 congress. Americans had difficulty making contacts with Soviet scientists and visiting Soviet facilities. One scientist reported that if an American asked to see a particular Soviet oceanographer, often he was told that the scientist was working at the institute; if at the institute an American posed the same question, the Soviet scientist was at the congress. One well-known Soviet scientist was Gleb Udintsev, who enjoyed some celebrity at the congress because of his spectacular new color atlas of the Pacific Ocean basin, which was on display at Moscow University for all to admire. A number of scientists, including K. O. Emery of Woods Hole, hoped to purchase a copy of the chart. Udintsev agreed to set aside a special room in which those interested in the chart could sign up to have it sent to them. About five scientists set out the following morning to find the room but were told upon their arrival that they were "off limits," and none of the Soviets present had any clue what they were talking about. Soviet oceanographer Konstantin Fedorov, then also secretary of the IOC, arrived and herded away the scientists, among whom was a group led by Scripps's Walter Munk that had met for a session in the adjacent room, telling them that they were on the wrong floor (they never did find the room).[89] A few days later two American scientists, D. S. Gorsline (University of Southern California) and Joseph Creager (National Science Foundation) set out to find copies of the chart at an address supplied by one of the Soviet interpreters. This address turned out to be World Data Center B, one of the central clearinghouses for internationally shared oceanographic data. Upon seeing the Americans, the incredulous staff demanded to know how they had discovered the center's location. They were told to leave immediately (the Soviets did take their names, to have the chart sent to them). They were informed that they should tell no one else where the center was, either by address or even general location. The Americans, naturally, became very suspicious about what could be in the Soviets' World Data Center that was not in the others.[90] No doubt such mishaps became an endless source of jokes and storytelling.

Although the Soviets organized some visits to scientific facilities, these were short and usually did not include visits to actual working laboratories. Hugh Bradner remarked that such visits "seemed to be a very open exchange until 12:00 noon, at which time the visitors suddenly turned into pumpkins and were rolled out the door."[91] American scientists complained about how little of the Institute of Oceanology they were permitted to see. They felt that many of the short visits to other institutions were a waste of time, and they guessed that the Soviets either must have been ashamed of the primitive state of their best facilities or must have been hiding something. Fred Phleger, who participated in a tour of the Black Sea after the congress, complained that he spent fewer than seven hours at oceanographic institutions over a period of eight days. Additional time was refused, and many of his specific questions went unanswered. "They did, however, show us youth camps, sanitaria and many botanical gardens—which none of us had the slightest interest in seeing. We all agreed that the trip was a complete failure and somewhat disagreeable because we were herded like cattle."[92]

Perhaps the majority of scientists who took the time to respond to NASCO's questionnaire were those who had some complaint to register. But certainly the congress served to reinforce, among some, negative stereotypes about the Soviet Union. At Moscow University, the scientists' movements were tightly controlled. Some of the corridors were lined with chairs meant to block entry into some of the rooms. The doors to the stairways were locked. On one occasion, Soviet scientist Aleksei A. Bogdanov took some scientists to visit a laboratory, only to tell them he had forgotten the key once they arrived. After he departed to retrieve the key, one of the visiting scientists simply decided to knock on the door: "He knocked and the door was opened.

A woman looked out, gasped, a shocked look came on her face, and she slammed the door." When Bogdanov returned, he explained that everyone had gone home and so they would not be able to visit the laboratory. There were many reports of this kind of "runaround" treatment. "As someone noted," wrote Stevenson, "all of the experiences we had would make the greatest Laurel and Hardy script of all time."[93]

A few Americans had more positive experiences, making the best of their opportunities. For example, Soviet biologist Lev Zenkevich and some other scientists led an unofficial tour of the biological facilities at the Institute of Oceanology, despite the remonstrations of the institute's director (this was Andrei S. Monin, who replaced Kort in 1965). Zenkevich was one of the best-known Soviet marine scientists, particularly among marine biologists, and he had clout enough to do this. His book on the biology of the Soviet Union's surrounding seas recently had been translated into English, and he was active in international activities.[94] When University of Miami biologist Gilbert Voss inquired why some Americans were having trouble visiting the facilities, the Soviet scientists accompanying him admitted to Voss that "it was by design as they did not care for us to go through the facilities." Voss noted, "This was partly because the facilities were not up to American and British standards, I understand, although for this I cannot vouch." Voss certainly was not impressed with the facilities he saw and wondered how so many papers could come out of such rudimentary facilities.[95]

Perhaps the Soviets feared spying by the American scientists. Some of the participants certainly thought they were under surveillance. At one point Stevenson, on one of the few facility tours at the Institute of Oceanology, was looking at a poster of two women arm-in-arm, one a Russian peasant and the other a Cuban soldier, on a wall about three feet from where he stood. It was then that he "saw this nondescript sort of fellow" watching him; the man stepped between him and the poster. At that point a few of the anonymous-looking men accompanying the group seemed less anonymous.[96] Scripps scientist Robert L. Fisher, who was accompanied by his wife, joked that "in Room 1317 of the Ukraine Hotel we daren't speak freely except by means of digestible notes written and passed beneath the bed sheets of a darkened room. Next time I'll learn Braille." Fisher actually did use his wife to ask other scientists to meet informally, and in that way he was able to meet with nearly everyone working in his field. He was also able to acquire some unpublished Soviet data that "we desperately needed in the so-called cooperative IIOE program."[97]

More than the inconveniences and strictures of the congress, the Americans were surprised at how little use it was to exchange scientific ideas with their Soviet colleagues. Stevenson's exhaustive report was the most explicitly critical of the Soviets; nevertheless many of the other respondents, however more diplomatic they may have been, generally followed the same line as Stevenson's:

> I learned nothing from the Russians; that is, nothing regarding the ocean, the sea floor, or life in the sea which I had not already known. The Russian papers to which I listened were mainly superficial. In some cases they were plainly ridiculous. In other cases, it was a pitiful performance and one felt sorry for the speaker.[98]

The Soviets spoke about religion, life in the Soviet Union and the United States, salaries, and culture—"But, when it came time to talk about Russian research vessels, Russian plans, about work that they themselves had done . . . the discussion was staggered, groped, or was guided into another line." Or suddenly, Stevenson wrote, the person who seemed to understand English perfectly well—when discussing food and the opera—no longer seemed to understand English well at all. Some may say, Stevenson continued, that despite the frustrations, it was a great benefit to exchange ideas on scientific subjects with colleagues from the East. To this Stevenson underlined the following words: *"Any such statements are pure and simply baloney."*[99]

The American attendees generally drew two conclusions from the Soviets' unwillingness to talk more in depth about science; they were incompetent and/or they were holding back. Many of the Americans in attendance would conclude that there was ample evidence for both. Scientists from thirty countries gave papers. Only 166 out of 1,200 Soviets in attendance were scheduled to give papers, compared to 175 by Americans. These two countries represented about 67 percent of the total, although scientists from twenty-eight other countries gave papers. But because the Soviets were hosting the event, many scientists were surprised that the Americans gave more papers than the Soviets. Some concluded that the Soviets were more eager to see what the rest of the world was doing rather than showcase their own results. This was particularly glaring in the subject of "Oceanic Instrumentation," in which Americans presented thirteen papers compared to only two by Soviets.[100] Many respondents to the NASCO survey were suspicious that the Soviets were just trying to glean as much information as possible from the West. Perhaps the Soviets' seeming lack of sophistication was no more than nondisclosure of significant results. One American participant, Margaret Robinson, wrote that the Soviets certainly learned more about American methods and results than the Americans learned from their Soviet counterparts. She noticed "that the Russians gave no papers in 'descriptive' oceanography, and since their theoretical papers do not describe the real ocean, I am wondering if they really are analyzing the data they collect."[101]

Western scientists were torn between judging the Soviets in terms of the science presented—which was universally thought of as poor—and assuming that the scientists simply were not allowed to share their most interesting results. Lamont geophysicist J. Lamar Worzel felt that the Soviets relied too heavily on topography, coring, and dredging and they made sweeping generalizations that reflected poor scientific standards. "The Russians started with a Rush [sic] into oceanography," he wrote, "but have not seemed to follow up—at least in geophysics."[102] Scripps scientist Fred Phleger described the Soviet papers as "very mediocre indeed. I have the impression that much of the work is the kind of thing which we were doing ten to twenty-five years ago."[103] Appalled by the quality of papers at the congress (not limited to Soviets), some felt that the Soviet organizers simply had accepted whatever papers had been proposed. Many agreed with T. H. van Andel, who said that for all countries' speakers, "an uncomfortable majority came out with trash or trivia," and that the quality "was so embarrassingly low, that one wonders whether we really rate the work and money needed for another Congress."[104] Criticism directed at the West typically derived from its reliance on summarizing past work rather than presenting fresh research, but the scorn directed at the Soviets came from the lack of quality. Americans could only assume that when Soviet marine geologist Panteleimon L. Bezrukov presented his chart of the sediment distribution for the entire Pacific Ocean, using only two thousand sediment samples, he must have been constrained by some authorities not to present his best data. With only two thousand samples ("A pittance! A nothing!" complained Stevenson), the results could not be considered valid.[105] Yet it was difficult to conclude that the Soviets were doing bad science, for it was clear to all in attendance that the Soviet national effort for oceanography was huge in scope. Germans, British, Australians, and Japanese alike expressed their bewilderment that, although the Soviet commitment often dwarfed the projects of their own countries, the Soviets' results appeared minuscule in comparison.[106]

The conclusions drawn served only to discredit Soviet science further. Although many concluded that the Soviets were holding back some of their best work, there was still the feeling that overall, Soviet oceanography was being misguided by the nature of the Soviet social and political system.

Fisher noted that he was pleased that the congress provided "the opportunity to do missionary work against the widespread practice in USSR geotectonics of beginning, rather than concluding, with the philosophy and interpretation."[107] A. E. J. Engel, who felt that the Soviets had done almost nothing of interest in oceanography, believed that there was a great potential for Soviet scientists to make genuine contributions. Nevertheless he added:

> But the Russians will have to abandon their veneration of dead and aging men of science, and clearly outmoded and unrealistic ideas, as well as the present institute setup. If the young men can break loose they will do big things—I would guess it will take several years for real progress, but with the odd, very excellent paper sandwiched between the innumerable bad reviews and windy, data-barren dialectical essays.[108]

These criticisms insinuated that the Soviet scientists were victims of their own society, dominated by the Communist Party and its supposed insistence on philosophical bases for scientific results. Stevenson went even further, making a clearer contrast between scientists from the East and their colleagues in the West. The individuals and the work were good in some areas, he admitted, but without sophistication. "What was obviously lacking," he wrote, "was *freedom*—freedom to pursue mental ruminations—freedom from external pressure for 'results'—freedom from political pressure for 'answers.'"[109] There was a substantive difference in some Americans' minds, tied closely to the central social and political contrasts of the Cold War, between oceanography done in the East and that done in the West. Recalling the frustrations of all the participants, Fisher summed up the conference well. "The local committees obviously had worked very hard to beat the 1959 congress but," he wrote, "except for the tremendous cultural events, were defeated by their own system, with its inflexibility and chain of command."[110] The congress managed to reinforce stereotypes of Soviet science, alienating the East and the West from each other rather than bringing them together.

The combination of scientific divergence and Cold War tensions only deepened after the congress. At the Montevideo Symposium on Continental Drift the following year (at which only two Soviets, geophysicist Aleksei Bogdanov and geochemist Alexei Tugarinov, were present), scientists made a resolution regarding the future direction of international research on the earth's crust. The Upper Mantle Committee was slated to end its tenure in 1970, so the participants made some recommendations about how to continue. Specifically, they resolved that the successor to the Upper Mantle Committee might focus its efforts upon investigating the problems associated with continental drift. Beloussov, who did not attend the symposium, opposed any plan with such a narrow focus. The study of the earth's interior, he wrote to Leon Knopoff, the Upper Mantle Committee's secretary general, "needs a wide-minded approach. . . . The study of the problem of whether the horizontal continental drift exists or not, covers only a part of the problems in that field. The relative size of the part depends on personal evaluations." Beloussov urged that they not neglect critical problems in other subjects such as the composition of the crust and upper mantle, the effects of high temperature and pressure, geophysical fields and deep structures, and the causes and mechanisms of poorly understood magma-related and metamorphic phenomena. He added that the international community should remember also to study "*different* types of tectonic movements" than those studied by the symposium participants (emphasis in original). An international body focused upon the study of one hypothesis, namely, continental drift, could not stand as a just successor to the Upper Mantle Committee.[111] Beloussov was revealing his own frustration with the West's obsession with the ocean and the strange characteristics of the seafloor. There was more to tectonics, he felt, than the oceans.

Beloussov suggested that the international scientific community should take up this topic at the upcoming International Geological Congress of the same year. Concurring, many Americans were excited to discuss ideas of continental drift with their Soviet colleagues, as some still felt that the clash of paradigms need not divide scientists along Cold War lines. One geologist who visited the Soviet Union in 1966 published a report on Soviet geology in *Geotimes* expressing his belief that dogma was not as prevalent among Soviets as Americans had assumed and that there was genuine hope for some collaboration. Although he thought "that the 'vertical tectonics' theories of V. V. Beloussov would be gospel in the USSR," he was surprised to find that several Soviets thought Beloussov was an extremist. There were some respected scientists, though not known in the West, "who can slide and jostle continents just as far and fast as the wildest of Western 'drifters.'"[112] The congress, to be held in the Eastern bloc, promised once more to bring scientists from the East and West together to discuss these scientific ideas.

The desire by Beloussov and others to bridge the gap between the geophysical ideas of East and West did not come to fruition at the congress. In fact, the Twenty-third International Geological Congress of 1968 could not have been a greater disaster in this respect. The fateful location of the congress was Prague, Czechoslovakia. Its opening on August 19, 1968, included welcoming remarks by the Czechoslovak prime minister and the mayor of Prague. About twenty-five ambassadors attended. It was a stately affair, including two selections by the Prague Symphony Orchestra.[113] In the spirit of scientific cooperation, the Americans intended to avoid politics. The leader of the American delegation, Thomas Nolan, conveyed to his fellow Americans the instructions he had received from the Department of State, namely, that they should not engage in embarrassing political conversation. "The thing to be avoided," Nolan reportedly said to his group, "was a shouting match with an unfriendly delegation."[114] It should not have been difficult to stick to the scientific or cultural tasks at hand. For the four thousand congress participants, there were field trips, good hotel facilities, and numerous sessions to keep the scientists busy talking for days about the constitution and dynamics of the earth.

The failure of the congress was obviously not the fault of the Czech organizers or of the Soviet scientists, who could not have foreseen Soviet tanks as a factor in disrupting the meeting. Not long after midnight on the morning of August 21, 1968, the Soviet Union invaded Czechoslovakia. Scientists awoke to the sound of gunfire as Czech citizens fought Soviet tanks on the streets of occupied Prague. The dozens of Soviets who had arrived at the conference, now no doubt deeply embarrassed, allowed their attendance to drop off sharply. Some scientists believed the Soviets had abandoned the congress, but actually nearly all of those remaining had simply removed their nametags, which listed national origin. Many congress members wore black strips on their nametags as a sign of protest, and for a brief while a congress sign was draped with a banner reading "Russian killers go home."[115] The Upper Mantle Committee was scheduled to meet August 22–23 in a special symposium titled "Deep-seated Foundations of Geological Phenomena." Although some papers were read, many were not, and attendance was low. Many of the attendees abandoned the congress and escaped the country if they could.[116]

The conversations of the international delegates were partly recorded in the notes taken by American scientist Hollis Hedberg. On Wednesday morning he came upon a special meeting of the bureau of the congress, "more or less by accident," and acted as an impromptu representative of the United States as the group decided how to proceed. The Czechs read the official statement of their Academy of Sciences, which protested the Soviet acts of aggression and voiced support for the Czech regime. Most of those present agreed that the congress should continue, at least until the following day.[117] On Thursday morning, the Czechs announced that the museum in Prague had been shelled and that the National Academy was blocked off. Although excursions in the area obviously would have to be canceled, there was no reason why the meeting could not continue.[118] The Soviets, who had not all departed, urged that the congress continue. The French delegation complained that it would be impossible to do so, as there was too much risk of machine-gun fire on the streets. Their embassy had instructed them to stay inside their hotels. The Unesco delegate proposed continuing on an ad hoc basis, with the understanding that it just might become physically impossible to carry on.[119] Hedberg admitted that, because public transportation out of the city was nearly impossible, they might as well try to make the most of the meeting. His notes reflect concern about his hosts:

> Question is how can we best support our Czechoslovak friends in this dire hour—by closing Congress or by continuing it. Museum has been shelled, Academy closed, and Charles University closed. The one thing that has been spared to date in the scientific life of the city is this Congress. Perhaps we should continue it as a refuge for Czech scientists. Perhaps we should continue it so that there will be some outside witnesses to what is going on. Fear that if we close it, things may go worse for our hosts.[120]

By Friday, August 23, although the Czech hosts announced that life in the city was improving, many people already had left the congress, and the others generally agreed that it had ceased to be a useful meeting. During the closing session, delegates from many countries made remarks. Hedberg himself tried to assure the Czech organizers that the congress had been valuable especially because of the "common bond of sympathy, deep feeling, and understanding which has been developed so strongly among all of us during the Congress and which I feel certain will always endure."[121] Indeed, the bond to which Hedberg referred cannot be denied. The disastrous congress and the mutual trauma of being there during the fighting probably enhanced the personal bonds between scientists of many nations.

The failed meeting in Prague, despite having brought a few members of the international scientific community together through the crucible of extreme circumstance, nevertheless seemed to foreshadow long-term damage to the coherence of the international scientific community, at least among geologists and geophysicists. It was an opportunity for earth science that disintegrated in the most physical manifestation of geopolitical conflict. As Nolan reported to the National Academy of Sciences, "there are direct intangible losses to the geological sciences. Several thousand geologists, knowledgeable of every continent of the earth and every subdivision of the subject, were given only a feeble chance for making the periodic review of the field, examining the evidence, and debating the differences so as to speed the flow of information and stimulate the world-wide growth of ideas."[122] For better or for worse, the Soviets would play a less and less important role in the development of geophysics. This was not a direct result of the Prague meeting; however, the congress would have been an ideal setting in which to discuss and debate the issues that thus far had fallen decidedly along political lines, as well as to formulate a plan for international action that met the wishes of all participants. The failed congress in Prague perhaps was the best symbol of the limits of international cooperation and of the solidifying boundaries around a scientific fraternity that was increasingly united in the face of embarrassing and unproductive actions of their Soviet colleagues. As one scientist, Linn Hoover, wrote to *Geotimes,* "the language of science is international . . . [b]ut Prague reminds us that science, too, is susceptible to the vagaries of politics."[123]

The focus of each community increasingly was different, with the West's most influential research deriving from studies of the ocean floor. The Soviets preferred to study continental tectonics. The planned agenda for the Prague meeting reveals this rather well; seven American and one British scientist comprised the whole subject "The Ocean Floors," whereas Soviet scientists presented their vision of global tectonics under the subject "Tectonics of the Continental Masses."[124] Rather than proponents of vertical (focusing on continental) tectonics such as Beloussov, it would be proponents of horizontal (focusing on seafloor) mobility such as Lamont seismologist Lynn Sykes who shaped the agenda of the Upper Mantle Committee in its final years. They took their cues from Bruce Heezen, who at the 1967 Upper Mantle Project Symposium on the World Rift System pointed out that current research so far indicated only that the earth's crust was expanding. He noted that although the creation of the ocean basin at the mid-ocean ridge largely had been accepted, an equally key concept of seafloor spreading—namely, the destruction of the crust at the trenches—had yet to be demonstrated.[125] Consequently, Lamont scientists conducted seismic studies that showed what happened to the crust as it dipped into the trenches, as seafloor spreading would have it. Sykes and his colleagues soon published papers revealing that different velocities of seismic waves indicated that a relatively cold, solid slab of crust was plunging deep into the molten mantle.[126] The same year, papers by other scientists at Lamont and Scripps synthesized disparate concepts such as fracture zones, mid-ocean ridges and rifts, magnetic anomalies, and seafloor spreading. These were the first papers in the new science of plate tectonics. The authors envisioned huge spherical caps, or blocks, moving horizontally over the earth's surface, created at spreading centers and destroyed at trenches.[127] Indeed Sykes and colleagues Jack Oliver and Bryan Isacks had prepared a paper to deliver at the Prague conference that showed how seismic work conducted by Lamont scientists had provided ample evidence for seafloor spreading, transform faults, and the underthrust of the crust. They summarized with conviction, "At present within the entire field of seismology there appear to be no serious obstacles to the new tectonics."[128]

There were no serious obstacles, certainly, except the Soviets themselves. The Soviets had proposed their own version of global tectonics, based upon vast continental structures called "platforms" whose mobility was characterized by "extreme conservatism"—a theory consonant with Beloussov's views, denying the validity of continental drift. But compared to the Americans' confidence that seismic data supported plate tectonics, what did platform tectonics have to recommend it? One Soviet abstract summarized, "To create a single and logically universally acceptable physical model of the evolution of the platform and especially in its historical aspect, is absolutely beyond the means of the authors at present."[129] This lack of certainty, from their perspective, was the very justification to pursue it further. But to Westerners they essentially were admitting that plate tectonics had no real competition. Granted, the ideas of the plate tectonics revolution would in subsequent years be debated and countered on the pages of Beloussov's international journal, *Tectonophysics,* and elsewhere. But there was a new paradigm for earth science, and the Soviets never were part of it.

What value were the Soviets, then, in international cooperation? Given the conceptual differences between East and West, it is not difficult to imagine why Western scientists put such little value in person-to-person contacts. Senior influential scientists seemed to block new ideas; Beloussov's reach extended beyond geophysics and included oceanography as well. In other fields not related to plate tectonics, such as physical oceanography, the Soviets were busy with surveys. The only tangible commodity the Soviets could offer the West in the scientific exchange was data, which they collected at great length and expense, much to the disdain of scientists such as

Deacon who preferred problem-oriented studies. Data from international programs kept Soviet scientists actively involved in international exchange and allowed them to publish without gaining approval from the government. Certainly it was for data alone that the Americans and British fostered the cooperative relationship with the Soviets; other alleged motivations—coordinating work, comparing techniques, or discussing new ideas—had little to recommend them by the late 1960s.

7 MARINE SCIENCE AND MARINE AFFAIRS

Western oceanographers widened cooperation in the 1960s to include not only Soviet scientists but also those least capable of carrying out research: countries of the developing world. This was part of American scientists' strategy of tying their work to economic exploitation at home and abroad. Often perceived as a necessary evil to finance large-scale research schemes, promoting marine science for its economic consequences yielded unforeseen (and some foreseen) problems. One problem already discussed here: physical oceanographers felt that their hands were tied. Projects had to gain the approval of a growing number of countries in the IOC that were less interested in conducting research than in attending to the political ramifications of research, meaning that scientific goals had to compete with each country's perceived need. In addition, tying projects to development left them vulnerable to interference by biologists and fisheries scientists. But even worse than these were expectations of the governments who paid for the work. If research was going to contribute to economic development, for example, why do it all the way in the Indian Ocean, when there are plenty of waters to investigate closer to home? And where was the evidence that such science really benefited anyone? These questions went naturally with the rising costs of oceanography, costs often felt most acutely in the smaller industrialized countries such as Britain and Norway.

The developing world did not treat these efforts as mere rhetoric. Scientists in Asia, particularly India, found themselves trying to work within the IOC's framework to establish permanent fixtures of marine science in their countries. But the weak coordination of the IIOE did little to convince them of the sincerity of scientists working on large projects. For their part, leading scientists had a difficult time trusting the competence of scientists in the developing world. The sense of community was false, which should come as no surprise because emphasizing the role of the developing world was simply, as one American put it, "for the good of the cause and the sake of the argument."¹ The cause, of course, was attaining more money and greater appreciation for scientific research: creating disciples of marine science. No one did this as well as the American oceanographers, who by the middle of the 1960s had convinced their government to create a Marine Sciences Council, led by the vice president of the United States, dedicated to putting science to use. These gains came with sacrifices, however; in particular was the diminished role of the scientists themselves. Marine science became highly politicized, tied inextricably to national interests and to some of the most difficult political problems at sea to be argued out by national delegations in an intergovernmental forum. Scientists' struggles to maintain support for basic research became more difficult than ever, with bureaucrats insisting that they make good on their promises of practical results. Considering their efforts to convince governments of the economic and social aspects of oceanographic research, the greater part of the blame for this can be placed on scientists.

DEATH AT THE BANQUET TABLE:
SCIENCE AND DEVELOPMENT

Connecting science to development occasionally met with considerable unease, and often opposition, from scientists. This was particularly true among some British and American scientists who felt the backlash of making these connections. But the support oceanography enjoyed in the 1960s resulted directly from American efforts to convince governments of the usefulness of oceanography for national defense and resource exploitation. British scientists knew that these connections were useful too, but they also were strongly apprehensive about the effects of making the connections so explicit. Britain's Ministry of Science met often during the early 1960s to discuss how the IOC was changing scientific research and its relationship to government. It was cognizant of some work in oceanography, through Solly Zuckerman and his role in the NATO Science Committee (and its subcommittee on oceanography), but the extent of the government's future role was unclear at the beginning of the decade. Mostly it watched for developments in the United States. A British scientific attaché in Washington wrote to the ministry in late 1959 that "it seems to me abundantly clear that oceanography is going to be one of the big subjects in science in the next few years and . . . the Americans are getting very active in this field."² By 1961, officials within the Ministry of Science were attempting to consolidate

Britain's governmental apparatus for support of oceanography to follow the lead of the United States. The Royal Society recommended an expansion in oceanography (like NASCO and the United States Navy had), but the means to do so were far from clear, and initially the Ministry of Science looked to the American ICO as a potential model.[3] But ultimately (much as in the United States) support for science in Britain would go through a number of changes during the 1960s before stabilizing. The debates were as fierce and complex as those in the United States, and often more so because there was considerably less money to go around.

When the IIOE was in its planning stages, British scientists in particular were wary of its implications, because some countries such as India were linking it very closely to fisheries development. "Some of those concerned," one official noted, "have reservations about such an approach which may be more commercial than purely scientific."[4] Despite Britain's commitment to the IIOE, it was having trouble recruiting scientists to work on its ship, the *Manihine,* because of a widespread conviction that such surveys were not particularly scientific. Because the studies would take on a routine character, requiring little imagination on the part of the observer, participation in the IIOE was unlikely to qualify students for a Ph.D. in a British university.[5]

Some felt that the Ministry of Science had missed the point that the Americans knew only too well, that such projects required a little bit of fanfare. Lacking any of the salesmanship that the United States had devoted to the cause of the IIOE, the British appeared to be less enthusiastic about it. Still, they were major participants, prompting one Admiralty official to wonder whether anyone realized that they were taking part. None of the newspapers and magazines, he complained, ever mentioned Britain, because as late as 1961, few scientists or officials in the country had made any public statements in support of the project. "I do not think it is good enough to be taking part," he complained to the Ministry of Science, "as I feel that we also ought to be seen to be taking part."[6] The reason for this lack of publicity, rather against the practice of the Americans, was the pervasive feeling that the IIOE would not produce valuable scientific results if tied so closely to economic development.

British scientists were slow to admit that the scale and scope of projects had a lot more promise when pursued under the banner of economic development, which could justify increased spending. But like the Americans, they too tried to take advantage of the opportunities. The British decided to involve not only scientists but also those whose interests lay predominantly in fisheries through its East African Marine Fisheries Research Organisation (EAMFRO), a body devoted to the fishing interests of the territories of Kenya and Tanganyika. Through the use of EAMFRO's ship, the *Manihine*, the British participation in the Indian Ocean could claim to be useful, because it would conduct some fisheries investigations on the far western side. Unfortunately for the British oceanographers, however, Britain recently had granted these countries (along with Uganda) greater independence, which meant that EAMFRO itself, although British, could exist only if it had some degree of local support. The local governments, however, decided that they had no interest, with their limited funding, in continuing to support what they considered a second-rate fisheries organization. This was severely demoralizing for the British oceanographers, as it promised to cut into their program considerably and deprive their program of its practical justification.

The situation pressured SCOR to take a firm position on the importance of oceanography for practical applications. Australian George Humphrey had taken over as the president of SCOR as the IIOE was getting under way. Unlike Roger Revelle, who was more than happy to link science to problems of society, Humphrey viewed such linkages with disdain. Nevertheless, Humphrey was faced with the reality that the IIOE was progressing with some success specifically because men such as Robert Snider had tried to demonstrate the connections to practical problems. The American fisheries scientist Wilbert Chapman, then director of the Van Camp Foundation, noted that the institution's troubles stemmed largely from SCOR's pious refusal to make such connections, a position that would let EAMFRO "wither on the vine for lack of support."[7] He complained to Humphrey that SCOR should court organizations that allocated funds for practical work, as did Unesco's Special Fund:

> If you want to do SCIENCE in capitals then you must get your money from sources that fund such operations. . . . On the other hand if you are prepared to admit for the good of the cause and the sake of the argument that your research in the Indian Ocean might accidentally lead to the development of fisheries, aid navigation, or predict rains that might help farmers in adjacent lands, then you will get an attentive ear at Special Fund.[8]

The administrators in the Special Fund were not so bad, wrote Chapman. They were efficient and flexible, and they did useful things with their money. "And," Chapman added, "they have a big pot of dough."[9] This reasoning applied not only to the Special Fund but also to all sources of financial support expecting practical results.

George Deacon, already ashamed of the lack of funds for his own National Institute of Oceanography, was annoyed by the potential closure of EAMFRO, which undermined his ability to tout the worthiness of the IIOE. He wrote to the Ministry of Science, "What a pathetic picture our country is presenting over the whole business!" Britain was positioned to make a rewarding contribution to the IIOE, "an outstanding milestone in international science," but problems of funding and unexpected opposition seemed equally positioned to halt them.[10] To another official he wrote that it was "dreadful" of Britain not to put a bit more determination into the oceans, and it "ought not leave everything to Japan, USSR or USA."[11] Britain found itself in the position of having men, a ship, but very little money to ensure the completion of the expedition. One National Institute of Oceanography official wrote, "All we lack is money, and when our needs are really so small and the dividends both in terms of progress in the science and national prestige are so large, it seems extraordinary that enough cannot be found."[12]

The British government was sympathetic to these complaints. The Ministry of Science took the position that the research off East Africa might have a major impact upon the fisheries of the region, being "of direct advantage, in the long term, to the East African territories bordering on the Indian Ocean." But when questioned on the issue, officials in the government in Kenya were appalled to find out that anyone cared about it. Their own understanding, officials explained, "was that EAMFRO had never been much good and had not done outstanding work," and further "had proved notably incompetent in running a research ship." The hull of the *Manihine* itself was covered with barnacles and it would require additional funds to ready the ship for the IIOE. With their limited money, there were far more advantageous avenues to explore with agriculture. In general, EAMFRO was a burden, and a second-rate burden at that.[13] However, after some negotiating, Britain convinced Zanzibar to make a small contribution, thus providing the semblance of local support, making it possible for the British to make a larger grant (under Britain's own regulations, such grants could be made only if a contribution was also made at the local level).[14] Thus EAMFRO and the *Manihine* were saved, thanks to the determination to promise economic dividends from scientific research.

By trumpeting the fisheries aspects, the British walked on treacherous political ground. Unlike the United States, the United Kingdom was already an active member of the International Council for the Exploration of the Sea (ICES), a scientific organization that also connected its work to fisheries, but that concentrated its activities in the North Atlantic. Now that the IIOE was getting under way, members of Parliament began to question why the new British ship, *Discovery*, was to be diverted from the Northern Hemisphere and sent, as Member of Parliament (MP) Hector Hughes put it, "with a fleet of foreign ships to explore the Indian Ocean."[15] This particular MP made it his business to hound the Admiralty about the IIOE and repeatedly inquired as to when the international fleet "will direct its scientific activities to exploring the North Sea and fishing grounds further north." He could not understand, despite repeated explanations of Britain's participation in the international effort, not to mention that the *Discovery* had not been designed for fisheries investigations, why more consideration had not been paid to "see that some return is made to Britain for those large financial contributions to this expedition." He was always answered to the effect that, although the expedition may find something of use to fisheries, it was a scientific expedition and not one of fisheries exploration.[16] The British found themselves in the position of having to contradict themselves, wanting the money that came with fisheries research without the responsibilities.

The *Discovery* was going to be British oceanography's flagship, conducting scientific research, while the *Manihine* was showcased as the practical aspect of the British contribution to the IIOE. But the *Manihine*'s troubles were only beginning. Initially its repairs were to have cost eight thousand British pounds, and then they were revised upward to ten thousand pounds. When British officials heard "disquieting rumours about not only the length of time they were taking but also the amount of money which they were consuming," they inquired into the matter and received the reply, "little short of shattering," that the estimates had risen to thirty-eight thousand pounds.[17] The Ministry of Science hurried to explain to the Treasury that the *Manihine*'s work in experimental fishing was a crucial point in defending the entire British participation in the IIOE as a practical effort aimed at economic ends. Without the *Manihine*, the fisheries connection would lose all credibility. There was no question of the *Discovery* doing this, as it was not equipped for fishing. The Ministry of Science threw its support behind the *Manihine*, without even attempting to account for the expansion in expenditure. "Much of the economic justification of the Expedition would be lost," one official explained, "if the research were not carried this step further toward development."[18]

The press ridiculed the British contribution to the IIOE. The *Daily Telegraph* did so by describing its present ship, the *Owen*, whose "research team" consisted of one scientist and two assistants. "This compares sadly," the article lamented, "with the efforts being made by the Americans, Russians and Japanese." Next to the dozens of scientists often aboard the other ships, this hardly seemed world-class. The *Owen* was about to be replaced, thankfully, by a new vessel, the *Discovery*. In the meantime, the *Manihine,* which was more than a half-century old, had taken two years to refit, and could house a whopping four scientists, "confidently expects spectacular results" from its cruise in the western part of the Indian Ocean.[19]

The ludicrously expensive struggle to overhaul the *Manihine* and put it into service simply to buttress the credibility of the IIOE's fisheries aspects was only one of Britain's many financial embarrassments. It highlighted the difficulties created with the United States and the Soviet Union on one side, the developing world on the other, and the rest of the world sandwiched in between. Britain ran into similar troubles later. Deacon even threatened not to attend one of the IOC's meetings because he needed to save the money "to buy a new armature for a fan motor in the ship."[20] His complaints merely annoyed British officials, who felt that the National Institute of Oceanography, next to many other institutions, did quite well in terms of funding. But Deacon was trying to keep in step with American institutions, which were expanding, and he was attempting to make his institute expand faster than British purse-strings could accommodate.[21]

In the middle of the IIOE, Deacon threatened to lay up the *Discovery* because of a lack of funds. This brought Edward Bullard into the affair, because scientists from Cambridge University had planned to conduct research aboard the ship. When British officials suggested he use his influence to obtain some money from the United States, Bullard indignantly refused. He was unwilling to ask the Office of Naval Research to fill in where the British government had left off. The government officials retreated from the idea. They agreed that it would be, for everyone, "most humiliating."[22] Deacon complained to his Norwegian colleague Håkon Mosby about the British oceanographers' perpetual financial problems, joking that his "next letter will probably be written from some debtor's prison." For him, the whole affair was another instance of the difficult time scientists had in persuading governments to take an interest in the oceans, despite its great importance compared to other fields. It was around this time that his daughter, Margaret Deacon, had begun to research the history of marine science during the seventeenth century. Familiar with her work, he wrote to Mosby: "In 1673 it was a question of whether Charles II should find money for an astronomical observatory or an automatic tide gauge, and, as usual, the Heavens won: no tide gauge till 1831."[23]

Deacon was increasingly frustrated during the 1960s because of what he believed was a miscalculation of priorities in big scientific programs, specifically the tendency to spend more on the atmosphere and outer space than the oceans. In order to obtain money and stimulate interest, large programs such as the IIOE were necessary, despite his obvious discomfort with the ramifications of touting practical implications.

Scientists in the developing world recognized the opportunism of industrialized countries. During the IIOE, American and Soviet survey ships spent a few days in Asian ports so that students could inspect them and learn some techniques. As one American said when planning to have a ship stop in Bangkok, "it would be a shot in the arm for Thai science."[24] However, as one Indian scientist complained, those same vessels did not bother to send weather messages to local ship-to-shore radio stations, "sending them instead back to La Jolla or to Moscow whence they may or may not emerge many months later." Such practices, exercising proprietorship over data collected and thus making no contribution to local weather forecasting, seemed to contradict the idea of helping local science and contributing to local development.[25] This cast doubt on the motivations of scientists from industrialized countries whose actions rarely measured up to their public proclamations. Indian scientists in particular often felt they were being used to serve Western designs, and thus they strove to achieve as much autonomous control of scientific activity as possible. Western scientists, however, had a difficult time trusting Indian scientists to do the job properly, creating an uneasy cooperative relationship.

By early 1964, the difficulties of trying to turn the IIOE into something truly cooperative became clear. One of the problems was the reticence of countries to clarify which of their expeditions were "declared" national programs, meaning that they intended to transfer data to World Data Centers and thus contribute to the international effort. The coordinators noted that very seldom were data submitted to World Data Centers within six months of cruises, as they previously had agreed. Partly this was due to the fact that scientists were reluctant to submit provisional data that had not yet been examined thoroughly. This was fine, because scientists seemed to be able to obtain data if they requested it personally. However, as already noted, a larger problem was that the International Meteorological Centre in Bombay was not receiving the data from research ships or data centers. Such data sharing was not prioritized by scientists. The most the IOC could do was to encourage national meteorological services to recognize that IIOE data were urgent and should be processed and submitted without delay.[26]

The Indians were keen to avoid the appearance of being led around by either the United States or the Soviet Union. Leading Indian marine scientist N. K. Panikkar often refused to go along with American recommendations. When Robert Snider first met him, Panikkar said he could do little to help the expedition, though Snider suspected this was not the case. Ultimately Panikkar became the leading Indian figurehead of the IIOE and subsequent oceanographic activities by India. In the course of the expedition, however, Woods Hole scientist John Ryther received some curt mail from Panikkar, who claimed that the Americans had recruited some Indian scientists to take part in American cruises without permission. As it turned out, Woods Hole had decided to include in the expedition some of its own graduate students, who happened to be Indian, without bothering to request permission from India. In Panikkar's mind, an Indian was an Indian, no matter where he lived or studied. His insistence on recognizing formalities alienated Americans and did little to foster a cooperative spirit, yet it also preserved Indian autonomy in the face of American dominance.[27]

Irritated, Americans wondered why they bothered to include the Indians at all. What was the point of Panikkar's resistance? Here was a leading scientist in India putting up roadblocks to cooperation. Most Americans did not consider cooperation with India an integral part of their work, and perhaps they felt that the Indians ought to be grateful for any role at all they might play in the project. Ryther wrote to the IOC that cooperation with India had become rather difficult and that he personally was "getting rather impatient with bending over backwards to be generous and accommodating, and getting kicked in the pants for my trouble."[28]

Despite such problems, Unesco thought it important to establish a permanent base for marine research activities in the region, to extend the spirit of the IIOE beyond its completion. It created a marine biological center in Cochin, India, to serve as a collection center for studies of the Indian Ocean. Enthusiasm for this international center, however, was not universal. On the one hand, it would be a nice symbol of cooperation and might even become a permanently useful institution in India. On the other hand, many scientists bristled at the implication that their biological samples might be ruined because of mishandling by incompetent Indians. The Americans particularly felt that the sorting of their samples should be the responsibility of scientists working in American institutions, particularly at the United States National Museum. John Ryther claimed that it had nothing to do with any disenchantment with India, but rather that the museum had offered to put up money, space, and staff to accomplish the task. "All of this doesn't do much for the international aspects of the Expedition," he admitted. They would still give plankton samples to the center, of course, though not their more important ones. Ryther wrote, "I know the Indians well enough to work around these problems without telling them to their face of my lack of confidence."[29]

The IOC refused to select an Indian to direct the center, believing this would undermine its credibility. It needed to be more than a trophy for the Indians and a symbol of international cooperation. There would have to be strict standards at Cochin, so scientists from other countries could have some faith in the reliability of its results. They would have to find someone who combined leadership qualities, scientific expertise, and a noncontroversial nationality—someone like IOC's first chairman, the late Danish scientist Anton Bruun. Also they would have to find someone who could get things done under very difficult circumstances. As Warren Wooster later said, they had to find a "mean Dane."[30] This they found in Dr. Vagn Hansen, who went to Cochin to act as curator of the collections. Some years later, when Hansen returned home to Charlottenlund, an IOC official praised the work he had done, despite the oppressive climate and depressing social conditions. Hansen and his assistant "established a fierce system of slavery in order to get more samples done... [and] established very high standards."[31]

The kind of work being done at Cochin was not always useful, but scientists were careful not to say so. In 1966, when an Australian scientist visited the center, an IOC official warned him not to hint that their work was unimportant. "Their whole lives are centered around [it] and they are terribly sensitive to any suggestions that net hauls may be of limited value."[32] It was important to offer the Indian sorters some encouragement. By mid-August 1963, Cochin had received nearly five hundred plankton samples, and more were to come. The greatest contributors were India and the United States, but contributions also came from Australia, Japan, South Africa, and the Soviet Union. Unesco helped pay for standard nets so that the plankton samples could easily be compared, but unfortunately little more than half of the samples were taken using the nets (many of the nets were lost in operation).[33]

The Indians, and particularly Panikkar, tried to combat the pervasive sentiment that the Indians could not be trusted with advanced scientific activity. He insisted that Indians should play as important a role as possible in the center. Although he was pleased to see international cooperation provide benefits to the science of India, he feared too much involvement by non-Indians. The IOC was aware of his sensitivity and hoped that Panikkar would not think that it wanted to dominate Indian marine science, but rather that he would feel free to use the IOC for his own goal of building up oceanography in India. By 1965, the IOC acquiesced to allow India to administer the center itself. Even then, however, only the *assistant* curator would be an Indian. As IOC secretary Konstantin Fedorov wrote, "scientific authority and imagination, goodwill and perhaps a lot of patience" were still urgently required by a non-Indian.[34] The IOC wanted the center's leader to be an eminent scientist from outside India, and it wanted to make sure non-Indians were seen to be in praise of it. In June 1965, the scientists at Cochin drafted an article to send to *Nature,* describing the work of the center. When Vagn Hansen sent a draft to the IOC to look over, he added his personal thoughts about authorship. He initially had included Panikkar and two other Indians as coauthors but decided against it because of the letter's optimistic tone and its overt praise of the Indian government. He should be the only author, he said, or else one of the non-Indian IOC officials should be coauthor. At the end of a letter written in English, Hansen added a single line in German: "Skeptische Personen würden mit indischen Mitverfassern misstrauisch sein" (Skeptical people would be mistrustful with Indian coauthors).[35]

By the end of the decade, Cochin showed mixed signs of usefulness. On the positive side, specialists visited the center often and the basic sorting phase was drawing to a close, with more detailed sorting planned for the coming years. It had handled nearly two thousand samples and separated them into more than sixty categories. On the negative side, the center had fulfilled many of Western scientists' unkind predictions about whether to trust the Indians with specimens. In particular, scientists at Cochin failed to ensure that the samples remained in good condition. In the summer of 1969, a significant part of the collection, specifically related to soft-bodied organisms and samples from some particular cruises, had undergone substantial deterioration. Containers were rusting, disintegrating, and leaking. Samples were exposed to dust and light. Some of the preservatives were ineffective, and the local tap water contained a great deal of bacteria. The center enjoyed support from India's National Institute of Oceanography and from Unesco in supplying equipment, but there were many delays and logistical difficulties. Humidity problems were rampant, with fungus growing uncontrollably on the optical equipment. To this was added the inadequacy of facilities for data processing and other tasks.[36] Ultimately the Cochin center became a permanent component of oceanographic research in India, standing as testimony to the linkages, fostered by the IOC and others, between scientific cooperation and building up science in developing countries. But it also stood to delineate the divide between North and South; acceptance of India as a major center of marine science was far from pervasive.

The IOC's emphasis on the developing world damaged its image as the coordinator of the highest levels of oceanographic research. It struggled to explain that its goals were supposed to apply to everyone, not just the developing world. Konstantin Fedorov wrote that "even among countries which are considered 'developed' there is a lot to do to improve communications, exchange of scientists and exchange of information."[37] As the IIOE was drawing to a close, Warren Wooster wrote to the editor of *Science* complaining about a recent article that gave short shrift to the scientific justification of the expedition. The Indian Ocean, he wrote, differed from all others because of the seasonal reversals of its surface winds. Because of this monsoon system, scientists recognized an opportunity to test theories about the relationship between winds and currents. "This feature of the Indian Ocean," Wooster wrote, "was a powerful attraction to physical oceanographers."[38] By the mid-1960s, IOC leaders were struggling to assure scientists that their aim was to support worthwhile research, not just to support the developing world.

Efforts to convince the world that the IOC's work was truly scientific underscored the illusory connections between science and development. The IIOE International Coordination Group acknowledged in 1964 that the odds were rather remote that the expedition could contribute anything at all to fisheries. Most of the existing local fishing activities, it concluded, were based not upon scientific data but upon trial and error, and this was unlikely to change. The only country in the region with the trained personnel to make genuine use of the scientific knowledge was Japan, but its fishing fleet was a domineering presence in the region with or without the IIOE.[39] The *Washington Post* carried an article the same year that tried to make the connection seem real. It recalled the story of the huge swath of dead fish, about six hundred miles long and more than a hundred miles wide, covering part of the Indian Ocean. Now, the newspaper reported, scientists had a new theory: the fish had suffocated from lack of oxygen. The monsoons caused a huge upwelling of fertile deep ocean water, carrying to the surface plants and organisms, which multiplied rapidly because of the effects of the sun. This depleted a huge amount of oxygen from the upper layers of the ocean, so when fish came to feed on the organisms, they found vast quantities of food but little oxygen. "And there the fish die," the *Washington Post* reported, "at the banquet table."⁴⁰ The article tied the IIOE neatly to its express purpose of helping to understand the problems of fisheries. It also created a nice metaphor for the fate of science seduced by fisheries money.

THE NATO SCIENCE COMMITTEE

Over the next several years, the IOC continued to pursue projects that were designed to attract governments and had little connection to specific scientific problems. It began to focus on regions, in order to simplify the process of coordinating and gathering government support. This yielded such unwieldy and vague project names as the Cooperative Investigations of the Northern Part of the Eastern Central Atlantic (CINECA). Other projects were regional as well, with no scientific goal other than data collection. The Cooperative Investigations of the Caribbean and Adjacent Regions (CICAR), for example, was not oriented toward a specific problem but, rather, initiated a "Survey Months Programme" to take various measurements and exchange the data. The important thing appears to have been not even the data but, rather, the cooperative act of exchanging the data. The most important result of CICAR was that it helped to interest countries in exchanging data and developing national infrastructures for studying the oceans. The achievements of CICAR, one author wrote, "appear to be more in the area of marine policy than in that of marine science." A similar program, proposed by the Soviet Union, focused on the Mediterranean Sea. The Cooperative Investigations in the Mediterranean (CIM) began officially in 1969, with Moscow as the regional data center, and included twenty-four countries (though not the United States). Another Soviet-backed plan, the International Coordination Group for the Southern Oceans (SOC), was to organize comprehensive studies of the waters around Antarctica. Neither of the two Soviet-inspired programs appears to have achieved much.⁴¹

The Soviet Union's vision once again seemed limited to survey work and data collection; it was not particularly interesting to scientists outside that country. Work in the developing world appeared equally bland. And yet the highest body devoted to oceanography (the IOC) was catering to the needs of both. Disgusted, some Western scientists doubted the prospects for the future of international cooperation in oceanography. In late 1968, George Deacon wrote to the Royal Society, "I really think that we shall not be able to work up much enthusiasm for a prominent, unified, programme—the sort we could give a big name to—just now." Cooperation ought to be confined to that between laboratories whose scientists were interested in the same subjects. He pointed to joint programs between Woods Hole and the National Institute of Oceanography and to joint studies with the French (in the Mediterranean) and the Germans (in the Norwegian Sea). "This is how the scientists are keen to work," he asserted, "and it is what they find most effective." These words hinted that it might be time to end the huge schemes of recent years.

Pandering to developing countries and the wishes of Soviet scientists led many oceanographers to seek to confine cooperation to Western countries, as had been the norm prior to the IGY. Even by the late 1950s, American oceanographers were looking to expand cooperative activities but to limit them to political allies whose scientists and patrons had common aims. Shortly after the launch of *Sputnik,* NATO established a Science Committee to address the perceived imbalances in scientific power between the West and the Soviet bloc. Through NATO, scientists hoped to achieve a broad base of cooperation that would reap the benefits of international cooperation in fields that could have unabashedly strategic value. As historian John Krige described it, the NATO Science Committee was "intended to create an 'international' elite, held together by ties of professional respect and friendship and by a political and ideological consensus, whose members could be counted on to promote the values of the Atlantic community."[42] At the same time, the Science Committee strove to promote itself as an organization with predominantly scientific goals unconnected to economic development, and not even necessarily connected to military objectives. This was partly because of the scientists' commitment to basic research, but it was also impractical to be explicitly defense oriented. Some feared that explicitly military NATO research would have the opposite effect to that intended: it could strip individual countries of their sovereignty in choosing their own defense research priorities. The end result was that the Science Committee funded basic research that, although it was not military research explicitly, would likely have an impact upon military technology or operations.[43] This turned out to be an ideal international forum for physical oceanographers disgruntled by the IOC and other international organizations.

Oceanography, with its importance for undersea warfare and long-range detection already demonstrated, was a natural avenue for NATO research. In late April 1959, NATO opened a new undersea warfare laboratory in La Spezia, Italy, financed through funds from the United States Department of State. Its top staff members were American physicists (Columbia University's E. T. Booth was the first director), all of whom had participated to some degree in Project Nobska a few years earlier. Despite its American funding, American staff, and the decidedly U.S. Navy–oriented research trajectory, the Americans hoped to project an image of international cooperation. The funding, after a few years, was to be shared among NATO countries, and British oceanographer George Deacon was invited to be the "senior oceanographer" to preside over the opening of the laboratory. Columbus Iselin felt that a non-American was needed to represent marine science. As he put it, "The laboratory is supposed to be truly international and the US is supposed to keep very much in the background."[44] The Americans who conceived of the laboratory were eager to implement it as an international undertaking.

To expand the oceanographic program and provide a more solidly international forum, the NATO Science Committee established a subcommittee on oceanography led by Norwegian oceanographer Håkon Mosby. Some of the most prominent figures in marine science in Western Europe and North America were members, including Columbus Iselin (United States), Anton Bruun (Denmark), George Deacon (Britain), and Marc Eyriès (France). The subcommittee focused on work that might have strategic implications but avoided specifically defense-related projects like those at La Spezia. The participation of NATO in oceanography, as Deacon explained to the British Admiralty, stemmed from its desire to have member countries work together on problems of common interest. They would work together on ships, or between ships, then go home and work in their own laboratories in "growing friendliness and cooperation." The initial program, begun in 1960, was modest. It consisted largely of sending the *Chain,* the largest research ship then in use by the United States, to prominent oceanographic institutions in NATO member countries. The *Chain* engaged in research and cooperated with other ships.[45]

The *Chain* became a symbol of what could be accomplished by broad international cooperative research between the top-notch research establishments of NATO countries. This meant that the world was watching its progress. The *Chain* visited Helsinki in July 1960, just in time to be available for visits by participants in the General Assembly of the International Union of Geodesy and Geophysics. But after visiting the *Chain* there, the naval attaché in Helsinki wrote to the chief of Naval Operations, Arleigh Burke, that it "was the most untidy ship I have ever seen, with gear literally strewn about the deck." The attaché also objected to the fact that the personnel were "uniformly unkempt, with soiled clothing and unshaven beards," and the wardroom was "sloppy."[46] This seemingly minor objection touched the highest levels of the NATO oceanographic community, because of the potential repercussions on the image of American leadership in oceanography. Woods Hole director Paul Fye wrote to Burke that he was "astonished and embarrassed" to hear of the report, because the vast majority of visitors to the ship had admired it and been impressed with the research program being undertaken. It certainly was a clean ship, he wrote, but because of the research program the ship was packed with equipment, much of it rusted from extended use at sea.[47] American scientists and naval leaders rallied to the cause, labeling the report as unjust and inaccurate, and Fye sent these on to Burke. Scripps scientist John Knauss wrote, "I don't suppose she was sent to Helsinki for propaganda purposes, but I'm sure she succeeded in impressing any of the 1800 scientists at the IUGG who came aboard her.... She was in the middle of a major research effort and the evidence of it was all around."[48] Influential scientists from Canada and France wrote letters of support; Canadian J. Tuzo Wilson wrote to Burke that "I think it was a very fine thing for the United States Navy to be represented by so impressive and workmanlike a ship. It emphasized the importance of the contribution being made by the United States to oceanography."[49] Burke gently suggested to Fye that American oceanographers make an effort to shelve some items and to keep some equipment hidden behind canvases. He wrote, "This matter of the *Chain* has received high level attention in the Navy Department primarily because we are all making every effort to give the best possible impression in foreign ports of American ships and the American way of life."[50] The chief of Naval Operations understood that the *Chain* was a working vessel and naturally would look like one, but he also knew that one part of the *Chain*'s purpose in 1960 was to kick-start the NATO program in oceanography and to showcase the preeminence of American marine science. Thus, appearances mattered.

The NATO efforts in oceanography sparked tensions among scientists in other countries. This was particularly true in Britain, after the program for 1961 went into the planning stages. Stimulated by American studies using direct measurement of currents, notably on the Cromwell Current in the equatorial Pacific, the program entailed a study of the transport of water in a strategic area of the North Atlantic, the Faeroe-Shetland Channel. Scientists from the United States, Britain, and Norway planned to operate ships and unmanned buoys to measure the velocity of currents, while also taking water samples and conducting other studies.[51] British fisheries oceanographer A. J. Lee objected to the NATO program on the grounds that it "contains no work of military importance, as far as I can see, and the pro-

Assistant Secretary of the Navy James H. Wakelin (left) and Paul Fye in 1960. Wakelin led the ICO in the early 1960s and was one of the principal "Potomac oceanographers." Courtesy Woods Hole Oceanographic Institution Archives

gramme could quite easily be carried out under the aegis of SCOR or ICES." Why, Lee asked, was NATO needed for such an operation that normally would have fallen under a broad international organization? To conduct such operations would severely undercut the spirit of international cooperation that had been built over many years. In particular, Lee warned, "to confine it to NATO countries and so exclude the USSR seems to us to run the risk of giving the USSR cause for withdrawal from such international organisations as ICES, ICNAF and SCOR, etc."[52] Lee was even more concerned with the fact that fisheries regulation was already very difficult to enforce in areas such as the Barents Sea, and without Soviet cooperation, it would be impossible. Only through ICES had any kind of progress toward fisheries agreements been made, and to Lee, the NATO program threatened to demolish it.

It was the lack of connection to fisheries work, however, that made scientists fierce guardians of NATO. Deacon wrote to Mosby, "Our Fisheries departments have gone very anti-NATO and will try to persuade yours to prevent you from taking an active part." Although fisheries ships would have been welcome, Deacon reasoned that their opposition should not stop NATO, which could succeed in its objective without their help. To answer Lee's objection, Deacon suggested to Mosby that "it might be a good thing if we could find a less proprietary name for our project than 'Faeroe-Shetland Channel,' which after all is only part of it." Deacon suggested something less geographically specific, and more scientifically oriented, such as "Interaction between Air and Water in the North Atlantic Ocean" or "Dynamics of Air and Water of the North Atlantic Ocean." In any case, he did not see the opposition from scientists in fisheries-oriented institutions as a real obstacle, and the problem with the Soviets was not insurmountable. Hinting at his resolve, he wrote to Mosby, "There is nothing like some opposition to sharpen the wits of the English."[53]

More than a year later, the same issue resurfaced in another of NATO's oceanographic plans, this time in the Irminger Sea, off Iceland. Lee objected strongly again, and now he had more reason for doing so, because ICES recently had been active in the area and had begun to discuss making a very similar survey. In addition, ICNAF had begun to plan a survey in the Irminger Sea with the cooperation of France, Norway, Portugal, the Soviet Union, and perhaps the United States and Canada. Lee did not object to the content of the scientific work to be done, because most of it was to be accomplished in the ICES and ICNAF programs. Instead, the "difficulty is caused by your programme being under NATO which excludes the USSR while the ICES–ICNAF programmes include her." To run two programs in the same area, one with the Soviets and one without, "would, in my opinion, be to court trouble from the USSR" in all cooperative organizations, from ICES and ICNAF to the IOC. Although NATO was a safe haven for physical oceanographers, it was creating a "clash of interests." Lee suggested that NATO ought to support laboratories and leave the cooperative programs to other more inclusive bodies.[54]

The conflict over NATO demonstrated the entanglement of two issues. From Lee's point of view, the problem was the mechanism for international cooperation and the political ramifications of limiting it to the Atlantic alliance. But to Deacon, Mosby, and many others, NATO offered a means to pursue physical oceanography without apology, for its own sake and not rhetorically linked to economic development. It reflected the frustrations of northern Europeans whose traditions in physical oceanography were quite strong, but whose funding for research was increasingly difficult to come by, especially compared to their colleagues in the United States and the Soviet Union. To Deacon, scientists were only just beginning to understand the processes of the deep ocean and the relative importance of wind and density differences in the sea. Physical oceanography needed to establish stronger theoretical premises and practical techniques with a far more concerted effort than was available in organizations whose focus was so dispersed. "Can ICES and ICNAF," he asked Lee, "having to devote most of their energy and facilities to fishery regulations and plankton, start what is almost a new science?" Although ICES at one time was a pioneer in both the physics and the biology of the oceans, he doubted whether the same could be said now. To Deacon, NATO offered a means to promote new ideas and techniques about the ocean's physical processes. He reproached Lee:

> You fishery scientists want to stop this because USSR might object to NATO countries joining up in any marine science outside ICES. There is no doubt that USSR, with or without its satellite countries, does a great deal of special work that never comes to ICES and that it would be glad to see NATO countries limit themselves to fishery research.[55]

Still, Deacon argued, such questions could not be argued out at their level. The British government supported the NATO Science Committee, and if the work occurred in the same areas as fisheries research, then "we ought to rejoice rather than be sorry."[56]

A number of prominent American oceanographers had come to share these sentiments, detesting survey projects that had objectives no more ambitious than making routine observations to "map" the oceans. Henry Stommel felt that oceanography needed to be more theoretically ambitious, and in 1962 he wrote to the president's science advisor, Jerome Wiesner, criticizing the leading institutions of American oceanography. "The situation in physical oceanography is much like that in mechanical engineering departments in universities some years ago: cook-books and tradition." For oceanography to take on its proper role as a science, with observations intended to test theories rather than simply to map the oceans, Stommel advised transforming the national effort in oceanography "by administering a strong dose of mathematical physics."[57] Much of this frustration came to the surface during the IIOE, with international cooperation taking on inconvenient dimensions for physical oceanographers, whose concern for helping developing countries was a distant second to their scientific objectives. Still, fisheries scientists were equally annoyed to have their toes stepped on by the physical oceanographers. Lee in particular, who played a major role in planning the ICNAF survey, attended the meeting of the NATO subcommittee in Reykjavík in April 1962 to argue against it. Ultimately, the subcommittee decided to continue the project but on a less intensive scale, given the major surveys already planned by ICNAF.⁵⁸ A year later, Deacon vented to NATO officials about the meddling behavior of the British Ministry of Science. "I rather think," he wrote, "they are browbeaten by the fisheries scientists, who really think they own the ocean, and probably there is something of the same trouble in Germany, Canada and other countries."⁵⁹

For physical oceanographers NATO became a forum in which to pursue their own research not only for their own academic interest but also for the strengthening of international ties without the drawbacks of an expansive organization like IOC. It required neither the diversion of funds to scientists in countries of the developing world nor tiresome negotiations with Soviet bloc scientists. Indeed, the rationale of the Irminger Sea project was not only fisheries research but also monitoring and determining the effects of Soviet nuclear tests.⁶⁰ This specific objective certainly was more appropriate for NATO, even if the scientific work essentially would be the same. As Deacon put it, such projects served the common commercial and defense interests of NATO countries. Helping to "cement the alliance" was a far easier and more vague justification for research than making promises of economic development, and it was usually good enough for the various defense establishments who funded the work.⁶¹ The subcommittee felt that oceanography was particularly suited for cooperation "to increase the efficiency of western science," by sponsoring projects that might have broad implications for navigation, weather forecasting, fisheries, and also defense. This was certainly the reason that NATO had not chosen a distant location, such as the Indian Ocean, as its object of study, nor had it made vague promises about vast underexploited food resources. Instead, it focused directly upon the physical interactions between the atmosphere and the oceans from the Straits of Gibraltar to the Arctic Ocean, the area of most direct interest to the NATO countries.⁶²

MARINE SCIENCE AND MARINE AFFAIRS IN THE 1960S

The retreat of many scientists to NATO promised to relieve them of a duty to promote research for its economic ramifications, but most scientists were caught up unavoidably in projects that tied oceanography to practical uses. The focus on exploitation of the sea made the future of oceanography in the United States the subject of debate at the highest levels during the 1960s. Its advocates in Congress insisted, as scientists had been doing for a number of years already, that the world of the oceans was at least as important to the nation's welfare as the world of space, which seemed to dominate, along with nuclear physics, not only the country's imagination but also the government's interest in science and technology. But despite the flurry of enthusiasm immediately following the first installment of the NASCO report in 1959, and sustained vocal support for marine science over the subsequent years, Congress had failed to pass comprehensive legislation to give American oceanography a radical overhaul. It made a few small steps: it lifted the jurisdictional limitations on the Coast Guard and the Coast and Geodetic Survey, it expanded the research component of the Bureau of Commercial Fisheries, and it granted authority to the Geological Survey to explore the ocean bottom. But oceanography still lacked the focus that many felt it needed. If oceanography was indeed more important than space, then oceanography deserved a kind of "wet NASA," codified by a marine science act.

Senator Warren Magnuson had been among the most vocal in this regard, and he was disappointed by the lack of progress in creating a comprehensive law. In 1965 he proposed another bill, this time for a National Oceanographic Council at the cabinet level, made up of several high-powered administrators and scientists, including the secretary of state. Magnuson insisted that the United States' future national security rested on a tripartite foundation, namely, "that we be strong in space, strong in the applications of nuclear energy for peaceful purposes and the national defense, and strong on, in, and under the oceans." The first two areas already had legislative bases from which to build strong national programs: the Atomic Energy Act of 1954 and the National Aeronautics and Space Act of 1958. These acts were significant not only in the financial and organizational structure they put in place but for the clear declaration of policy and purpose contained in them. Magnuson argued:

> It is similar because the policies declared in those two acts and the pending [marine science] bill are national policies; the purposes are national purposes, and these national policies and purposes express the determination of the Congress to strengthen our security, economy, and welfare.[63]

The policy and purpose of American oceanography, as set forth by Magnuson, would be first of all to contribute to the expansion of human knowledge of the marine environment. Second, it should be "the preservation of the role of the United States as a leader in oceanographic and marine science and technology." Other goals (he enumerated ten) included the advancement of education, coordination among the various agencies interested in oceanic research, and last on the list, the cooperation of the United States with other nations in research "when such cooperation is in the national interest."[64]

Magnuson's bill echoed the various attempts by members of Congress in the early 1960s to implement a national policy that would lead to developing a long-range, coordinated program for oceanography in the United States. The closest they came was in 1962, with a bill that would have given the Office of Science and Technology (OST) supervision over other oceanography-oriented agencies. The bill passed both houses, but was pocket vetoed by President Kennedy for a number of reasons, largely because it would have placed the OST between the president and the chiefs of other agencies.[65] Moreover, although both Presidents Kennedy and Johnson were supporters of oceanography, they could not draw upon any clear consensus among the oceanographic community about what shape the national infrastructure should take. Generally, those who opposed a "wet NASA" did so because of fears of overcentralization and the emergence of an oceanographic "czar" in Washington who would strip institutions of their autonomy. At the same time, doing nothing at all appeared equally undesirable, as the importance of oceanography seemed in need of some direction at the national level.[66] The enormous surge in support for fundamental research after the launch of *Sputnik* calmed considerably during the mid-1960s, and some feared that oceanography had missed its opportunity to gain the kind of support afforded to space and nuclear sciences.

Oceanography retained some importance because of its explicit focus upon the practical benefits of scientific research, not merely to the developing world but also to national interests. The vehicle for oceanography's legislative resurgence was the 1964 Convention on the Continental Shelf. This international agreement, adopted under the auspices of the United Nations Conference on the Law of the Sea, granted rights to the resources of the continental shelf to the adjacent country; such rights extended far away from the coast until the water reached two hundred meters depth. In the United States, these areas ranged from ten to three hundred miles from the coast. According to Alaska senator Bob Bartlett, the convention gave the United States sovereign rights "over the most extensive territory since the Louisiana Purchase."[67] Even more important, the wording of the convention suggested that sovereignty might be extended further if a country could demonstrate its ability to exploit deeper areas. The thought intrigued Congress: here was a tangible way in which science and technology could serve the national interest. It appeared to be the first time that one nation's rights to resources were directly dependent upon that nation's technological prowess. Already marine technology seemed capable of exploring and exploiting beyond the two hundred–meter line, and the technological constraints of the present, many reasoned, would be overcome in the years ahead. In a speech before the Senate, Bartlett mused:

> Although there may be a question as to whether the United States under the Continental Shelf Convention could legally homestead the mid-Atlantic ridge, there should be no doubt but that the present trend . . . is to associate sovereign rights over resources with technological capacity to exploit them.[68]

Bartlett's plan to establish a federal commission to exploit the continental shelf gained support from a number of influential senators, including Robert F. Kennedy.

In the House of Representatives, Massachusetts congressman Hastings Keith was equally enthusiastic about the continental shelf as a specific object of scientific research. Surely, he reasoned, the Soviet Union would be pursuing the same objective in the oceans. In a letter to the chairman of the House Committee on Merchant Marine and Fisheries, he wrote that "the recent international agreement has, willingly or not, entered us in what might be termed an oceanspace race."[69] Both Bartlett and Keith, as well as many others, insisted that this race was far more important than the space race. The continental shelf promised immediate economic and military benefits far greater "than our costly probes of the desolate and limitless reaches of outerspace."[70] Bartlett, in closing his speech in favor of establishing a federal commission for the exploitation of the continental shelf, echoed this sentiment while adopting the catchphrase of the Johnson administration. "A Great Society," he said, "cannot be blind to any potential resource development," particularly when the resources of the oceans represented an area nine times greater than that of the moon.[71]

A federal commission, whose express purpose was the exploitation of the resources at the bottom of the sea, offered an attractive merger of scientific and national interest, promising a deepened role of the federal government in supporting the nation's oceanographic effort. To Woods Hole director Paul Fye, the twin objectives of the legislation, namely, providing federal coordination of the country's oceanographic effort while also helping to expand the nation's claims to the resources beyond the two hundred–meter depth line, seemed laudable. The commission, by supporting an active research program in both exploration and engineering, would go far to provide a solid justification for American diplomats to make such claims.[72] As Keith wrote to Fye, the United States had "more reason to pursue the objectives of this bill than we have in attempting to put a man on the moon—legally, militarily, scientifically and economically."[73]

The bill that finally made its way through Congress blended the federal control advocated by Magnuson and the focus on the continental shelf envisioned by those in Congress motivated by the sea's economic potential. President Lyndon Johnson signed it into law in 1966, as the Marine Resources and Engineering Development Act. It established the National Council on Marine Resources and Engineering Development; this body was known more generally as the Marine Sciences Council, though its true name is worth remembering, as it made no mention of science. It put marine science and technology at the president's fingertips and enjoyed cabinet-level status. It assigned to the president himself the responsibility of establishing new initiatives for the effective use of the sea. It was chaired by the vice president of the United States, who at the time was Hubert Humphrey, an active promoter of marine science. Organized thus, the council attained a level of prestige previously unknown to oceanography, largely through Humphrey's influence and the benediction of the president. Its meetings frequently were attended by the secretary of State, secretary of the Navy, secretary of the Interior, and other influential individuals in Johnson's administration. Marine affairs assumed greater importance during the late 1960s than ever before or since in the United States.[74]

Although the delays in proposing an acceptable piece of legislation resulted from the typical political and bureaucratic wrangling in Washington, they also reflected diverging needs of marine scientists and government. The NASCO report had galvanized Congress, and when President Kennedy was elected, the executive broke away from its shyness toward oceanography. Kennedy spoke out in favor of an expanded oceanographic research budget and insisted on the importance of the seas to the United States' economic and military well-being.[75] Oceanographers found disciples of marine science throughout government. Nevertheless, as oceanography gained political support, few of its new enthusiasts adopted the idea that basic research, as opposed to applied research, should be pursued on a large scale. Questions about the relevance of projects plagued oceanographers now far more than they had with the Navy. For years, oceanographers had nursed the notion of basic research through Navy patronage, arguing that research with no specific technological implications was the best way to ensure the growth of science and the strength of the country. But the purse strings of

Vice President Hubert Humphrey with Woods Hole Director Paul Fye, 1967. Courtesy Woods Hole Oceanographic Institution Archives

defense organizations were relatively immune to popular dissent, particularly if the research was classified secret; convincing the Navy of a project's value usually was the scientist's only serious obstacle. In the political arena, scientists had to convince the politicians not only that research was needed but also that voters would agree. Civilian politicians were to prove less accepting than the Navy, and the relative strengths of basic and applied research were not as important to leaders who had to show something for their money. The first executive secretary of the Marine Sciences Council, Edward Wenk, Jr., wrote of the basic versus applied conundrum, "Semantics aside, the actual dichotomy broke into open conflict between the academic oceanographers and their government sponsors." This conflict, Wenk observed, was often characterized by mutual disrespect and misunderstanding.[76]

Wenk presided over an explicit abandonment of the scientists' rhetoric of basic research, and his distaste for the scientists' point of view is manifested clearly in his two books about his years on the Marine Sciences Council. The new legislation went far beyond the limited objectives of NASCO and the Interagency Committee of Oceanography, both of which promoted oceanographic research and tried to coordinate the various agencies conducting it. Tired of scientists' unwavering commitment to basic research for its own sake, and cognizant of the socially oriented goals of the president, Wenk adopted a very different strategy for promoting the oceans. By law, he argued, the Marine Sciences Council had to devote its energies not only to the "means" for research but to the "ends" of research as well. The council began to adopt a new term to reflect its new focus. "Applying knowledge of the oceans to public purposes impelled a deepening blend of science and society," he later wrote. "We were deliberately forcing a transition from what previously had been termed 'oceanography'... to a completely different universe of activities that we termed 'marine science affairs.'" To mark the transition, the first report of the council bore this term as its title, reflecting not only basic research but also a host of connections between science and society, with science clearly identified as a means to tangible ends.[77]

The formation of the council marked a major shift from science toward technology and exploitation. The Interagency Committee on Oceanography had been a coordinating body to assess and negotiate the needs of both producers and consumers of scientific knowledge. The Marine Sciences Council tried to embed tangible benefits more thoroughly into its project ideas. This was partly due to Wenk's conviction that one role of the council should be to ensure the effective utilization of science. But more important, it was a strategic move calculated to attract the attention of President Johnson, whose ambitions to create a Great Society made him receptive to programs that seemed to serve social needs. Johnson was cool toward scientists, according to Wenk, because of the scientific community's opposition to the war in Vietnam. While continuing to show support as Kennedy had, he was also far more concerned than his predecessor had been about what specific benefits would emerge from supporting science. Consequently, the Marine Sciences Council deliberately stopped arguing that science ought to be supported for its own sake.[78]

How did Wenk's attitude differ from Roger Revelle's in the 1950s? Both painted a broad picture of marine science and pointed to its social consequences. But Revelle was seeking disciples, converting patrons to the view that if oceanography had support, the social ramifications *potentially* could be great. Revelle's vision was above all a means to support science. With a deft appropriation of the same rhetoric, Wenk and others were prepared to downgrade science to its function as a means of pursuing technological or policy goals. The Marine Sciences Council made no secret of this transformation. All of the council's proposals, Wenk claimed later, had been worded to appeal to Johnson's agenda for a strong social policy. In addition, Wenk had the committed support of the vice president, who gave the council his personal endorsement, presided over its plenary sessions, and actively helped to pursue its recommendations. Wenk came to appreciate Hubert Humphrey as the nation's "Chief Oceanographer."[79] Humphrey met with him to discuss the underlying purpose of the council. They agreed that science and technology should be viewed as means to social ends, and not as ends for their own sake. They looked forward to a problem-oriented council, which would address issues such as pollution and waterfront decay, nuclear arms in the ocean, and the sticky international issue of the resources of the seabed. As Wenk later wrote of his meeting with Humphrey, "This was more than a meeting of minds; it was a meeting of souls."[80]

The Marine Sciences Council's importance was fleeting. With the change in administration in 1969, the council lost all of its influence. Although Wenk was reappointed as executive secretary by President Nixon, he did not have the president's ear, and the council lost a number of minor yet meaningful privileges, such as the physical location of its office in relation to the White House. More important, the rapport that Wenk had shared with Humphrey was not renewed when Spiro T. Agnew became Nixon's vice president. There was little personal contact at all, the council's activities were delegated to a subcabinet committee, and Agnew took such little interest in marine affairs that the council's only business was in carrying out the programs begun during Johnson's administration. The Nixon administration's lack of interest in new programs, and particularly what Wenk felt to be Agnew's professional incompetence, ultimately caused Wenk to resign in 1970. The council was dissolved in 1971 after Nixon begrudgingly formed the longer-lasting National Oceanic and Atmospheric Administration (NOAA) after incessant pestering from pro-oceanography members of Congress. Although NOAA, founded in 1970, provided a focus for proposing new civilian-based oceanographic initiatives, it did not have the same influence that the Marine Sciences Council briefly enjoyed. After 1971, Wenk recalled, marine affairs never again merited enough importance to be included in a major address or speech by a president.[81]

During the Johnson years, however, oceanography saw the zenith of its recognized importance for international relations and the economic well-being of all mankind. Johnson released the report of the President's Science Advisory Committee (PSAC), *Effective Uses of the Sea*, in 1966. Its title reflected what PSAC considered the ultimate objective of oceanic research, namely, its use "by man for all purposes currently considered for the terrestrial environment: commerce, industry, recreation and settlement; as well as for knowledge and understanding." The advisory committee assigned the highest priority to those applications of the oceans dealing with national security. The report was released at the commissioning ceremony for the U.S. Coast and Geodetic Survey's new ship *Oceanographer*. Giving a short speech at the ceremony, Johnson affirmed his commitment to oceanography, saying that in the months ahead, the United States would establish its priorities and follow them, "just as we have followed an orderly and relentless program for the exploration of space."[82] Missing from *Effective Uses of the Sea*, members of NASCO noted, was "the important matter of the formulation of policy, planning and organization of international programs for the investigation and effective use of the sea." This was one of the few discrepancies between the PSAC report and a report issued by NASCO the same year titled *Oceanography 1966—Achievements and Opportunities*.[83] It appeared that the Americans' renewed effort in oceanography, along with its focus on exploitation, was going to be limited to the United States, with little commitment to the international community.

Johnson, however, was not going to forget about the international arena. In fact, he made international cooperation the cornerstone of his marine affairs agenda. Rather than treat it as an "oceanspace" race, as Congressman Keith termed it, the president tried to avoid turning the oceans into yet another avenue in which to pursue global competition with the Soviets. According to Wenk, Johnson viewed international cooperation as a potential means to promote goodwill abroad, to blunt the negative effects of the war in Vietnam. The Marine Sciences Council capitalized upon the president's view by proposing a number of initiatives that would appeal to Johnson's desire to promote programs that could make a positive impact on society. In doing so, the Marine Sciences Council gave short shrift both to Cold War competition and to basic science, favoring instead projects with practical applications for the benefit of all mankind, just as Unesco had been doing since the 1950s. One brainchild of the council was the International Decade of Ocean Exploration (IDOE), which it conceived as a worldwide effort to study the sea and develop its resources. Wenk pitched the IDOE to Johnson, and keeping the president's priorities in mind, he "skipped arguments on contributing to basic knowledge and focused on the pragmatic." He described the IDOE as a way to have all countries of the world cooperate on mutual problems such as using fish proteins to feed a rapidly expanding population, preventing coastal deterioration, controlling pollution of the coasts and the ocean, and increasing the world's capacity to exploit oil, gas, and minerals.[84] The strategy worked. Johnson was enthusiastic. In his

1968 State of the Union address, Johnson announced that he would propose that the United States launch "with other nations an exploration of the ocean depths to tap its wealth and its energy and its abundance."[85]

To the Johnson administration, the IDOE seemed an ideal scientific project to pursue a peaceful foreign policy initiative to help offset the negative effects of the war in Vietnam and to use science as a means to assist less developed countries. Both of these were foreign policy aims of the administration, and Wenk's council wisely used them to rationalize the IDOE. It was not much of a departure from the rhetoric that scientists themselves had been using for years, to gain support for their own research. The difference was that the designers of the IDOE wanted to start with the premise of economic development, rather than merely point vaguely toward it to justify scientific objectives. The IDOE would not be a single project, but a series of multinational projects devoted to the sustained exploration and study of the sea on an international basis, which Wenk felt improved upon what he considered to be the sporadic efforts of the past.[86]

The announcement of a major oceanographic project by President Johnson was a great but uneasy victory for American marine scientists. To the members of the Marine Sciences Council, Johnson's inclusion of it in a State of the Union address "crowned council achievements . . . this presidential imprimatur was like winning the brass ring."[87] But to others, notably the scientists within NASCO, it was a further step in taking the initiative for the country's national efforts and for international cooperation out of their hands. At its June 1968 meeting, NASCO members tried to grasp what the IDOE was to mean for them. Fye was afraid that the president had preempted the role of the IOC in taking it for granted that such a program would be accepted by the international community. He urged NASCO to avoid lending credence to what appeared to be an already fixed and permanent organizational structure for the IDOE. The chairman of NASCO, John Calhoun, however, had to remind him that the organization was already "an accomplished fact," despite there having been little input by either American scientists or anyone outside the United States. Members of NASCO were not witnessing the birth of an idea but instead were seeing the project, more or less, as fully grown.[88]

Scientists viewed the developments leading up to the IDOE with caution. The creation of the Marine Sciences Council and the issue of *Effective Uses of the Sea* by PSAC both portended an increased emphasis on ocean exploitation over scientific investigation. The role of marine scientists in the formulation of future projects and policies seemed uncertain. Many scientists, in assessing the future of NASCO, urged that this body continue to focus on science, in order to provide the most credible sources of information on the subject, independent of government pressures. Some of them were alarmed at the explicit focus by the Marine Sciences Council on technology, a concern that was doubled by the fact that the recent report by PSAC, rarely a friend to the oceans, seemed to place the advisory locus squarely in the executive branch of government. The role of NASCO was threatened both on the policy-planning and science advisory fronts. Scripps scientist John Isaacs felt that NASCO would be strengthened by refocusing its energies on science itself and leaving the policy implications to the other bodies. It should have scientific panels to report and to advise on the progress of basic scientific research. Just a few of these, Isaacs wrote to NASCO's Richard Vetter, "would be greatly superior to a number of panels reporting superficially on practical implications, methodology or politics."[89] Gordon A. Riley, the director of Dalhousie University's Institute of Oceanography, felt that PSAC was "trying to run with the ball," without the competence to do so. He called *Effective Uses of the Sea* an "ignorant and amateurish document," resulting from the work of nonspecialists. He thought that NASCO needed to be kept alive to ensure that there was a high-level group of oceanographers who stood a chance to steer clear of the government pressures that were sure to plague the other agencies.[90]

The IDOE's birth in Washington gave rise to uncertainty about the role of oceanographers themselves. At a meeting in October 1967, the National Academy of Sciences discussed the situation in American oceanography and the role of its oceanography committee for the future. Its president, Frederick Seitz, said that although it was now clear that the United States government was committed to oceanography, the focus of activity would be the continental shelves and other avenues of economic exploitation. For its part, NASCO should help to determine what would aid the cause of science itself. One component of this would be in continuing its concern for the relationship between oceanography and international affairs, particularly by trying to address the international constraints on the progress of science such as restrictions on access to study the continental shelf off the coasts of foreign countries. Because NASCO was itself the United States National Committee to SCOR, this role would be a natural one. Paul Fye noted that oceanographers often had to struggle to convince other scientists that oceanography was not a second-rate science. By renewing its focus on the promotion of science, NASCO could set the record straight. Others at the meeting, while agreeing upon the necessity to pursue science for its own sake, urged their colleagues not to flee toward their proverbial ivory towers. It was imperative, National Research Council member John Coleman urged, that scientists not be satisfied with science, but endeavor also to be good politicians. Atomic Energy Commission scientist Arnold Joseph agreed, recalling that under Harrison Brown, NASCO had gone out of its way to sell itself and its findings to the public. The result had been an unprecedented show of support from Congress. Oceanographers should be prepared to do the same in the future.[91]

The National Academy of Sciences swallowed its pride and, along with the National Academy of Engineering, submitted a report to the council incorporating what Wenk would later describe as a "new, enlightened stance." In other words, the scientists caved in. Their 1969 report, *An Oceanic Quest*, pointed explicitly to national goals of exploiting the seabed and managing the marine environment.[92] But despite this concession, the structure of the IDOE opened up sores that had festered among oceanographers throughout the 1960s. In the past, obviously, scientists had played a far greater role in planning, and now it seemed that the projects would be controlled at a governmental level, with no autonomy for the scientists. The best programs, Fye complained, had been developed in the past not within government but among scientists, and they had been argued out in international scientific forums. Now, NASCO's only role was to develop a study to provide preliminary advice on the United States' participation, which seemed a backward way to go about things. Scientific planning ought to come first, not after the project had already been announced, fully conceptualized, by the president of the United States.[93]

The Marine Sciences Council viewed it differently. The IDOE was an international venture, to be sure, but unlike previous efforts it was to be executed through international agreements. Thus, the American delegation had to represent not merely American scientists but the United States as a whole. Governments were more than capable of identifying which problems were the most pressing and which projects would serve national and international interests best. The scientists were needed only to help outline the specific investigations. They were to have no policy role. The project was not for science; science simply would contribute to the project. Trying to be sensitive to the scientific community, Wenk offered to appoint a scientist from the National Academy to a government post in order to take part in the delegation, but the National Academy rejected the idea. As Wenk admitted, "the nervous gavotte between the government and science was never totally in step."[94]

In trying to assert some control over the IDOE, NASCO had failed. Wenk's own view was that NASCO never understood the political compromises that were necessary to get the IDOE off the ground. Members of NASCO threatened to declare support only if the project were handed over to them to plan it themselves, on a sound scientific basis. But in Wenk's view, this would have alienated the president, who had praised the IDOE's social basis. Losing the president, he felt, would have been worse than losing the scientists. Wenk had little but disdain for the scientists' point of view, which he viewed as all complaints and no solutions. "Few cared to get embroiled in the politics," Wenk wrote of them. "It was easier to take potshots from outside." He compared the relationship between NASCO and the Marine Sciences Council to courtship among animals. In planning the International Decade, "the ballet of approach and rejection must have been equally entertaining. The result, however, was never fully successful."[95] The National Academy of Sciences accepted the task of drawing up specific recommendations, fully aware that the result would be a report for Wenk's use, not a plan of action with explicit marching orders. Thus *An Oceanic Quest* acknowledged the IDOE's emphasis on the uses of the sea, not just on science for its own sake. Although this report can be seen as a capitulation by scientists, a kinder view would be that it was a last-ditch effort to inject the ideas of scientists into an agenda that had come to them from above.[96]

Fresh from this failed struggle with the Marine Sciences Council, no oceanographer of national stature was willing to occupy a government post to oversee planning of the IDOE. The task fell upon Wenk, who included only one nongovernment scientist, Garth Murphy of the University of Hawaii, in his planning group. Later, when he tried to gain a broader base of support for the IDOE among nongovernment scientists, Wenk again met with stubborn resistance. Licking its wounds, NASCO was trying to squeeze what it could from the IDOE, by insisting on increased funding for the project. If the council wanted scientists' blessing, it would come with a high price. To Wenk, this maneuver was frustrating, as it was clear that the scientists were going to use the IDOE as a means to press for a major increase in funding for science, with few specific scientific priorities and little indication as to their social value. His irritation was fired even more when he discovered that at the same time, scientists were advising PSAC that the council was already requesting more money than the scientists could utilize productively. Summing up his annoyance toward these high-level scientists, Wenk wondered, "Could that community be schizophrenic?"[97]

In the end, the IDOE took place, but neither the Marine Sciences Coun-

Marine Science and Marine Affairs

cil nor the National Academy of Sciences controlled it. As already noted, the council itself disintegrated during the Nixon administration and was replaced by the National Oceanic and Atmospheric Administration (NOAA). The creation of NOAA resulted from the recommendations of a commission formed under the same 1966 legislative mandate that had created the Marine Sciences Council. Like the council, the Commission on Marine Science, Engineering, and Resources (COMSER), to which Massachusetts Institute of Technology president emeritus Julius A. Stratton was appointed chairman in early 1967, concerned itself not only with scientific and technological issues but environmental, educational, industrial, and legal ones as well. In 1969 it released its report, *Our Nation and the Sea*, recommending that most federal agencies supporting oceanography should be consolidated under a single independent agency. After another two years of political wrangling by COMSER and the Marine Sciences Council, President Nixon created the new agency, but it was significantly weaker than its advocates had hoped. Although NOAA provided a civilian base for oceanographic science, it was not independent and, missing the Coast Guard, it lacked a robust component of marine science that would have attracted major funding. It was not quite the "wet NASA" that many had envisioned. The Marine Sciences Council, perceived as redundant with the creation of NOAA, was killed in favor of its less influential successor (in its favor, however, NOAA was an operating agency, which the Marine Sciences Council was not). Nixon took the oversight of the IDOE, by far the council's greatest achievement, out of the new governmental marine affairs infrastructure altogether and made the National Science Foundation responsible for it. By 1971, when NOAA was born, oceanography had a new but weaker voice in Washington, and international science was dumped back onto the scientists.[98] The late 1960s was a frustrating and uncertain time for marine science in the United States. Scientists' efforts to gain support by connecting science to economic exploitation succeeded a bit too well, with the creation of "marine affairs" bodies that threatened not only to diminish the importance of basic science but also to take scientists out of the science policymaking process for national and international projects.

IMPLEMENTING THE INTERNATIONAL DECADE OF OCEAN EXPLORATION

The IDOE got off to a painful start because at the turn of the decade of the 1970s, political considerations dominated marine affairs at the expense of scientific cooperation. Scientists began to complain in the 1960s about the potential research constraints of the United Nations Convention on the Law of the Sea, which required scientists to request permission to conduct research in coastal waters of a foreign country. Although in theory the coastal states were expected to grant permission on a routine basis, in practice it gave them far more control over scientific activity than they had possessed previously. Enraged by the potential political limitations on research that they had always conducted with freedom in the past, most scientists from countries with traditionally strong oceanographic programs (such as the United States, the Soviet Union, and Britain) took on "freedom of the seas" as a new rallying cry. Scientists were not there to exploit the areas, they claimed; what they discovered would be published and available for all. Developing countries, however, viewed it quite differently. Warren Wooster noted that the IOC had a reputation as a "rich man's club," and now these other members were asserting their sovereign rights and exercising considerable—some might say undue—influence in the IOC.[99] These countries reasoned that if local populations were not in a position to exploit their own resources, they were also unlikely to be able to police them, and "freedom" of research simply made it easier for other nations to explore and exploit where the coastal state could not.

The United Nations was taking a greater role in the ocean's scientific, economic, and legal affairs. This only increased the bureaucratization of oceanography already taking place in the IOC, where national delegations represented more interests than just those of science. In the late 1960s, other sectors within the United Nations began to take an accelerated interest in the oceans. In 1966, the United Nations Economic and Social Council (ECOSOC) put the oceans on its agenda, initiating a survey of knowledge of the nonfish resources of the sea beyond the continental shelf. This action in the United Nations had a significant effect on the focus of work in the IOC. At the fifth session of the IOC, Unesco's assistant director-general for science, A. N. Matveyev, gave an address that reflected the focus on marine resources. He pointed to the United Nations, with its 1966 resolution "Resources of the Sea," as having "put the Commission and its activities in a so-to-speak international limelight." Since that time, Unesco had called upon the United Nations and its specialized agencies to look to the IOC for advice on all international cooperative matters regarding the ocean. Matveyev spoke of this responsibility in terms of "concerted international actions for exploration of the ocean," not necessarily the scientific focus of that exploration.[100]

Further complicating oceanography's growing politicization were new ocean-related proposals in the United Nations. At the Twenty-second Session of the United Nations General Assembly, in August 1967, the representative from Malta, Arvid Pardo, set forth a novel idea. He proposed that the seabed be demilitarized and its resources internationalized. The exploitation of these resources could be managed by the United Nations itself, and the proceeds could go toward programs in less developed countries. The result of Pardo's proposal, in addition to stirring up interest among developing countries in the riches of the seabed, was the creation of a study committee on the peaceful uses of the seabed under the United Nations itself, not under the IOC, and thus not under the banner of science.[101] These developments convinced officials in the State Department and the White House that the United States needed to come up with a constructive alternative to propose at the international level if it hoped to prevent the Pardo proposal or something similar from taking hold. Fortunately, the Marine Sciences Council had already begun to conceptualize the IDOE; these international political considerations were added to Johnson's own motivations, all of which made it a major initiative for both science policy and foreign policy.[102]

Gaining support from other countries was a slow process, with each country wishing to know the commitments of others before committing themselves. When trying to stir up support in Europe, Wenk found that most countries shared similar problems as the United States, particularly the struggle between scientific growth for its own sake and scientific contributions to society. But his observations of the differences reveal some perceived national characteristics:

> West Germany alert to new opportunities; the United Kingdom uncomfortable when the boat was rocked; Norway agonizing over its small size; the USSR eager to match the competition.[103]

Even more interesting than these ad hoc characterizations is the fact that Wenk saw things on an explicitly national level. Gone were the personalities of individual scientists, or even of institutions; it seemed that finally international scientific cooperation had attained this purely diplomatic character.

One reason the Soviets were eager to participate in the IDOE, other than the Soviet scientists' desire for intergovernmental coordination, was the same growing politicization of what the Americans were now calling "marine affairs." Because the Marine Sciences Council occupied a high place in the

United States government, the Soviets understood immediately that the IDOE might have ramifications beyond science alone. In light of recent proposals at the United Nations with regard to the seabed, the Soviets suspected that the IDOE's express purpose was to take the initiative with robust cooperative proposals in the ocean, to mute the effect of the Pardo proposal and prevent the United Nations from exploring that path. Neither the United States nor the Soviet Union wished to put the sea's resources under international control. The Pardo proposal, which neither superpower welcomed, helped convince the Soviets to back the IDOE, despite its being yet another U.S.-proposed international project. According to Wenk, the Soviets hoped that the United States and the Soviet Union together could convince other nations that Pardo's proposal was premature and that the IDOE was a genuine way to promote international cooperation in the uses of the sea.[104] This, combined with the preference of Soviet officials for intergovernmental arrangements and top-down scientific planning, virtually guaranteed the enthusiasm of the Soviet Union for the IDOE.

The idea of the IDOE was completely in step with what the Soviets wanted to do. Recalling his efforts to gain international support for the IDOE, Wenk wrote that he was surprised that "the shadow most feared, cold war politics, never materialized."[105] But really this is not much of a surprise because the proposed structure of the IDOE was precisely what the Soviets had been trying to accomplish since the formation of the IOC. It was also what many influential American and British scientists had resisted tooth and nail. The IDOE, as formulated by the Marine Sciences Council, was to be government-to-government, not scientist-to-scientist, with objectives planned from above. Because the projects would be the result of negotiation between national delegations, there would be little of the flexibility so prized by many scientists, and there would be more mechanisms to compel scientists to conform to an overall plan. The Soviets had been clamoring for this for years. They initially were suspicious of the intentions of the IDOE, but Wenk met with senior government officials in the Soviet Foreign Ministry and with government science and technology advisors, convincing them that all nations would benefit from a concerted approach to marine problems facing everyone. The Soviet Academy of Sciences, unlike its American counterpart, was perfectly comfortable with the need for intergovernmental apparatus, and when the Soviets realized the scope and structure of the IDOE, they were thrilled to take part.[106] The Marine Sciences Council, which had come to differ sharply with academic oceanographers, found itself in line with the Soviet view. Like the Soviets throughout the 1960s, Wenk and his

Marine Sciences Council looked at cooperation and saw its quantitative potential: "together we accounted for roughly 70 percent of the global capability."[107] This point of view took into consideration numbers of ships, numbers of scientists, national commitment, and thus the volumes of data to be acquired and area to be surveyed. It placed far less emphasis upon specific scientific problems, scientific originality, or novel ideas.

On the political level, the United States and the Soviet Union tended to agree that the growing influence of the developing world was a cause for solidarity among superpowers. In an article in *Foreign Affairs*, Arvid Pardo criticized "the great oceanic maritime powers" for their silence or vagueness on the question of all countries participating in the economic exploitation of the seabed. He also noted that there was a deep divergence between those who favored a complete demilitarization of the ocean floor and those who wanted to take limited, more cautious steps. To prevent massive undersea territorial claims, uncontrolled exploitation, and unregulated use of the sea as a waste disposal site, Pardo called for a strong international regime that would provide some share of the ocean floor's resources for everyone. To present the different sides of the debate, he did not focus on East-West groupings; instead, Pardo's divisions clearly separated countries according to technological and economic development, which made the United States and the Soviet Union allies.[108]

Ultimately, no hard negotiations took place in the IOC about whether to pursue the International Decade, despite the fact that scientists were being handed a project that they had no role in producing. The incorporation of the IDOE into the IOC's agenda was dictated by the United Nations. Its General Assembly in December 1968 adopted a resolution requesting that the IOC intensify its activities in coordinating the scientific aspects of a "long-term and expanded programme of world-wide exploration of the oceans and their resources of which the International Decade of Ocean Exploration will be an important element." It added that the purpose of this expanded program would be to increase the knowledge of the ocean "with the goal of enhanced utilization of the ocean and its resources for the benefit of mankind."[109] The IDOE was incorporated into the IOC agenda as a supposedly international initiative, the Long-Term and Expanded Programme of Oceanic Exploration and Research (LEPOR). This was itself a diplomatic gesture by IOC, designed in Wenk's view "[t]o suppress US authorship, which seemed curiously necessary in international political circles."[110] The IDOE was treated as a major component of a broadly conceived international program to establish a long-term basis for the study and exploration of the seas.

In 1969, the IOC formed a special working group for the LEPOR, and it stated that its purpose was to increase knowledge of the oceans and its subsoil, the processes therein and interactions with surroundings, for its practical benefits.[111] It called for greater collaboration with existing international organizations such as ICES and ICNAF and for close contact with the newly formed United Nations Committee on Peaceful Uses of the Seabed and the Ocean Floor Beyond the Limits of National Jurisdiction. It identified existing projects as good places to begin: the Cooperative Study of the Kuroshio and Adjacent Regions (CSK) and similar expedition-style investigations projected for the Caribbean, Mediterranean, and other areas.

The IDOE, as an "acceleration phase" of a more general expanded program of research, may well have marked the end of the style of international cooperation in oceanography as described in this book. It did not identify a specific region or problem for which scientists wanted to pool their resources. The IDOE was rather different, providing a host of projects with the intention of simply increasing the level of research on the oceans, as broadly as possible. The only requirements were that the projects be multinational, that they entail reporting of data, that they be peaceful, that they somehow "accelerate" knowledge, that scientists from other countries be encouraged to take part, and that they take place between 1970 and 1980. These were rather wide parameters. Hans Roll wrote later that the new strategy "leaves more freedom for ideas, initiative and action to the scientific community than the former co-operative investigations which were started, organized and co-ordinated by the IOC. Perhaps this is the reason why IDOE had proved so successful."[112] Certainly it provided greater flexibility, which was a positive thing for scientists. At the same time, increasing the level of research and accelerating knowledge acquisition were rather vague requirements that did not necessitate specific scientific objectives. The political support enjoyed by the IDOE in the United States was unprecedented in its preparatory phase (though not necessarily during its implementation), and when it was transferred to the National Science Foundation, the scientists perhaps received more than they had hoped for—money, a major long-term international program, and management by a science-oriented body.

Still, the oceanographic community saw much of the initiative for projects taken out of its hands. International oceanography had moved irrevocably into the realm of social implications. The Committee on Peaceful Uses of the Seabed and the Ocean Floor Beyond the Limits of National Jurisdiction, resulting from Arvid Pardo's proposal about the exploitation of the seabed for the benefit of developing countries, gave the IOC some jurisdic-

Marine Science and Marine Affairs

tional competition and held more political clout (and potential long-term ramifications) than the IOC could hope to attain. As the United Nations gained more members during the 1960s, as a result of newly independent nations, so did the IOC. By the 1970s, promoting oceanography without tying it to some tangible social benefit seemed impossible. The United States delegation to the IOC set forth this philosophy clearly:

> International co-operation in science is not an end in itself, but serves other ends. It should be undertaken only when the sum of scientific, political, and economic benefits exceeds the cost. These benefits cannot be separated in practice. Scientific co-operation will not be effective for any purpose unless it is good science, yet governments are unlikely to support it unless it serves economic and political purposes, as well as scientific ones.[113]

This certainly reflected the attitude taken by the Marine Sciences Council as formulated during the Johnson years. Other countries, notably the Soviet Union, agreed with it. "Comprehensive world ocean study," a Soviet position paper proclaimed, "must be closely linked with practical use of the knowledge obtained." With the help of programs such as the IDOE, the IOC seemed poised to achieve what Soviet scientists had long dreamed of: the standardization of observational methods, top-down planning, and the close coordination of large-scale oceanographic surveys.[114]

At its eighth session, in 1973, the IOC further associated its scientific programs with human problems. It identified four main areas of concentration for the LEPOR: environmental forecasting, the quality of the marine environment, the resources at the bottom of the sea, and the living resources of the sea. The purpose of the IDOE would be to accelerate all these areas within the LEPOR. The IOC adopted an official "philosophy" toward IDOE projects, namely, that they all be of significant size and scope, not only in terms of science but also of international cooperation, "to 'accelerate' our knowledge and understanding of the ocean more rapidly than if individual programmes were conducted separately and at a normal rate."[115] Many of the IOC's existing programs were simply incorporated into this accelerated phase, with their components spread among the four major categories.

In addition, a host of new projects were envisioned, some of which outpaced previous efforts at cooperation and extended scientific knowledge of the sea. Under the rubric of environmental forecasting, the North Pacific Experiment (NORPAX) was the most ambitious. It would evaluate the large areas of abnormally hot or cold sea surface temperatures, thought to have a major effect upon the weather and climate of the entire North American continent, and attempt to set up an international environmental monitoring system throughout the North Pacific Basin. One of the projects for environmental quality was the Global Baseline Data Project, part of the Geochemical Ocean Sections Study (GEOSECS). This survey of physical and chemical properties was not really an experiment with a specific goal but, rather, an effort to develop a baseline data set to establish parameters for future quantitative studies of marine geochemistry. In the third category, assessing the resources of the seabed, the most renowned would be a coordinated study of the Mid-Atlantic Ridge by France, the United States, the Soviet Union, the United Kingdom, the Netherlands, and Iceland. Labeled FAMOUS, this project would utilize submersible craft to explore the ocean floor and to study the most seismically active areas, in order to investigate the formation of new crust and further understand a crucial mechanism driving the motion of the earth's plates.[116] Many of these projects were successful because scientists managed to regain control of them, despite the original conception of the IDOE.

To the relief of many scientists in the West, the era of extensive oceanographic surveys seemed to be coming to an end. New technologies were poised to head off the Soviet demand for more data collection projects and the demands for practical information by developing countries, leaving scientists to focus on more problem-oriented work, on expeditions and in the laboratory. For many oceanographers, one of the most promising of these technologies was the development of a permanent system to provide oceanographic data in "real time," similar to that already typical with meteorological data. In 1968, the IOC convened the first working committee on an Integrated Global Ocean Station System (IGOSS). This concept promised to eliminate the problem of countries withholding data that had not yet been "worked up." Also, in the past, oceanographic research was based upon historical data, stored and exchanged between data centers throughout the world. Now, with the recent development of a number of advanced oceanographic services throughout the world, and with the cooperation of the World Meteorological Organization, "real time" data seemed possible. The 1968 working committee concluded that IGOSS would "for the first time, provide truly synoptic data (and charts) over considerable ocean areas which will allow the development of research in many ways that have hitherto not been possible." The scientists initially used the term "real time," but despite its appeal, they had to admit that it was not quite accurate; still, the amount of time needed to make data available would be measured in hours rather than months or years.[117]

These major innovations and projects were, once again, tied closely to practical uses. In 1972 countries used IGOSS to report data from expendable bathythermographs with great success, and scientists felt that the uses of such a data network had the potential to revolutionize scientific investigation while creating a powerful tool for weather and climate forecasting. After the first year of testing, the IOC envisioned numerous uses in oceanography, meteorology, and fisheries research. "No doubt other justifications will appear in due course," one report proclaimed, "but even when these are not dependent upon real-time (i.e. operational time) oceanographic reports, we consider that real-time reporting is the best practical method of informing interested parties throughout the world, unless delays of months or years are deemed acceptable."[118] The IGOSS data network, and any others like it, promised to negate the endemic problem of gaining access to data that some nation or another allowed to rest hidden behind its own borders for political or bureaucratic reasons. Instead of collecting, amassing, and storing data for later use by interested scientists, the IOC hoped to develop a system of data exchange that could facilitate ocean and atmospheric forecasting in the short term. The IOC's chairman, Rear Admiral W. Langeraar, recognized this by 1971 and devoted his speech at the opening of the IOC's seventh session to the subject of weather forecasting and environmental monitoring, which would be a beneficial by-product of the scientific investigation of the sea during the IDOE. The IOC "must prepare itself and gird its loins to become—more than ever—an important part of an international machinery that will guard and protect the integrity and health of the human environment."[119] This focus on monitoring and protection would grow in importance throughout the decade.

The focus on economic exploitation and other practical pursuits, for developing countries and industrialized ones alike, had been the cornerstone of oceanographers' decision in the early 1960s to connect international oceanography to problems of society. In grappling with controversial issues—the environment, the relationship of oceanography to the nutritional needs of the world, freedom of scientific research in coastal waters, the legal status of unmanned stations, and control of the resources of the seabed—the tasks of the IOC in the coming decades were to become more and more problematic for scientists. The 1970s would be an era of renewed focus on social problems and increasing politicization of science in the international context. These were all brought about largely by the scientists themselves, who in their efforts to create disciples of marine science in government brought difficulties upon themselves, for better or for worse, that never would go away. American scientists in particular had helped to push things in this direction, in the hope that it would mean more money and opportunities for basic research in oceanography. Social benefits were a key justification to pursue science by the beginning of the 1960s. A decade later, they were not a vague justification but were instead the focus of activity. The IOC and its powerful sponsoring governments had made the utilization of the sea its primary objective, with science merely as a means to that end.

8 CONCLUSION

After three decades of creating disciples of marine science, oceanographers had established a massive support infrastructure for themselves that spanned national and international organizations but, at the same time, had opened new problems to be negotiated and new threats to their autonomy. By the early 1970s, international cooperation had undergone both a change in emphasis and a change in cast. Most of the scientists who advocated international cooperation after World War II and who saw their efforts come to fruition during the late 1950s and 1960s no longer exercised a strong voice in such cooperation. Though many remained active, some retired and others died. At the seventh session of the IOC, in 1971, the international oceanographic community began its meeting by mourning the recent deaths of a number of prominent figures in marine science, among them British vice admiral Archibald Day, Soviet marine biologist Lev Zenkevich, and three Americans—former Woods Hole director Columbus Iselin, Milner Schaefer, and Wilbert Chapman. The last, Chapman, had been a persistent voice in urging oceanographers to abandon their cherished notions of "pure" science and to take fisheries money during the IIOE in order to further their cause. All of the scientists being commemorated had spent parts of their careers promoting cooperative projects that crossed the political frontiers of the Cold War, and it might have seemed fitting to have the IOC acknowledge them. But it is doubtful that any of these figures would have felt at home in the IOC of the 1970s. In fact, after paying homage to the deceased, that very meeting got bogged down in doing precisely what all of them had claimed they had wanted *not* to do during the preceding decades. The Romanian delegation brought up questions of representation and stated that it would recognize only the People's Republic of China and the Provisional Revolutionary Government of South Vietnam as legitimate delegations of their respective countries and that it deplored the absence of representatives from East Germany, North Vietnam, and North Korea. Other delegates voiced their own sentiments, each lodging objections about various governmental disputes of the ongoing Cold War.[1] Such political noise added to the already controversial nature of the issues at stake in the IOC: ownership of ocean resources, freedom of scientific inquiry at sea, protection of the environment, not to mention fulfilling the needs of scientific communities with increasingly divergent priorities.

Easing political tensions, however, had never been the primary motivation for international cooperation in oceanography. The ideas that gave rise to the major international cooperative efforts in oceanography during the first two decades or so of the Cold War were a blend of opportunities provided by military funding, the rhetorical power of science aiding in development, and the attractiveness to scientists of the international community—with its combined efforts, coordination of plans, and openness of scientific results. Oceanography presented a paradox: it was both a major outlet for military research funds and a model field of inquiry in which all nations could cooperate. Although "easing tensions" may have been one of the stated international benefits of cooperation during the 1950s, it was not a genuine impetus for any significant ventures. International cooperation, far from being a "feel good" aberration from oceanography's otherwise nationalistic endeavors, was a significant component of the American scientific effort during the Cold War.

Although marine science was connected to foreign policy from the early 1950s, it was through military patronage that major expeditions began and the first major cooperative venture beyond North America and northern Europe succeeded. And despite trying other strategies, oceanographers had the most success with the Navy in acquiring funding, in jointly identifying goals, and in exploring scientific ideas without much interference. Recent studies of oceanography after 1945 have begun to emphasize, quite rightly, that the relationship between scientists and the Navy was the critical interaction of oceanography during this period. It was critical not only because of the great boost that Navy sponsorship gave a host of scientific endeavors related to the oceans but also because of the position oceanography came to occupy in the scientific and technological struggles with the Soviet Union during the Cold War. Historian Paul Forman showed what military goals and funding lay "behind" quantum electronics.[2] The same was true for oceanography, from ocean dynamics, geology, meteorology, zoology, acoustics, to the modern theory of plate tectonics. Funding for oceanography, beginning during World War II and reaching an intense period dur-

Conclusion

ing the 1960s, originated in the need to provide the armed services, particularly the Navy, with a superior understanding of the ocean environment. International cooperation was useful to the Navy for the same reasons that it was useful to scientists: it increased effort without increasing cost. Or, if costs rose to accommodate an international endeavor, the participants judged that sharing data made it more than cost-efficient. The Navy came to support international research, to gather strategic data without having to double or triple its efforts in oceanography.

Scientists shared these motivations and added some of their own: cooperative projects allowed them to circumvent the problem of classification for some projects, to coordinate activities, and to maintain their international reputations while still conducting a great deal of classified military work. International cooperation was not directed against the military, which usually put up a lot of money for such projects. Oceanographers worked closely with the military and often considered their goals to be similar. International cooperation was a part of this relationship, and it would be naïve or misleading to separate their activities between "warlike" military work and "peaceful" international cooperation. Roger Revelle once recalled that the most productive event toward preserving peace was not the IGY, or the formation of SCOR or the IOC, or in fact any international cooperative endeavor. Instead, he saw his greatest contribution to preventing a third world war as the 1956 decision, by scientists during Project Nobska, to help the Navy develop the capability to launch nuclear missiles from beneath the sea, giving rise to *Polaris*-equipped submarines. Revelle reflected that "the *Polaris* submarines and their successors, with their terrible weapons, have probably more to do with keeping the peace between the Soviet Union and the United States in the past 20 years than all the diplomats and politicians put together."[3] There is a certain pride of accomplishment in that statement, and it speaks volumes of the strength of the bond created between oceanographers and the Navy during the early 1950s and carried through the entire period covered by this book.

The development of ideas combining data acquisition with openness of results culminated in the first cooperative project that included scientists from the Soviet bloc: the International Geophysical Year. Scientists sought to convince their patron that although the free flow of data would benefit all countries, it would benefit Americans first. Put to test amid the shocks of the IGY (the Soviet Antarctic expedition, the extensive oceanographic program, and most of all the launch of *Sputnik*), this idea faltered when it became clear that the Soviet Union stood to gain as much by it as the United States. The IGY, regardless of having crossed political boundaries by bringing together scientists of many nations, showcased the Cold War and turned the oceans into a competitive enterprise much like the space race. After the IGY, oceanographers had to adopt new methods for promoting their research.

In the international arena, scientists dropped "easing tensions" and focused more upon the social ramifications of research. The Scientific Committee on Oceanic Research phrased its agenda in terms of practical goals from the beginning, and the International Indian Ocean Expedition was sold to the developing world as a possible means to study the impact of the ocean processes on food resources. The establishment of the Intergovernmental Oceanographic Commission carried this salesmanship further and added to it the trappings of bureaucracy; with national delegations, the research programs had to be negotiated not just with science in mind but with a blend of scientific, political, and economic considerations as the basis for making decisions. Nevertheless, countries of the developing world were themselves becoming disciples of marine science, appreciating more and more that perhaps oceanography could contribute to their nations' economic health and the value of their scientific communities. Americans were selling more than the idea of economic benefits; they were selling the concept of "marine science," a field so broadly defined and with applications so vaguely promised that anyone could be a practitioner and, more important, anyone could be a patron.

In the United States, the new Soviet oceanographic programs convinced many marine scientists to look inward and to achieve a "first," to combat enemy propaganda. Although the impact of the launch of *Sputnik* upon marine science has not received nearly the amount of attention as the impact upon space science and science education, American oceanographers benefited handily from the competitive spirit engendered by Soviet scientific and technological successes. The anxiety sparked by *Sputnik* touched many fields of science, oceanography included. American national support for oceanography began to escalate dramatically after the IGY, and the newly created National Academy of Sciences Committee on Oceanography initiated a propaganda campaign to find potential disciples of marine science in Congress, which they did with astounding success, capitalizing on what seemed to be oceanography's equivalent of the "missile gap." The scientists of the Mohole project refused to transform their work into an international endeavor, fearing that the Soviets would somehow turn it into another great success for the communist world. Many of the efforts to promote national programs were good ideas, leading to more money for science; others were self-defeating, leading to resentment in international forums. Focusing on the Mohole project even led to the abandonment of initiative in spearheading cooperative projects, which the Soviets soon did with the Upper Mantle Project. Even in the IOC, deciding on international programs turned into a battle for influence between the United States and the Soviet Union, neither of which could ignore the Cold War politics dividing them.

The creation of a viable international infrastructure for supporting oceanography depended upon creating disciples of marine science among the scientists themselves. Unesco's training programs aimed to cultivate interest in oceanography in countries in the developing world, to help them participate in big projects and to expand the community of scholars who worked on matters connected to the oceans. In industrialized countries, not all scientists were convinced that oceanography, broadly defined (and thus requiring the negotiation of research agendas), would be healthy for their individual branches of science. This was particularly true for physical oceanographers, who feared being dominated by fisheries scientists; after all, if they promised to do work that might lead to understanding fisheries, it followed that the problems of physical oceanography might not be addressed when it came to deciding research priorities. But even the skeptics agreed that endorsing a broad definition of the field and becoming a marine scientist—rather than a physical oceanographer or a marine zoologist, for example—was a way to act in concert and provide unified recommendations to governments to support international activity. This, however, led to protests against the bureaucratization of scientific activity. How could scientists work on their own problems, maintaining their intellectual autonomy, when projects required the negotiation and approval of dozens of nations? In addition, the thought of an intergovernmental body trying to compel action was anathema to many scientists who prized intellectual autonomy over joint surveys planned in high-level committees.

Even when scientists ignored intergovernmental organizations and considered only purely scientific bodies, they found that they did not necessarily wish to belong to an all-inclusive scientific community. The presence of Soviet scientists in these international forums was an embarrassing reminder that scientists, despite the self-proclaimed universality of their subject, were as impaired by political divisions as their governments were. One might be tempted to say that Western scientists, by rejecting Soviet views and consistently opposing Soviet proposals in international forums, were betraying the "republic of science." But as Michael Polanyi noted in his 1962 article "The Republic of Science," a republic is not a democracy. "The more widely the republic of science extends over the globe, the more numerous become its members in each country and the greater the material resources at its command," Polanyi wrote, "the more clearly emerges the need for a strong and effective scientific authority to reign over this republic."[4] Certainly American oceanographers expended considerable energy to become and to remain this authority, and the Soviet effort to do so can be viewed only as a dismal failure.

Some marine scientists were squeamish about aggressively finding disciples of marine science in government. The IGY's justification had been scientific, with its practical benefits so vague—easing tensions—that it had not been necessary to formulate detailed explanations about what social good might come of oceanic research. But in the years that followed, the IOC defined its work in terms of feeding the world's hungry, predicting climate change, and finding places in the ocean for deadly waste products. Marine scientists in all fields found themselves promising that their work undoubtedly would lead to some future exploitation of the sea. The practice was criticized as disingenuous by some marine biologists and as dangerous by others who felt that selling projects for their fisheries applications would undermine the concept of supporting science for its own sake. This worry proved well founded, particularly in the United States. By the mid-1960s, bureaucrats in government had established a council to harness the whole spectrum of national interests at sea: military, economic, diplomatic, and scientific, usually in that order. Instead of the broad and powerful conception "marine science," American oceanography had become a component of "marine affairs," leaving scientists relatively unimportant in the policy-making process. The most ambitious cooperative project covered in this book, the International Decade of Ocean Exploration, sprang forth from the American marine affairs bureaucracy, not from American scientists, and certainly not from the international community. Like so many of the efforts of scientists to create "disciples of marine science" throughout the 1950s and 1960s, the IDOE resulted in a lot of money for scientific projects, and it can be judged a successful effort to induce a major government to appreciate oceanography. At the same time, that appreciation politicized science, threatened scientists' autonomy, and took the initiative for shaping the international scientific community out of the hands of scientists themselves.

The process of stirring up interest required a lot of effort on the part of oceanographers, who saw their community competing with more glamorous

Conclusion

ideas such as atomic bombs, outer space, and even Antarctica. Often historians are tempted to ask whether scientists were manipulated or even coerced into doing certain kinds of science, for economic or military reasons. In assessing the Cold War, with so much science supported by the military, this question is unavoidable. This book has tried to make clear that in the case of oceanography, scientists made such choices themselves, and they did so consciously. They did not, moreover, simply react passively to military courtship or to other financial opportunities. Oceanographers actively sought out military patronage, chose to work closely with the military, chose to help the military to define its goals, and enjoyed the benefit of seeing their field flourish on military funding. But even further, leading American oceanographers went out of their way to tailor their scientific community to appeal to a broad spectrum of patrons, including but not limited to military ones. Their goal, as the title of this book suggests, was to create disciples of marine science wherever they could find them. International cooperation provided the vehicle to do this, with all the scientific and professional advantages of basing research on international foundations. Along the way, however, scientists appear to have lost their anchor, particularly as marine science drifted in a sea of political expectations. These political problems were natural fruits of their decades-long strategy of trying to bring as many parties on board the international marine science ship as possible. If there is a lesson to be learned, it is partly that scientists themselves created the problems from which they later suffered. But more important, the story of international oceanography provides a striking case of how active, how dynamic, and how influential scientists were in defining their own roles in the Cold War.

NOTES

INTRODUCTION

1. Johnson's speech, and the comments of one government think tank, can be found in Technical Analysis Office, Hughes Aircraft Company, "Study of Oceanography for Reducing International Tensions in an Arms Control and Disarmament Environment, suggested to Advanced Warfare Systems Division, Naval Analysis Group, Office of Naval Research," 27 Apr 1964, OSTSF, box 391, folder "Oceanography 1964."

2. Ibid.

3. Many of the most prominent oceanic scientists of the twentieth century were deeply involved in both military work and international cooperation. See Menard, *The Ocean of Truth*; Wertenbaker, *The Floor of the Sea*; Shor, *Scripps Institution of Oceanography*; and Schlee, *On Almost Any Wind*.

4. Mills, "The Historian of Science and Oceanography after Twenty Years," 14.

5. Rainger, Oreskes, and Weir have made substantial contributions to this literature, pointing out such personal and institutional connections. See Rainger, "Constructing a Landscape for Postwar Science"; Rainger, "Patronage and Science"; Oreskes, *"Laissez-tomber"*; and Weir, *An Ocean in Common*.

6. See Richardson, "The Benjamin Franklin and Timothy Folger Charts of the Gulf Stream."

7. Deacon, *Scientists and the Sea*, 366–77.

8. See Emery, "The *Meteor* Expedition"; and Mills, "'Physisches Meereskunde,'" 51–52.

9. Pernet, *International Science*, 153–61. The best source on ICES's history is Rozwadowski, *The Sea Knows No Boundaries*. For a general discussion of the early years of ICES, also see Schlee, *The Edge of an Unfamiliar World*, chap. 6.

10. Schlee, *The Edge of an Unfamiliar World*, 261–65.

11. Raitt and Moulton, *Scripps Institution of Oceanography*, 12–22; and Revelle, "The Oceanographic and How It Grew."

12. On marine biology before the war, see Mills, *Biological Oceanography*. Bigelow is quoted in Bigelow, *Oceanography*, 3. The *Discovery II* expeditions in the Antarctic Ocean during the 1920s and 1930s were useful in providing physical oceanographic data of the most southerly seas. See Coleman-Cooke, *Discovery II in the*

Antarctic. On physical oceanography in the United States, see Mills, "The Oceanography of the Pacific"; and Raitt and Moulton, *Scripps Institution of Oceanography*.

13. Marine geology, because of technical constraints, was limited to shallow waters until about mid-century. See Schlee, *The Edge of an Unfamiliar World*, chap. 4.

14. See Mills, "Oceanography, Physical."

15. References to easing tensions can be found in many sources. See, e.g., Kistiakowsky, "Science and Foreign Affairs." For a discussion, see Skolnikoff, *Science, Technology, and American Foreign Policy*.

16. The classic study of the postwar movement for social responsibility in science is Smith, *A Peril and a Hope*.

17. Henry Stommel to Paul Fye, 13 Feb 1961, ODWHOI, box 30, folder WHOI Policy Review 2 of 2.

18. J. N. Carruthers, "Unesco and Oceanography," Casablanca, 1961. UNESCOR, box 551.46 A06 (64) "61."

1 / BEGINNINGS OF POSTWAR MARINE SCIENCE AND COOPERATION

1. Robert S. Dietz to Director and Staff, Scripps Institution of Oceanography, 25 May 1953, SIOSF, box 7, folder 31.

2. Recent work on the role of scientists in the international arena during the early postwar years suggests that scientists and policymakers saw various, often conflicting, roles for science in foreign relations. See Doel, "Scientists as Policymakers, Advisors, and Intelligence Agents." By the 1960s, the relationship between science and international relations had received far more attention. Representative works from that era, covering science more broadly defined than atomic physics, include Skolnikoff, *Science, Technology, and American Foreign Policy;* Gilpin and Wright, *Scientists and National Policy-Making;* Haskins, "Technology, Science and American Foreign Policy"; Kistiakowsky, "Science and Foreign Affairs"; and Brode, "National and International Science." Historians have focused on the role of atomic energy in American politics and foreign relations, notably in Hewlett and Duncan, *Atomic Shield, 1947–1952;* and Hewlett and Holl, *Atoms for Peace and War, 1953–1961*. An analysis of President Harry Truman's foreign policy goals, including the role of the atomic bomb, can be found in Leffler, *A Preponderance of Power*.

3. Many of the most prominent oceanic scientists of the twentieth century were deeply involved in both military work and international cooperation. See, e.g., Menard, *The Ocean of Truth;* Wertenbaker, *The Floor of the Sea;* Shor, *Scripps Institution of Oceanography;* and Schlee, *On Almost Any Wind*.

4. Polanyi, "The Republic of Science."

5. See Chambers, "Does Distance Tyrannize Science?"; and Pyenson, *Cultural Imperialism and Exact Sciences*. For a critique of Pyenson's views, see Palladino and Worboys, "Science and Imperialism." Science in Japan, under various local and foreign influences, is analyzed in Bartholomew, *The Formation of Science in Japan*. The

origins of the scientific networks in the Pacific region are treated in Rehbock, "Organizing Pacific Science." Basalla is quoted in Basalla, "The Spread of Western Science," 613.

6. Rehbock, "Organizing Pacific Science," 208–12. For the Australian Congress, see Macleod and Rehbock, "Developing a Sense of the Pacific," 209–26. The committees were "Oceanography of the Pacific," "Coral Reefs of the Region," "Volcanic Rocks of the Central Islands" (the fourth was "Pacific Anthropology"). Elkin, *Pacific Science Association*, 28.

7. Essays exploring the unique historical role of science in the Pacific Ocean (not limited to the twentieth century) can be found in Macleod and Rehbock, *Nature in Its Greatest Extent*.

8. Elkin, *Pacific Science Association*, 32.

9. Carl Hubbs to Laura Hubbs, 30 May 1929, CLHP, box 18, folder 64.

10. For example, although Indonesians of non-European descent held some research positions, Dutch-born scientists or children born of Dutch parents were at the top of the scientific hierarchy. See Pyenson, "Assimilation and Innovation in Indonesian Science."

11. Biological and medical sciences were particularly strong in Japan. See Bartholomew, *The Formation of Science in Japan*.

12. Carl Hubbs to Laura Hubbs, 29 Jun 1929, CLHP, box 18, folder 64.

13. Dennis, "Historiography of Science," 5–6.

14. Elkin, *Pacific Science Association*, 43.

15. See Mills, "'Physisches Meereskunde.'"

16. See Macleod, "'Kriegsgeologen and Practical Men'"; and Little, "Natural Resources in Their Relation to Military Supplies."

17. Oreskes, "Weighing the Earth from a Submarine," 63–64.

18. Ibid., 66.

19. On physical oceanography in Germany during the 1930s, see Mills, "'Physisches Meereskunde.'" On Sverdrup at Scripps, see Mills, "The Oceanography of the Pacific." On oceanography and meteorology in Norway, see Friedman, *Appropriating the Weather*.

20. "Interview with Sir Edward Crisp Bullard of Cambridge University," at U.S. Naval Institute, Annapolis, Maryland, 12 Dec 1969, by Glen B. Ruh, ECBP, folder A.8.

21. M. Hill, "Geological and Geophysical Investigations of the Floor of the Ocean and of Neighbouring Shallow Seas Undertaken by the Department of Geodesy and Geophysics, Cambridge University," 1959, ECBP, folder B.22.

22. See Sverdrup, Johnson, and Fleming, *The Oceans*. The new focus on physical oceanography is emphasized in Rainger, "Patronage and Science."

23. The experience of oceanographers during World War II has been analyzed by a number of scholars. The work of Gary Weir has been particularly useful in understanding the details of a growing partnership between scientists and the Navy. Good treatments include Weir, *Forged in War*; Weir, *An Ocean in Common*; and (less com-

prehensive) Rainger, "Patronage and Science." Also see Sapolsky, *Science and the Navy*.

24. "Interview with Sir Edward Crisp Bullard of Cambridge University," at U.S. Naval Institute, Annapolis, Maryland, 12 Dec 1969, by Glen B. Ruh, ECBP, folder A.8.

25. B. C. Browne to J. Anderson, Superintending Scientist, Portland Naval Base, 11 Jul 1946. Quote is from Edward Bullard to Sir Charles Darwin, National Physical Laboratory, 1 Mar 1946, ECBP, folder F.85.

26. Torben Wolff, "The 50 years' anniversary of the Galathea Deep-Sea Expedition 1950–1952," 5–9.

27. M. Hill, "Geological and Geophysical Investigations of the Floor of the Ocean and of Neighbouring Shallow Seas Undertaken by the Department of Geodesy and Geophysics, Cambridge University," 1959, ECBP, folder B.22.

28. On the expeditions to the Marshall Islands, see Rainger, "Science at the Crossroads," and Weir, *An Ocean in Common*, 198–209.

29. Maurice Ewing to Edward Bullard, 12 Jun 1946, ECBP, folder J.29. Quote from Edward Bullard to Maurice Ewing, 26 Jun 1946, ECBP, folder J.29.

30. A. E. Maxwell to Edward Bullard, 20 Nov 1949, ECBP, folder D.406.

31. Roger Revelle to Edward Bullard, 15 Oct 1950, ECBP, folder J.118.

32. See McEvoy and Scheiber, "Scientists, Entrepreneurs, and the Policy Process."

33. See Rainger, "Patronage and Science."

34. Sverdrup, "New International Aspects of Oceanography," 75–78.

35. Curti and Birr, *Prelude to Point Four*, 216. See also Kay, "Rethinking Institutions." For the impact of the Rockefeller Foundation's funding upon scientific communities, see Jonas, *The Circuit Riders*. On the case of Latin America, see Cueto, "Visions of Science and Development."

36. H. L. C., "Mapping the Valley of Caracas," in *Caracas Journal*, 5 Sep 1949 [clipping, no page number visible], HHHP, box 6.

37. Harry Hess to Victor M. Lopez, Ministerio de Fomento, Estados Unidos de Venezuela, 18 Nov 1941, HHHP, box 6.

38. See Pyenson, *Cultural Imperialism and Exact Sciences*.

39. See Greenaway, *Science International*.

40. See Calluther, "Development? It's History." For a discussion of Truman's policies for containing communism in the context of the Cold War as a whole, see Gaddis, *Strategies of Containment*.

41. Quoted in Rostow, *Eisenhower, Kennedy, and Foreign Aid*, 78–79.

42. Alcalde, *The Idea of Third World Development*, 120.

43. Acheson, *Present at the Creation*, 265–66.

44. Charles A. Thomson and Walter Walkinshaw, "Observations on Unesco Operations in South Asia," 12 May 1950, IDABR, box 14, folder "Unesco."

45. Department of State to "Certain Diplomatic Officers," confidential memorandum (unsigned), 16 May 1951, IDABR, box 14, folder "Unesco."

46. The budget for the expert's activities and training program would be shared

Notes to Pages 16–21

between the Institute for Inter-American Affairs and the government of the receiving country. See Tickner, *Technical Cooperation*, 94–95.

47. Charles B. Wade to Chief, Fishery Mission to Peru, 5 Jan 1953, IIAACF, box 112, folder "Annual Report: US Fisheries Mission, Lima, Peru, 1952."

48. Tickner, *Technical Cooperation*, 24–25.

49. Charles B. Wade to Chief, Fishery Mission to Peru, 5 Jan 1953, IIAACF, box 112, folder "Annual Report: US Fisheries Mission, Lima, Peru, 1952."

50. Ibid.

51. N. D. Jarvis, "Annual Report," n.d. [covers period 17 Oct 1952–31 Dec 1952], IIAACF, box 112, folder "Annual Report: US Fisheries Mission, Lima, Peru, 1952."

52. Robert O. Smith, "Report on Activities of S.O.Y.P.," 27 Feb 1952, IIAACF, box 125, folder "Survey of Oceanographic and Fishery Service—Uruguay."

53. Robert O. Smith, "Information Relative to a Program of Technical Aid in Fisheries for Columbia," 9 Jun 1953, IIAACF, box 35, folder "Technical Aid in Fisheries for Colombia, Special Report, Robert O. Smith."

54. McMahon, *The Cold War on the Periphery*, 4. Also Brands, *The Specter of Neutralism*, 105.

55. "British Commonwealth Scientific Official Conference, Report of the Oceanography and Fisheries Committee," n.d. [1946], ADM 213/791.

56. A. A. W. Landymore to W. D. Allen, 21 Oct 1955, FO 371/116942, "Question of Admission of Japan and Afghanistan to and French participation in Colombo Plan, 1955."

57. For details on these events, see FO 371/111908, "Question of admission of Japan and Afghanistan to Colombo Plan; increase in Canadian aid; French participation in Colombo Plan, 1954."

58. Richard N. Johnson to W. A. Harriman, 16 Feb 1951, IDABR, box 5, folder "Government: The White House IDAB."

59. Isaacson and Thomas, *The Wise Men*, 732.

60. Acheson, *Present at the Creation*, 266. Others maintain that Point Four programs, despite their ineffectiveness, were still significant because they were "instrumental in the rise of multilateral development assistance and the emergence of the United Nations as a champion of world development." See Alcalde, *The Idea of Third World Development*, 207.

61. Black, *The Strategy of Foreign Aid*, 14.

62. For a discussion of the changing purpose of science aid in developing countries, see Skolnikoff, *The Elusive Transformation*.

63. Waggoner, "State Department Urges U. S. Office of World Science."

64. Charles B. Wade to Chief, Fishery Mission to Peru, 5 Jan 1953, IIAACF, box 112, folder "Annual Report: US Fisheries Mission, Lima, Peru, 1952." Warren Wooster informed me that José Barandiaran, a Peruvian, had participated in the Northern Holiday expedition of 1951, and he helped to facilitate the Shellback expedition to Peru in 1952.

65. Carl L. Hubbs to Claude ZoBell, 22 Oct 1952, and Claude E. ZoBell to Harold J. Coolidge, 18 Nov 1952, SIOSF, box 7, folder 29.

66. Harold J. Coolidge to Claude E. ZoBell, 12 Jan 1953, SIOSF, box 7, folder 29.

67. Ibid.

68. Robert S. Dietz to Director and Staff, Scripps Institution of Oceanography, n.d., SIOSF, box 7, folder 29.

69. See McEvoy and Scheiber, "Scientists, Entrepreneurs, and the Policy Process," 393–406.

70. Rainger, "Patronage and Science," 58–59.

71. On Revelle's conflicts with marine biologists, see ibid.

72. C. N. G. Hendrix, ONR Liaison Officer, to Claude ZoBell, 25 Feb 1953, SIOSF, box 7, folder 29.

73. C. N. G. Hendrix, ONR Liaison Officer, to Director, Office of Naval Research Branch Office, Pasadena, 24 Apr 1953, SIOSF, box 7, folder 30.

74. Rainger, "Patronage and Science," 78–79.

75. Robert S. Dietz to Director and Staff, 25 May 1953, SIOSF, box 7, folder 30.

76. Ibid. Dietz had not been familiar with the emperor's scientific works.

77. Roger Revelle to All Hands, 30 Sep 1953, SIOSF, box 7, folder 35.

78. This story comes from Warren Wooster, in WWI-01 and in his comments to me regarding a manuscript of the present book.

79. Warren S. Wooster to Kazuhiko Terada, 2 Jul 1953, SIOSF, box 7, folder 33.

80. T. R. Folsom to Captain Hale, 26 Jun 1953, SIOSF, box 7, folder 32.

81. Kanji Suda to Roger Revelle, 18 Sep 1953, SIOSF, box 7, folder 35.

82. Roger Revelle to Ryohei Morimoto, 29 Dec 1953, SIOSF, box 7, folder 36. The work described was eventually published in Kuno, Fisher, and Nasu, "Rock Fragments and Pebbles Dredged near Jimmu Seamount, Northwestern Pacific."

83. Roger Revelle to Kanji Suda, 19 Jan 1954, SIOSF, box 7, folder 36.

84. Roger Revelle to various American, Canadian, and Japanese scientists, 8 Apr 1954, SIOSF, box 7, folder 43.

85. Joseph Reid's responses to questionnaire by Mike Connelly, United Press Staff Correspondent, n.d., SIOSF, box 7, folder 44.

86. Roger Revelle to various Japanese scientists, 15 Nov 1954, SIOSF, box 7, folder 43.

87. Roger Revelle to various American and Canadian scientists, 15 Nov 1954, SIOSF, box 7, folder 43.

88. "To Discuss Results of Largest Ocean Survey," Press Release, Oceanographic Publications, 29 Jan 1956, SIOSF, box 7, folder 46.

89. The Japanese scientists accepted the American proposal at a January 1954 meeting and designed a national program of about fifteen ships. Koji Hidaka to Joseph Reid, 31 Jan 1955, SIOSF, box 7, folder 43. The ten additional ships are noted in another letter in the same folder, Kanji Suda to Joseph Reid, 31 Jan 1955.

90. Roger Revelle to Kanji Suda, 18 Jun 1955, SIOSF, box 7, folder 44.

Notes to Pages 29–33

91. Joseph Reid to Gordon G. Lill and Arthur Maxwell, Geophysics Branch, Office of Naval Research, 29 Dec 1955, SIOSF, box 7, folder 45.

92. Roger Revelle [by Joseph Reid] to Koji Hidaka, 16 Dec 1955, SIOSF, box 7, folder 45.

93. "To Discuss Results of Largest Ocean Survey," Press Release, Oceanographic Publications, 29 Jan 1956, SIOSF, box 7, folder 46.

2 / OCEANOGRAPHY'S GREATEST PATRON

1. One should note that many internationally minded scientists shared a defense-oriented view after World War II, especially because they felt that the imposition of communist ideas on scientific practice in the Soviet Union had made internationalism unrealistic. For an analysis of American efforts to gather intelligence on Soviet science during the late 1940s, see Doel and Needell, "Science, Scientists, and the CIA." For a treatment of many scientists' pro-military stance during the mobilization for war in Korea, see Kevles, "K1S2."

2. The unsuccessful efforts to put government support for science more firmly into the hands of civilians have been chronicled in analyses of the genesis of the National Science Foundation. See Reingold, "Vannevar Bush's New Deal for Research"; and J. Wang, "Liberals, the Progressive Left, and the Political Economy of Postwar American Science." For an overview of these events, see England, *A Patron for Pure Science*.

3. In specifically analyzing marine scientists, sociologist Chandra Mukerji concluded that government patronage can be explained in terms of its desire to maintain a highly skilled elite reserve labor force, useful for expertise as much as for scientific information. Certainly there is some truth to this, but one cannot ignore the fact that military and civilian patrons alike expected tangible results for their money, whether in the short or long run. In discussing the Navy, historian Gary Weir emphasizes the importance of key individuals who spoke the languages of the Navy and of science and served as "translators" between the two cultures, each with its own set of values, expectations, and motivations. Weir's approach echoes that of Allan Needell, whose analysis of Lloyd Berkner centered on that man's efforts to balance the needs of science with those of the State Department. Such outlooks are most valuable in understanding what the boundaries between communities were, and how differences were addressed and cooperation ensured. Too often one is tempted, however, to perceive the motivations of each group as inherently different; in the case of science and the Navy, one runs the risk of missing just how important scientific research became to the Navy as a whole. See Mukerji, *A Fragile Power;* Weir, *An Ocean in Common;* Needell, *Science, Cold War, and the American State.*

4. The United States Navy had a long history of periodic support for science, particularly in exploration and survey projects. See, e.g., Rothenberg, "'In Behalf

of the Science of the Country.'" The Navy's willingness to fund research became especially clear during the interwar period, as described in McBride, "The 'Greatest Patron of Science'?"; Van Keuren, "Science, Progressivism, and Military Preparedness"; and Oreskes, "Weighing the Earth from a Submarine."

5. Wertenbaker, *The Floor of the Sea*, 42–44. For a more detailed analysis of the development of wartime research, see Weir, *An Ocean in Common*, chaps. 7–9.

6. RRI-84, 6, 11. For reflections on Woods Hole's wartime experience with the Navy, see also JLWI-96, 124–37.

7. See Rainger, "Science at the Crossroads." The cooperative arrangement between science and the military existed not only for oceanography but also for other scientific disciplines, in what became known as the era of "Big Science." This term was used by Derek Price to describe the exponential growth of science since the seventeenth century; see Price, *Little Science, Big Science*. Despite such steady growth, few can dispute that the definitive factor in the growth of American science during and after World War II—often termed "Big Science"—was government (largely military) involvement in supporting basic research. See Dupree, "The Great Instauration of 1940"; and Capshew and Rader, "Big Science: Price to the Present."

8. The dilemma between social responsibility and making a living marked the experience of scientists throughout the Cold War; some, however, felt less anxiety about it than others. See Roland, "Science and War."

9. Ridenour, "Military Support of American Science, a Danger?"

10. Huxley, "A Positive Program of Research for Peace"; Wiener, "The Armed Services Are Not Fit Almoners for Research"; and Einstein, "The Military Mentality."

11. Bush, "Dangers to Research, If Recognized, Can Be Avoided."

12. RRI-84, 51.

13. Waterman and Conrad, "Office of Naval Research Discusses Ridenour's Views."

14. RRI-84, 4, 18.

15. Ridenour, "Military Support for American Science, a Danger?" 222.

16. Bowen, *Ships, Machinery, and Mossbacks*, 137.

17. Ibid., 354.

18. Greenberg, *The Politics of Pure Science*, 273.

19. Sapolsky, *Science and the Navy*, 60.

20. Zachary, *Endless Frontier*, 340–41.

21. Reynolds, *History and the Sea*, 186.

22. Potter, *Admiral Arleigh Burke*, 320–27. For a treatment of this episode, with particular regard to the role of aircraft carriers, see Barlow, *Revolt of the Admirals*.

23. Director, Naval Intelligence, to Director, Strategic Plans, 19 May 1952, SPDR, box 274, folder "A16–8 Antisubmarine Warfare Operations."

24. Goldstein, *A Different Sort of Time*, 97–103.

25. Kevles, "KIS2," 327.

26. Potter, *Admiral Arleigh Burke*, 330.

27. Ibid., chaps. 22–24.

28. Frank Akers, Assistant Chief of Naval Operations (Undersea Warfare), "Undersea Warfare Newsletter No. 2–51," 15 Nov 1951, SPDR, box 264, folder "A16–6 Submarine Warfare Operations." For a discussion of LOFAR, see Weir, *An Ocean in Common*, 292–315.

29. Chief of Naval Operations to Chief of Bureau of Ships, Chief of Naval Research, and Hydrographer, Top Secret memorandum, 6 Jun 1952, SPDR, box 272, folder "A1 Plans, Projects and Development." For a discussion of Jezebel and Caesar, and LOFAR generally, see Weir, *An Ocean in Common*.

30. Chief of Naval Research to various recipients, memorandum, 9 Mar 1951, SPDR, box 264, folder "A16–6 Submarine Warfare Operations."

31. Director of Naval Intelligence to Director, Strategic Plans, 19 May 1952, SPDR, box 274, folder "A16–8 Antisubmarine Warfare Operations."

32. Minutes of Meeting of Anti-Submarine Plans and Policies Group, 5 Feb 1952, SPDR, box 271, folder "A16–8 Antisubmarine Warfare Operations," 1–3, 6.

33. Rosenberg, "American Naval Strategy in the Era of the Third World War," 245.

34. JLWI-96, 294.

35. Troebst, *Conquest of the Sea*, 66. JLWI-96, 128.

36. Revelle and Maxwell, "Heat Flow through the Floor of the Eastern North Pacific Ocean."

37. RRI-84, 41.

38. Ibid.

39. Menard, *The Ocean of Truth*, 52. See chap. 4 for accounts of the expectations and findings of these expeditions.

40. The alleged boundary between pure and applied science was reinforced by Bush's report. See Kline, "Construing 'Technology' as 'Applied Science,'" 220.

41. ASI-89, 96.

42. G. Deacon, "Ocean Waves and Swell." Also see Schlee's chap. 8, "Oceanography and World War II," in *The Edge of an Unfamiliar World*.

43. Sverdrup, "New International Aspects of Oceanography."

44. Fleming, "The International Scientific Unions," 123.

45. Hunsaker, "International Scientific Congresses," 128.

46. Millikan, "The Interchange of Men of Science," 132.

47. Schlee, *The Edge of an Unfamiliar World*, 196–97.

48. Nierenberg, "Harald Ulrik Sverdrup," 357. Indeed, one of Sverdrup's first acts as director of the Norwegian Polar Institute in Oslo was to organize the 1949–52 Norwegian-British-Swedish expedition to Antarctica.

49. Schlee, *The Edge of an Unfamiliar World*, 193–95.

50. Assistant Chief of Naval Operations (Undersea Warfare) to Director of Naval Intelligence, 1 Mar 1951, SPDR, box 261, folder "A1(2) Navy Research (Agenda Items, Minutes of Meeting)."

51. Harold J. Coolidge to Claude ZoBell, 12 Jan 1953, and C. N. G. Hendrix to

Claude ZoBell, 25 Feb 1953, SIOSF, box 7, folder 29. See this folder for other correspondence regarding the TRANSPAC expedition.

52. VBI-64, 272.

53. Op-03D3 to Chief of Naval Operations, 12 May 1953, SPDR, box 289, folder "EF-61 Russia."

54. Potter, *Admiral Arleigh Burke*, 369.

55. Director, Strategic Plans, to Deputy Chief of Naval Operations (Operations), 24 Apr 1953, SPDR, box 279, folder "A1 Plans, Projects, and Development."

56. Deputy Chief of Naval Operations (Air) to Deputy Chief of Naval Operations, 22 Jul 1953, SPDR, box 279, folder "A1 Plans, Projects, and Development."

57. Director, Strategic Plans, to Director, Fleet Operations, Top Secret letter, 25 Feb 1954, SPDR, box 302, folder "Hydrography."

58. Ibid.

59. Director, Fleet Operations, to Deputy Chief of Naval Operations (Operations), 16 Mar 1954, SPDR, box 302, folder "Hydrography."

60. Op-33 to Op-03, memorandum, 13 Apr 1954, SPDR, box 302, folder "Hydrography."

61. Lill, "Office of Naval Research Laboratory of Oceanography and Hydraulics Laboratory."

62. Op-33 to Op-03, memorandum, 13 Apr 1954, SPDR, box 302, folder "Hydrography."

63. Potter, *Admiral Arleigh Burke*, 377.

64. Robert S. Dietz to Commanding Officer, Office of Naval Research, Secret letter, 11 Aug 1954, SPDR, box 302, folder "Hydrography."

65. Planning Group, for Employment of a Chartered Research Vessel, to Director, Strategic Plans, 16 Nov 1954, SPDR, box 302, folder "H1 Hydrography."

66. Ibid., and Enclosure (1), "Brief of Background Correspondence," n.d., SPDR, box 302, folder "H1 Hydrography."

67. Director, Strategic Plans, to Deputy Chief of Naval Operations (Operations), 3 Mar 1954, SPDR, box 302, folder "Hydrography."

68. Op-605 to Op-533, Secret memorandum, 5 Oct 1954, SPDR, box 302, folder "Hydrography."

69. Op-533 to Op-33, Secret memorandum, 1 Nov 1954, SPDR, box 302, folder "Hydrography."

70. Planning Group, for Employment of a Chartered Research Vessel, to Director, Strategic Plans, 16 Nov 1954, SPDR, box 302, folder "H1 Hydrography."

71. Gordon G. Lill to Commander E. B. Rankin, Op-605D1, Secret memorandum, 16 Sep 1954, SPDR, box 302, folder "Hydrography."

72. Hevly, "The Tools of Science," 226.

73. VBI-64, 279.

74. Goldstein, *A Different Sort of Time*, 102.

75. J. Wang, "Science, Security, and the Cold War," 248, 267.

76. For an analysis of the effect of anticommunism upon the academic community, see Schrecker, *No Ivory Tower*.

77. Shils, *The Torment of Secrecy*, 178.

78. VBI-64, 269.

79. Ridenour, "Secrecy in Science," 8.

80. Menard, *The Ocean of Truth*, 111.

81. J. D. H. Wiseman to Menard, 5 Nov 1953, and Menard to Wiseman, 11 Jan 1954, HWMP, box 2, folder 24 "Correspondence 1954."

82. Menard, *The Ocean of Truth*, 111.

83. Hess to Captain Hobbs, Hydrographic Office, 14 Sep 1951, HHHP, box 5, folder unlabeled.

84. Roger Revelle to Hess, 19 Apr 1952, HHHP, box 5, folder unlabeled.

85. Hess to Revelle, 24 Apr 1952, HHHP, box 5, folder unlabeled.

86. Hess to Captain Joseph Cochrane, USN, Hydrographer, 13 Jul 1953, HHHP, box 5, folder unlabeled.

87. Earl Droessler to Revelle, 26 Mar 1952, HHHP, box 5, folder unlabeled.

88. Berkner, "Is Secrecy Effective?" 68.

89. Rear Admiral Felix Johnson, Memorandum of Information, 27 Feb 1950, SPDR, box 255, folder "Intelligence."

90. Chief of Naval Research to Chief of Naval Operations, 8 Dec 1952, SPDR, box 269, folder "A1 Plans, Projects, and Developments."

91. Director, Strategic Plans, to Director, New Developments and Operational Evaluation, 12 Nov 1952, SPDR, box 269, folder "A1 Plans, Projects, and Developments."

92. Hess to Captain Joseph Cochrane, USN, Hydrographer, 13 Jul 1953, HHHP, box 5, folder unlabeled.

93. Detlev Bronk to Hess, 3 Mar 1954, and Hess to Bronk, 10 Mar 1954, HHHP, box 22, folder "National Academy of Sciences."

94. Shils, *The Torment of Secrecy*, 186.

95. VBI-64, 329.

96. Berkner, "Is Secrecy Effective?" 68.

97. Athelstan Spilhaus to Donald A. Quarles, Assistant Secretary of Defense for Research and Development, 28 Dec 1953, HHHP, box 5, folder unlabeled. For some scientists' perceptions of Eisenhower's attitude toward classification, see Menard, *The Ocean of Truth*, 63–64.

98. Siple, *90 South*, 126–27.

99. Rosenberg, "American Naval Strategy in the Era of the Third World War," 247. Defense Secretary Charles E. Wilson caused many Navy programs to be cut in an effort to implement Eisenhower's New Look strategy, which sought to reduce defense expenditures by relying less on conventional weapons and more on the threat

of nuclear retaliation. See Geelhoed, *Charles E. Wilson and Controversy at the Pentagon*.

100. Hess to Senator Styles Bridges, 10 Feb 1960, HHHP, box 22, folder "NAS–NRC Committee on Oceanography Ocean-Wide Survey Panel January–June 1962."

3 / THE INTERNATIONAL GEOPHYSICAL YEAR, 1957–1958

1. BNCIGY, NGY/17 (54), Correspondence relating to the International Geophysical Year, 11 Aug 1954, including Dwight Eisenhower to Chester I. Barnard, Chairman of the National Science Board, National Science Foundation, 24 Jun 1954, GERDP, folder G5/1, "British National Committee for the International Geophysical Year, Correspondence and Papers, 1953–54."

2. For an analysis of the conflicting obligations of an internationally minded scientist in the years after the launch of *Sputnik*, see Doel, "Evaluating Soviet Lunar Science in Cold War America." The many effects of the Soviet satellite on American science are discussed in Geiger, "What Happened after Sputnik"; Divine, *The Sputnik Challenge*; and McDougall, *The Heavens and the Earth*; Aaserud, "Sputnik and the 'Princeton Three'"; Damms, "James Killian, the Technological Capabilities Panel, and the Emergence of President Eisenhower's 'Scientific-Technological Elite'"; and Z. Wang, "American Science and the Cold War." For a discussion of attitudes toward international cooperation by top-level science advisors during these years, see the classic memoirs of the first three chairmen of the President's Science Advisory Committee, created after the launch of *Sputnik*: Killian, *Sputnik, Scientists and Eisenhower*; Kistiakowsky, *A Scientist in the White House*; and Wiesner, *Where Science and Politics Meet*.

3. Bullis, *The Political Legacy of the International Geophysical Year*, 52, 55–62.

4. Kistiakowsky, *A Scientist in the White House*, 59.

5. For the earlier period, see Baker, "The First International Polar Year."

6. See Needell, *Science, Cold War, and the American State*.

7. Bullis, *The Political Legacy of the International Geophysical Year*, 7–8.

8. Ibid., 35.

9. Gould, "The History of the Scientific Committee on Antarctic Research (SCAR)," 48.

10. Senator Thomas Reid, Chairman of the International Pacific Salmon Fisheries Commission, to W. C. Herrington, Special Assistant to the Under Secretary of State, 12. Aug 1957, IGYC, drawer 12, folder "Oceanography 1957."

11. Sullivan, *Assault on the Unknown*, 29–32.

12. Quote is from U.S. Embassy, minute, 25 Oct 1954, FO 371/108781, "International Geophysical Year." Several documents related to Soviet activities in Antarctica can be found in this folder.

13. P. A. Wilkinson, British Embassy, Washington, to M. C. G. Man, Foreign Office, 15 Dec 1954, FO 371/108781, "International Geophysical Year."

14. R. A. Hibbert, minute, 2 Feb 1955, FO 371/108781, "International Geophysical Year."

15. Dr. Roberts, minute, 15 Jul 1955, FO 371/113960, "International Geophysical Year."

16. J. S. Whitehead, minute, 17 Oct 1955, FO 371/113963, "International Geophysical Year."

17. Dufek, *Operation Deepfreeze*, 189.

18. Extract from *Pravda*, "In Academy of Sciences of USSR before Departure of Antarctic Expedition," sec. B, 21 Nov 1955, and extract from *Soviet Fleet*, "American Expedition to Antarctic," sec. A, 30 Nov 55, FO 371/113963, "International Geophysical Year."

19. Robert S. Dietz to Edward Smith, 7 Jun 1956, ODWHOI, box 22, folder "Act. ONR-Washington."

20. G. E. R. Deacon to Edward H. Smith, 6 Jun 1955, IGYC, drawer 12, folder "Oceanography 1954–1955."

21. G. Laclavère to E. H. Smith, 13 May 1955, IGYC, drawer 12, folder "Oceanography 1954–1955."

22. Robert S. Dietz, Technical Memorandum, "Program to Measure Deep Ocean Currents at NIO," 9 Mar 1955, ODWHOI, box 22, folder "Act. ONR-Washington."

23. Edward Bullard to D. C. Martin, n.d., enclosed with D. C. Martin to George Deacon, 24 Mar 1953, GERDP, folder G5/1, "British National Committee for the International Geophysical Year, Correspondence and Papers, 1953–54."

24. On the role of ICES in the IGY, see Rozwadowski, *The Sea Knows No Boundaries*.

25. Henry Stommel to George Deacon and M. S. Longuet-Higgins, Trinity College, Cambridge, 4 Sep 1953, and George Deacon to Henry Stommel, 9 Sep 1953, GERDP, folder D12/1, "International Geophysical Year, 1957–58: Preparation, Operations and Results, Correspondence, 1953–54."

26. George Deacon to Archibald Day, 30 Jun 1954, GERDP, folder D12/1, "International Geophysical Year, 1957–58: Preparation, Operations and Results, Correspondence, 1953–54."

27. Edward Smith to Athelstan Spilhaus, 30 Sep 1955, ODWHOI, box 24, folder "Individuals: Spilhaus, A. F."

28. Edward Smith to George Deacon, 2 Dec 1955, GERDP, folder D12/4, "Correspondence, 1957." This letter evidently was misfiled.

29. George Deacon to Günther Böhnecke, 8 Feb 1955, GERDP, folder D12/2, "Correspondence, 1955."

30. Edward Smith to Alan T. Waterman, 28 Mar 1956, ODWHOI, box 22, folder "Act. National Science Foundation, 1 of 4."

31. Robert Sinclair Dietz, "USSR Program in Oceanography for the International Geophysical Year," Technical Report ONRL-91–55, 28 Sep 1955, RSDP, box 5, folder "Dietz, Robert S., ONR Technical Reports, 1955," 1–2.

32. Soviet scientists used the terms "oceanology" to describe the science of the sea and "oceanography" to describe its measurement; however, often these terms were used interchangeably. See ibid., 4.

33. Ibid.

34. Edward H. Smith to Georges Laclavère, 22 Mar 1955, IGYC, drawer 12, folder "Oceanography 1954–1955."

35. Roger Revelle to Joseph Kaplan, 22 Feb 1955, IGYC, drawer 12, folder "Oceanography 1954–1955."

36. Gordon Lill to G. E. R. Deacon, 7 Mar 1955, IGYC, drawer 12, folder "Oceanography 1954–1955."

37. John L. Farley to Detlev Bronk, 9 Aug 1956, NASCPF, folder "Earth Sciences Committee on Oceanography: Proposed 1956."

38. Rear Admiral R. Bennett to Detlev Bronk, 14 Aug 1956, NASCPF, folder "Earth Sciences Committee on Oceanography: Proposed 1956."

39. Ibid.

40. S. D. Cornell to William R. Thurston, 19 Nov 1956, NASCPF, folder "Earth Sciences Committee on Oceanography: Proposed 1956."

41. Paul Weiss to Frank L. Campbell, 12 Nov 1956, NASCPF, folder "Earth Sciences Committee on Oceanography: Proposed 1956."

42. Lloyd Berkner to S. D. Cornell, 24 Oct 1956, NASCPF, folder "Earth Sciences Committee on Oceanography: Proposed 1956."

43. Engineer Alfredo Obiols Gomez, President, IGY National Committee for Guatemala, to Gordon Lill, 22 Dec 1956, IGYC, drawer 12, folder "Oceanography 1956."

44. John Goodwin Locke to W. Maurice Ewing, 15 Mar 1956, IGYC, drawer 34, folder "International Co-operation Administration (ICO)."

45. Roger Revelle to Gordon Lill, 14 Oct 1957, SIOOD, box 49, folder 1, "Japan 1954–1958."

46. Gordon G. Lill to Rear Admiral F. A. Studds, Director, United States Coast and Geodetic Survey, 23 Mar 1955, IGYC, drawer 12, folder 28, "Oceanography 1954–1955."

47. Sullivan, *Assault on the Unknown*, 357–60.

48. Gordon G. Lill to Rear Admiral W. D. Leggett, Jr., 29 Mar 1955, IGYC, drawer 12, folder 28, "Oceanography 1954–1955."

49. Roger R. Revelle, "Deep Sea Research as a Cooperative Enterprise," 1956, RRP, AC 6A, box 63, folder 16.

50. Harry Hess to H. W. Menard, 22 May 1968, HHHP, box 11, folder "H. W. Menard," and Harry Hess to Immanuel Velikovsky, 8 Mar 1957, HHHP, box 12, folder "Velikovsky."

51. Sullivan, *Assault on the Unknown*, 356.

52. Ibid., 352–54.

53. Menard, *The Ocean of Truth*, 40.

54. Wertenbaker, *The Floor of the Sea*, 146–47.

55. H. W. Menard to Bruce C. Heezen, 20 Mar 1957, and H. W. Menard to Bruce Heezen, 6 Jan 1958, HWMP, box 69, folder 6, "[For Ocean of Truth,] Heezen 1957–1984."

56. Hugh Odishaw and Stanley Ruttenberg, "Meeting of Technical Panel on Oceanography," 31 Oct 1956, IGYC, drawer 12, folder 31, "Oceanography 1955–1958."

57. Chief of Naval Research to Chief of Naval Operations, 18 Apr 1957, IGYC, drawer 12, folder 26, "Oceanography 1957."

58. Zenkevich, "Importance of Studies of the Ocean Depths," 67–70.

59. Sullivan, *Assault on the Unknown*, 362.

60. Lisitzin and Zhivago, "Marine Geological Work of the Soviet Antarctic Expedition, 1955–1957."

61. Edward Smith to George Deacon, 19 Aug 1955, and George Deacon to Edward Smith, 24 Aug 1955, GERDP, folder D12/4, "Correspondence, 1957." These letters evidently were misfiled.

62. Günther Böhnecke to George Deacon, 3 Dec 1955, GERDP, folder D12/3, "Correspondence, 1956."

63. Edward Smith to Georges Laclavère, 21 Aug 1956, GERDP, folder D12/3, "Correspondence, 1956." On a copy sent to George Deacon, Smith noted his doubts that anyone would want to join the Soviet plan.

64. Vladimir Beloussov to G. R. Laclavère, 14 Mar 1956, SIOSF, box 24, folder 55, "International Geophysical Year, 1955–1956."

65. G. Laclavère to the Chairman of the National Committees for the IGY, of the nations participating in Antarctic Operations during the IGY [Argentina, Australia, Chile, France, Great Britain, Japan, New Zealand, Norway, Union of South Africa, United States, and Soviet Union], 5 Aug 1956, SIOSF, box 24, folder 55, "International Geophysical Year, 1955–1956."

66. G. E. R. Deacon to G. Laclavère, 23 Aug 1956, SIOSF, box 24, folder 55, "International Geophysical Year, 1955–1956."

67. Hugh Odishaw to Walter M. Rudolph, Office of the Science Advisor, Department of State, 13 Sep 1957, IGYC, drawer 12, folder 26, "Oceanography 1957."

68. See Sullivan, *Assault on the Unknown*.

69. N. N. Sysoev to G. E. R. Deacon, 31 Mar 1956, SIOSF, box 24, folder 55, "International Geophysical Year, 1955–1956."

70. Hugh Odishaw and Stanley Ruttenberg, "Meeting of Technical Panel on Oceanography," 31 Oct 1956, IGYC, drawer 12, folder 31, "Oceanography 1955–1958."

71. Zaklinskii, "Deep Circulation of Water in the Indian Ocean."

72. Kaufman, *Trade and Aid*, 18.

73. Ibid., 52–53.

74. Rostow, *Eisenhower, Kennedy, and Foreign Aid*, 111–12.

75. John B. Hollister to Joseph Kaplan, 18 Jun 1956, IGYC, drawer 34, folder "International Co-operation Administration (ICO)."

76. Joseph Kaplan to John B. Hollister, 10 Jul 1956, IGYC, drawer 34, folder "International Co-operation Administration (ICO)."

77. Extract from minutes of Tenth Meeting of the United States National Committee, 13 Jul 1956, IGYC, drawer 34, folder "International Co-operation Administration (ICO)." Lloyd Berkner to Hugh Odishaw, 11 May 1956, IGYC, drawer 34, folder "International Co-operation Administration (ICO)."

78. Lloyd Berkner to Herbert Hoover, Jr., 17 Jul 1956, IGYC, drawer 34, folder "International Co-operation Administration (ICO)."

79. D. A. Fitzgerald to Lloyd Berkner, 21 Aug 1956, IGYC, drawer 34, folder "International Co-operation Administration (ICO)."

80. Kaufman, *Trade and Aid,* 63–68, 73.

81. H. S. Young, Ministry of Defence, Division of Scientific Intelligence, to Edward Bullard, 23 Mar 1956, ECBP, folder E.115.

82. Edward Bullard to H. S. Young, Ministry of Defence, 17 Apr 1956, ECBP, folder E.115.

83. Robert S. Dietz, "Meeting of the 13th International Limnological Congress with Notes on Finnish and Russian Oceanography," Technical Report ONRL-98–56, RSDP, box 5, esp. pp. 4–5.

84. G. L. Turney, Ministry of Defence, to George Deacon, 11 May 1956, and George Deacon to G. L. Turney, 14 May 1956, GERDP, folder D12/4, "Correspondence, 1957."

85. Robert S. Dietz, "Meeting of the 13th International Limnological Congress with Notes on Finnish and Russian Oceanography," Technical Report ONRL-98–56, RSDP, box 5, esp. pp. 17, 20–21.

86. BNCIGY, NGY/45(56), Lecture to the Royal Society of New Zealand (Wellington Branch) by Professor V. G. Kort of the Russian ship "OB" (Dr. E. Marsden FRS in the chair), 23 May 1956, GERDP, folder G5/3, "British National Committee for the International Geophysical Year, Correspondence and Papers, 1956."

87. Joseph Kaplan to Admiral Arleigh Burke, 1 Feb 1957, and Admiral Arleigh Burke to Joseph Kaplan, 8 Mar 1957, SIOSF, box 24, folder 55, "International Geophysical Year, 1955–1956."

88. Stanley Ruttenberg, Coordinating Secretary to Technical Panel on Oceanography, memorandum, 29 Mar 1957, SIOSF, box 24, folder 55, "International Geophysical year, 1955–56."

89. England, *A Patron for Pure Science,* 304.

90. Alan T. Waterman to Roger Revelle, 21 Feb 1957, and Alan T. Waterman to Joseph Kaplan, 21 Feb 1957, SIOSF, box 24, folder 56.

91. Roger Revelle, "The Scientist and the Politician," 1957 Charter Address, University of California, Riverside, 22 Mar 1957, RRP, MC 6, box 29, folder 1, esp. pp. 1, 7.

92. For details on Chinese participation, see FO 371/127420, "Chinese Participation in International Geophysical Year." Also see Bullis, *The Political Legacy of the International Geophysical Year,* 24.

93. Press Intelligence, Inc., "Soviet Ship Not Welcome at Brazil Port," *Buffalo*

Courier-Express, 11 Jan 1958 [clipping, no page number visible], IGYC, drawer 34, folder "USSR–Zarya Cruise."

94. Hugh Odishaw to J. A. Armitage, Soviet Desk Officer, Department of State, 30 Jan 1958, IGYC, drawer 34, folder "USSR–Zarya Cruise."

95. Richard H. Fleming to Roger Revelle, 17 Nov 1958, IGYC, drawer 12, folder 25.

96. See Divine, *The Sputnik Challenge.*

97. Eisenhower, *Waging Peace,* 209–10.

98. "Soviet Bloc International Geophysical Year Information," Department of Commerce, 25 Apr 1958, SIOSF, box 25, folder 23.

99. Wilson, *IGY: The Year of the New Moons,* 212–13.

100. Marian Oakleaf, National Academy of Sciences, memoranda to Files, 11 Dec 1958, 3 Feb 1959, 10 Mar 1959, IGYC, drawer 34, folder "Central Intelligence Agency." These memoranda summarize telephone conversations between Oakleaf and an unnamed CIA representative.

101. A. H. Shapley to James Morgan, Executive Director, USNC-IGY Offices, 1 Apr 1958, IGYC, drawer 34, folder "Central Intelligence Agency."

102. Sir Graham Sutton, report, "Visit to Moscow July 29th–August 10th 1958," FO 371/135399, "Soviet Participation in International Geophysical Year."

103. "Preliminary Report on Role of the Federal Government in International Science," National Science Foundation, Dec 1955, HPRC, folder 10.7.

104. Joseph Kaplan to Neil McElroy, Secretary of Defense, 31 Dec 1958, IGYC, drawer 34, folder "Department of Defense."

105. Ibid.

106. H. W. Menard to N. Zenkevich, 11 May 1959, HWMP, box 3, folder 2.

107. For these topics see McDougall, *The Heavens and the Earth,* and Geiger, "What Happened after Sputnik."

108. Bullis, *The Political Legacy of the International Geophysical Year,* 39.

109. United States Congress, *United States Postinternational Geophysical Year Scientific Program,* 20.

110. Ibid., 23.

111. Ibid., 28.

112. Ibid., 29.

113. Richard H. Fleming to Roger Revelle, 17 Nov 1958, IGYC, drawer 12, folder 25.

114. London, "Toward a Realistic Appraisal of Soviet Science," 176.

115. Wenk, *The Politics of the Ocean,* 62.

116. Gould, "The History of the Scientific Committee on Antarctic Research (SCAR)," 54.

117. Memorandum of Discussion at the Department of State–Joint Chiefs of Staff Meeting, Washington, D.C., 24 Jan 1958, in Glennon, *Foreign Relations of the United States, 1958–1960,* vol. 2, *United Nations and General International Matters,* 467–70.

118. Preliminary Notes on the Operations Coordinating Board Meeting, Washington, D.C., 10 Dec 1958, in ibid., 2:514.

119. Report of the Operations Coordinating Board, 21 Jan 1959, in ibid., 2:522, 530–32.

4 / THE NEW FACE OF INTERNATIONAL OCEANOGRAPHY

1. Ovey, "Forward," 1.
2. Wiseman, "International Collaboration in Deep-Sea Research," 3.
3. Wolff, "The Creation and First Years of SCOR (Scientific Committee on Oceanic Research)," 338–39.
4. Unesco, "The Unesco Marine Science Programme, an Outline of Activities," 13 Aug 1957, UNESCOR, Unesco/NS/OCEAN/64. The other members were D. V. Bal (India), Luis Howell Rivero (Cuba), and D. J. Rochford (Australia).
5. George Deacon to M. N. Hill, 29 Jun 1956, GERDP, folder H3/2, "Scientific Committee on Oceanographic Research: SCOR; Correspondence and Papers, 1956–58."
6. Unesco, "The Unesco Marine Science Programme, an Outline of Activities," 13 Aug 1957, UNESCOR, Unesco/NS/OCEAN/64.
7. Unesco, "Meeting of Experts to Consider the Terms of Reference and Mode of Operation of the International Advisory Committee on Marine Sciences," Rome, FAO's Headquarters, 9–10 May 1955, Communication Received from Dr. G. E. R. Deacon, Director of the National Institute of Oceanography of the United Kingdom, 29 Apr 1955, UNESCOR, Unesco/NS/OCEAN/15.
8. George Deacon to Georges Laclavère, 29 Apr 1955, GERDP, folder H1/1a, "Unesco International Advisory Committee on Marine Sciences, Correspondence 1952–61."
9. Unesco, International Advisory Committee on Marine Sciences, First Session (1956), Letter from Norway in Response to Circular Letter Inviting Proposals, Oslo, 11 Sep 1956, UNESCOR, Unesco/NS/OC/40.
10. Unesco, International Advisory Committee on Marine Sciences, Circular N. 3 to Members, 26 Jun 1956, UNESCOR, Unesco/NS/OCEAN/CIRC/3.
11. Unesco, Marine Sciences Programme, "Report of Dr. Anton Fr. Bruun, University of Copenhagen, and Member of the International Advisory Committee on Marine Sciences, on a Voyage to the Countries around the South China Sea and Adjacent Seas, surveying their efforts in Marine Research," 6 Jun 1957, UNESCOR, Unesco/NS/OCEAN/61.
12. Ibid.
13. Ibid.
14. Unesco, "The Unesco Marine Science Programme, an Outline of Activities," 13 Aug 1957, UNESCOR, Unesco/NS/OCEAN/64.
15. Yoshida to Anton Bruun, 5 Jun 1957, GERDP, folder H1/1b, "Unesco International Advisory Committee on Marine Sciences, Correspondence 1952–61."
16. J. N. Carruthers, "Unesco and Oceanography," Casablanca, 1961, UNESCOR, box 551.46 A06 (64) "61."

17. Ibid.

18. Unesco, International Advisory Committee on Marine Sciences, Third Session, 24–30 Sep 1958, Annotations to the Provisional Agenda, 18 Sep 1958, UNESCOR, Unesco/NS/OCEAN/86. Unesco, Marine Sciences Programme, "Outline of Programme Proposed by IACOMS for a Major Project in the Marine Sciences," 20 Mar 1958, UNESCOR, Unesco/NS/OCEAN/77.

19. Unesco, Marine Sciences Programme, "Proposed International Oceanographic Vessel," 5 Jun 1958, UNESCOR, Unesco/NS/OCEAN/78.

20. Ibid.

21. Unesco, Natural Sciences Department, "Special Long-Term Project for Biological and Physical Oceanographic Investigation of South East Asian Seas," 8 Jan 1959, UNESCOR, Unesco/NS/OCEAN/88.

22. G. E. R. Deacon, "Proposal for Co-operation in Marine Science in Southeast Asia," 6 Mar 1959, GERDP, H1/1C, "Unesco International Advisory Committee on Marine Sciences, Correspondence 1952–61."

23. See Rainger, "Constructing a Landscape for Postwar Science."

24. L. Zenkevich to Yoshida, Mar 1956, "Proposals for Marine Biological Investigations during the International Geophysical Year," GERDP, folder H4/2, "Recommendations by IAPO to IACOMS meeting in Lima, 1956." For Revelle's comments, see M. N. Hill, "Abstract of the Discussion at the First Meeting of the Bureau of the Special Committee on Oceanic Research," 18 Jan 1957, NASCPF 1957–61, folder "ADM: IR: IU: ICSU: SCOR: Bureau: Meetings: 1st."

25. Unesco, International Advisory Committee on Marine Sciences, Third Session, 24–30 Sep 1958, 9 Sep 1958, UNESCOR, Unesco/NS/OCEAN/83.

26. Ibid.

27. George Deacon to B. Kullenberg, 3 Dec 1956, GERDP, folder H4/1, "International Association of Physical Oceanographers, IAPO: Correspondence, 1956–67."

28. George Deacon to C. H. Mortimer, Scottish Marine Biological Association, 18 Feb 1957, GERDP, folder G3/1, "British National Committee for Geodesy and Geophysics, Subcommittee for Physical Oceanography (Physical Science of the Ocean Subcommittee), Correspondence and Papers, 1953–79." Also see Doodson to George Deacon, 11 Jan 1957, and George Deacon to D. C. Martin, 22 Jan 1957, GERDP, folder H3/2, "Scientific Committees on Oceanographic Research: SCOR; Correspondence and Papers, 1956–58."

29. For Coulomb's comments, see Unesco, International Advisory Committee on Research in the Natural Sciences Programme of Unesco, Fifth Session, Moscow, 6–8 May 1958, Report of the Secretariat, UNESCOR, Unesco/NS/IAC/79, and Sixth Session, Giessen, 20–24 Apr 1959, Draft Report of the Secretariat, UNESCOR, Unesco/NS/IAC/91.

30. M. N. Hill, "Abstract of the Discussion at the First Meeting of the Bureau of the Special Committee on Oceanic Research," 18 Jan 1957, NASCPF 1957–61, folder "ADM: IR: IU: ICSU: SCOR: Bureau: Meetings: 1st."

31. Sir Harold Spencer Jones to L. V. Berkner, 4 Mar 1957, NASCPF 1957–61, folder "ADM: IR: IU: ICSU: SCOR: Beginning of Program."

32. L. V. Berkner to W. W. Atwood, Jr., Office of International Relations, National Research Council, 8 Mar 1957, NASCPF 1957–61, folder "ADM: IR: IU: ICSU: SCOR: Beginning of Program."

33. L. V. Berkner to Sir Harold Spencer Jones, 25 Mar 1957, NASCPF 1957–61, folder "ADM: IR: IU: ICSU: SCOR: Beginning of Program."

34. Roger Revelle to M. N. Hill, 12 Feb 1957, GERDP, folder H3/2, "Scientific Committees on Oceanographic Research: SCOR; Correspondence and Papers, 1956–58."

35. These statements come from three letters by Deacon. See George Deacon to A. Shavitsky, 27 Feb 1957, George Deacon to M. N. Hill, 20 Mar 1957, and George Deacon to R. Fraser, 2 Apr 1957, GERDP, folder H3/2, "Scientific Committees on Oceanographic Research: SCOR; Correspondence and Papers, 1956–58."

36. Wolfle, *Renewing a Scientific Society*, 216–18.

37. Anonymous, "Khrushchev, Stressing Peace, Plans No Military Aides on Trip," 6.

38. Anonymous, "US Extends Ban on Atomic Tests," 1.

39. RRI-84, 104.

40. Wolfle, *Renewing a Scientific Society*, 219–20.

41. RRI-84, 103.

42. Harrison Brown to Detlev W. Bronk, 28 Jul 1959, NASCPF 1957–61, folder "IR: International Congresses: Oceanography: First NY 1959."

43. R. C. Vetter to S. D. Cornell, 29 Jul 1959, NASCPF 1957–61, folder "IR: International Congresses: Oceanography: First NY 1959."

44. Arifin Bey to Dr. Atwood, 25 Sep 1959, NASCPF 1957–61, folder "IR: International Congresses: Oceanography: First NY 1959."

45. Behrman, *Assault on the Largest Unknown*, 12. More details about the origin of the project, the Indian Ocean's characteristics, and the scientific results of the IIOE may be found in Rao and Griffiths, *Understanding the Indian Ocean*.

46. Ronald Fraser, "SCOR: Report by the Administrative Secretary," Mar 1958, NASCPF 1957–61, folder "ADM: IR: IU: ICSU: SCOR: Reports: Report to ICSU."

47. T. S. Satyanarayana Rao to Yoshida, 9 Jan 1958, GERDP, folder H1/1c, "Unesco International Advisory Committee on Marine Sciences, Correspondence 1952–61."

48. Behrman, *Assault on the Largest Unknown*, 16.

49. Robert G. Snider, untitled report, 8 Oct 1959, NASCPF 1957–61, folder "SCOR 1959."

50. Behrman, *Assault on the Largest Unknown*, 92.

51. RRI-84, 95. The next line in the interview transcription is "Not very bright."

52. Robert G. Snider, journal entry for Japan, 5 Feb 1960, NASCPF 1957–61, folder "SCOR: International Trip to Survey Participants 1960."

53. Snider, journal entry for Ceylon, 17 Feb 1960.

54. Snider, journal entry for Pakistan, 24 Feb 1960.

Notes to Pages 123–135　　287

55. Snider, journal entry for Japan, 5 Feb 1960.
56. Snider, journal entry for Indonesia, 11 Feb 1960.
57. Snider, journal entry for Singapore, 11 Feb 1960.
58. Snider, journal entry for Ceylon, 17 Feb 1960.
59. Snider, journal entries for Indonesia and Ceylon, 11 Feb 1960 and 17 Feb 1960.
60. Snider, journal entry for India, 17 Feb 1960.
61. Ibid.
62. Snider, journal entry for USSR, 11 Mar 1960.
63. Behrman, *Assault on the Largest Unknown*, 37.
64. Noyes, "Do We Need a Foreign Policy in Science?" 237.
65. Barron, "Why Do Scientists Read Science Fiction?"
66. Greenberg, "Science and Foreign Affairs," 123.
67. RRI-84, 105.
68. Ibid., 104–8.
69. Brode, "National and International Science," 736.
70. Stevenson, "Science, Diplomacy, and Peace," 405.
71. Rusk, "Building an International Community of Science and Scholarship," 625, 626.
72. Brode, "The Role of Science in Foreign Policy Planning," 274.
73. See Kistiakowsky, "Science and Foreign Affairs."
74. Gilbert L. Voss to John Ryther, 28 Nov 1960, NASCPF 1957–61, folder "SCOR: IIOE: USNC: Working Groups: Biology."
75. Ibid.
76. Behrman, *Assault on the Largest Unknown*, 34.
77. Lev Zenkevich to Robert Snider, 3 May 1960, GERDP, folder H3/19, "Scientific Committees on Oceanographic Research: SCOR: Correspondence on the Indian Ocean Expedition, 1957–67."
78. G. E. R. Deacon to Marine Sciences Panel, Royal Society Unesco Committee, 18 Mar 1959, GERDP, H1/1C, "Unesco International Advisory Committee on Marine Sciences, Correspondence 1952–61."
79. G. E. R. Deacon to Günther Böhnecke, 3 Jun 1959, GERDP, H3/3, "Scientific Committees on Oceanographic Research: SCOR; Correspondence and Papers, 1959–67."
80. Gene LaFond, letter, *Indian Ocean Bubble*, no. 4 (13 Jul 1959), ODWHOI, box 24, folder "Pers: Stommel, Henry."
81. G. E. R. Deacon to D. C. Martin, Royal Society, 20 Apr 1961, GERDP, G1/1, British National Committee for Oceanic Research, Correspondence 1959–69.
82. Kazuhiko Terada, Chief, Marine Division, Japan Meteorological Agency, to Warren Wooster, 23 Jun 1962, UNESCOR, box 551.46 A02 IOC "-66," pt. 2.
83. Unesco, Preparatory Meeting of the Intergovernmental Conference on Oceanographic Research, 21–29 Mar 1960, Research Programme, 21 Mar 1960, UNESCOR, Unesco/NS/OCEAN/92.

84. Edward Miles, "COSPAR and SCOR: The Political Influence of the Committee on Space Research and the Scientific Committee on Oceanic Research," 133–52, esp. 141–42.

85. Warren Wooster to D. V. Villadolid, Biological Research Center, National Institute of Science and Technology, Manila, 25 Jul 1961, UNESCOR, box 551.46 A02 IOC "-66," pt. 1.

86. Intergovernmental Oceanographic Commission, *Intergovernmental Oceanographic Commission (Five Years of Work)*, 5.

87. Item 5(f), Paris, 26 Oct 1961, translated from the Spanish, UNESCOR, Unesco/NS/IOC/Res.

88. Intergovernmental Oceanographic Commission, *Intergovernmental Oceanographic Commission (Five Years of Work)*, 6.

89. W. H. Menard to Harrison Brown, 11 Mar 1963, and Lawrence C. Mitchell to H. W. Menard, 26 Mar 1963, HWMP, box 16, f. 3, "OST 1962–1966 Russia [Office of Science and Technology]."

90. RRI-84, 96.

91. Ibid., 89.

92. Intergovernmental Oceanographic Commission, *Intergovernmental Oceanographic Commission (Five Years of Work)*, 11.

5 / COMPETITION AND COOPERATION IN THE 1960S

1. "The Next Ten Years in Oceanography (A Survey of the Growth Potential at Existing Institutions)," by Gordon G. Lill, A. E. Maxwell, and F. D. Jennings, ODWHOI, box 22, file "ONR."

2. Arleigh Burke, Chief of Naval Operations, to Distribution List, 1 Jan 1959, ODWHOI, box 22, file "ONR."

3. "The Next Ten Years in Oceanography (A Survey of the Growth Potential at Existing Institutions)," by Lill, Maxwell, and Jennings, ODWHOI, box 22, file "ONR." The other existing institutions described in the report were University of Miami, University of Washington, Texas Agricultural and Mechanical College, Chesapeake Bay Institute, Narragansett Marine Laboratory, Oregon State College, and New York University.

4. Ibid.

5. Ibid.

6. ASI-89, 83, 85.

7. Ibid., 100.

8. Wenk, *The Politics of the Ocean*, 51.

9. National Academy of Sciences–National Research Council, *Oceanography 1960 to 1970: A Report by the Committee on Oceanography,* chap. 10, "International Cooperation," 1–8.

10. NASCO called these "Extremes of Estimate Increases in Human Population

by 1980" (in millions): Canada, 6–11; Europe (excluding Soviet Union), 60–87; Soviet Union, 85–100; Middle America, 43–57; tropical South America, 72–95; temperate South America, 12–15; Asia (excluding Asian parts of Soviet Union), 710–990; Oceania, 7–8. The total for the world was 1,160–1,590. See National Academy of Sciences–National Research Council, *Oceanography 1960 to 1970*, chap. 3, "Ocean Resources," 2.

11. Ibid.

12. Ibid., chap. 1, "Introduction and Summary of Recommendations," 27–28.

13. Ibid., chap. 2, "Basic Research in Oceanography during the Next Ten Years," 19–21.

14. Ibid., chap. 1, "Introduction and Summary of Recommendations," 4.

15. Wenk, *The Politics of the Ocean*, 65. Wenk's assessment was based upon his conversations with Kistiakowsky.

16. Hubert M. Humphrey, "Importance of Studies in Oceanography," 17 Feb 1959, *Congressional Record, Proceedings and Debates of the 86th Congress, First Session*, ODWHOI, box 23, folder "Act. Subcommittee on Oceanography, House of Representatives."

17. Ibid.

18. U.S. Congress, *Oceanography in the United States*, 3–5.

19. U.S. Congress, *Education in the Field of Oceanography*, 35.

20. Athelstan Spilhaus to Wakelin, 18 Aug 1960, OSTSF, box 182, folder "Letter to Dr. Wakelin, Jr., from Athelstan Spilhaus re his review of the national program in oceanography 8/18/60." See also Wenk, *The Politics of the Ocean*, 66.

21. Wenk, *The Politics of the Ocean*, 59. The Merchant Marine and Fisheries Committee conducted preliminary hearings on the NASCO report, under the auspices of its Special Subcommittee on Oceanography. Eventually it retained jurisdiction. But in 1959 and 1960 the Science and Astronautics Committee tried to put oceanography under its control. Wenk claimed that in 1966 he spoke with senior Soviet officials, only to discover that the Soviet Union had been far from committed to a major thrust in oceanography at this time (see ibid., 532 n. 27).

22. Ibid., 55.

23. Paul M. Fye to Gordon Lill, 29 Oct 1958, ODWHOI, box 22, folder "Dir: Activities: ONR, Washington."

24. Paul M. Fye to F. Joachim Weyl, Research Director, Office of Naval Research, 9 Dec 1958, ODWHOI, box 22, file "ONR."

25. William S. von Arx to Director and Members of the Scientific Policy Committee, Jun 1960, ODWHOI, box 30, folder WHOI Policy Review 2 of 2.

26. The Director [Paul M. Fye], Memorandum for the Executive Committee, "Policies for the Woods Hole Oceanographic Institutions in the 60's," 20 Jan 1961, ODWHOI, box 30, folder WHOI Policy Review 1 of 2.

27. Ibid.

28. William S. von Arx to Paul M. Fye, 2 Feb 1961, ODWHOI, box 30, folder WHOI Policy Review 2 of 2.

29. Allyn C. Vine to Paul Fye, 20 Mar 1961, ODWHOI, box 30, folder WHOI Policy Review 2 of 2.

30. E. Bright Wilson, Jr., to Paul Fye, 8 Feb 1961, ODWHOI, box 30, folder WHOI Policy Review 2 of 2.

31. Henry Stommel to Paul Fye, 13 Feb 1961, ODWHOI, box 30, folder WHOI Policy Review 2 of 2.

32. Ibid.

33. Ibid.

34. Gordon Lill to Jerome B. Wiesner, 18 Dec 1962, OSTSF, box 161, folder "Oceanography—Title Folder."

35. U.S. Congress, *Oceanography in the United States*, 44–53.

36. U.S. Congress, *Marine Science*, 97–98.

37. U.S. Congress, *Oceanography in the United States*, 131.

38. U.S. Congress, *Education in the Field of Oceanography*, 34.

39. U.S. Congress, *Oceanography in the United States*, 170.

40. On the genesis of the Interagency Committee on Oceanography, see Wenk, *The Politics of the Ocean*, 62–66.

41. U.S. Congress, *Oceanography 1961—Phase 1*, 2.

42. Chairman, Ocean Survey Advisory Panel, to Chairman, ICO, 13 Jul 1961, and "United States National Oceanographic Program Fiscal Year 1963," n.d., OSTSF, box 181, folder "Oceanography 1961–1962."

43. Jerome B. Wiesner to Cecil H. Green, 9 Nov 1962, OSTSF, box 161, folder "Oceanography—Title Folder."

44. ASI-89, 95.

45. Ibid., 96. When asked if he knew how W. Maurice Ewing felt about such international projects, Spilhaus responded, "I never knew what Maurice Ewing thought. I thought he was a good scientist and a different person and I tried to keep from talking to him as much as possible."

46. U.S. Congress, *Marine Science*, 84.

47. WHMI-98, 34.

48. U.S. Congress, *Marine Science*, 85.

49. WHMI-97, 1.

50. WHMI-98, 62.

51. London, "Toward a Realistic Appraisal of Soviet Science," 169–73, 176.

52. Ivanov, "The Work of the Non-magnetic Vessel *Zarya* in the Pacific Ocean."

53. Ibid.

54. Bogorov, "Marine Scientific Organizations in the Indian Ocean."

55. *Ocean Science News* 3 (10 Mar 1962): 3, OSTSF, box 180, folder "ICO Corr-1962 File #1."

56. Turkevich, "Soviet Science Appraised," 493.

57. ASI-76, 57.
58. Harry Hess to George Kistiakowsky, 15 Feb 1960, HHHP, box 12, folder "AMSOC-MOHOLE General Correspondence 1960."
59. Willard Bascom to Editor, *New York Times,* 18 Sep 1961, HHHP, box 12, folder "AMSOC-MOHOLE General Correspondence 1961, Jul–Dec."
60. William L. Petrie to Gordon Lill, 27 Jun 1961, HHHP, box 12, folder "AMSOC-MOHOLE General Correspondence 1961, Jan–Jun."
61. Anonymous, "Soviet Considers 5 Earth Debates," 13.
62. Harry Hess to G. Medina, 18 Sep 1961, HHHP, box 12, folder "AMSOC-MOHOLE General Correspondence 1961, Jul–Dec."
63. Willard Bascom to Editor, *New York Times,* 18 Sep 1961, HHHP, box 12, folder "AMSOC-MOHOLE General Correspondence 1961, Jul–Dec."
64. Harry Hess to Merle Tuve, 14 May 1958, HHHP, box 12, folder "American Miscellaneous Society."
65. Gordon Lill to Editor, *Science,* 10 Jul 1961, HHHP, box 12, folder "AMSOC-MOHOLE General Correspondence 1961, Jul–Dec."
66. Willard Bascom to Richard Vetter, 9 Nov 1961, HHHP, box 12, folder "AMSOC-MOHOLE General Correspondence 1961, Jul–Dec."
67. Anonymous, "Race to Inner Space," 15–17.
68. Ralph S. O'Leary, "Project Mohole," *Houston Post,* 22 Jul 1962, [clipping, no page number visible], MPC, folder "ES: AMSOC Com: Mohole Project: Soviet Competition 1961–1963."
69. Willard Bascom to Richard Vetter, 9 Nov 1961, HHHP, box 12, folder "AMSOC-MOHOLE General Correspondence 1961, Jul–Dec."
70. Bascom, *A Hole in the Bottom of the Sea,* 324.
71. Detlev Bronk to Harry Hess, 27 Nov 1961, HHHP, box 12, folder "AMSOC-MOHOLE General Correspondence 1961, Jul–Dec."
72. Willard Bascom to Richard Vetter, 9 Nov 1961, HHHP, box 12, folder "AMSOC-MOHOLE General Correspondence 1961, Jul–Dec."
73. William Petrie to Almirante Octacillo Cuñha, Presidente, Conselho Nacional de Pesquisas, 15 Feb 1962, HHHP, box 12, folder "AMSOC-MOHOLE General Correspondence 1962, Jan–Mar."
74. White House press release, 19 Mar 1964, OSTSF, box 391, folder "Oceanography 1964."
75. Greenberg, *The Politics of Pure Science,* chap. 9.
76. Robert J. Uffen, "Canadian Scientific Committee for the Upper Mantle Project," n.d., OSTSF, box 353, folder "International—Upper Mantle 1964."
77. International Council of Scientific Unions, "Upper Mantle Project," 14 Oct 1963, NASCPF 1962–65, folder "ADM: IR: IU: Geodesy and Geophysics: Com on Upper Mantle."
78. G. D. Garland, "Report on Upper Mantle Project," 11 Feb 1965, NASCPF

1962–65, folder "ADM: IR: IU: Geodesy and Geophysics: Upper Mantle Project: Report: Report to ICSU."

79. Minutes of the Meeting of the Upper Mantle Committee, 29 Feb 1964, OSTSF, box 353, folder "International—Upper Mantle 1964."

80. "Relations of IUGG with Developing Countries," [no author], 1964, NASCPF 1962–65, folder "ADM: IR: IU: Geodesy and Geophysics 1964."

81. William L. Petrie to George P. Woollard, 19 Apr 1962, MPC, folder "ES: AMSOC Com: Mohole Project: Soviet Competition 1961–1963."

82. R. Rollefson to Frederick Seitz, 8 Jan 1963, MPC, folder "ES: AMSOC Com: Mohole Project: UMC."

83. M. A. Tuve to J. B. Wiesner, 30 Jan 1963, MPC, folder "ES: AMSOC Com: Mohole Project: UMC."

84. Alan T. Waterman to Jerome B. Wiesner, 29 Jan 1963, MPC, folder "ES: AMSOC Com: Mohole Project: UMC."

85. NAS-NRC Geophysics Research Board to Members of the U.S. Upper Mantle Committee, 12 Dec 1963, MPC, folder "ES: AMSOC Com: Mohole Project: UMC."

86. Minutes of the Meeting of the Upper Mantle Committee, 29 Feb 1964, OSTSF, box 353, folder "International—Upper Mantle 1964."

87. Johnson is quoted in Technical Analysis Office, Hughes Aircraft Company, "Study of Oceanography for Reducing International Tensions in an Arms Control and Disarmament Environment," OSTSF, box 391, folder "Oceanography 1964."

88. Turkevich, "Soviet Science Appraised," 492.

89. WHMI-97, 104–5.

90. "Soviet Oceanographic Fleet," n.d., OSTSF, box 180, folder "Oceanography-ICO."

91. Ibid.

92. Harris B. Stewart, Jr., to Paul Fye, 12 Aug 1963, ODWHOI, box 36, folder "Act. ICO 2 of 3." Stewart, a Coast and Geodetic Survey official, used this term to describe the ICO members and others in Washington with a policy-oriented interest in marine science.

93. Intergovernmental Oceanographic Commission, First Session, Paris, Unesco, 19–27 Oct 1961, Proposals of the United States of America, UNESCOR, NS/IOC/INF-6.

94. Statement of Dr. Arthur E. Maxwell, Head, Geophysics Branch, Office of Naval Research, and Chairman of the Interagency Committee on Oceanography Panel on International Programs, before the Subcommittee on Oceanography of the House Committee on Merchant Marine and Fisheries, 1 Mar 1962, ODWHOI, box 32, folder "Act. [ICO] 1962 3 of 3."

95. A. E. Maxwell to Members of PIPICO, 8 Feb 1962, ODWHOI, box 32, folder "Act. [ICO] 1962 3 of 3."

96. H. Arnold Karo to Vernon Brock, Coordinator, Tropical Atlantic Investigation, 1 Mar 1962, ODWHOI, box 31, folder "Act. Equalant I and II."

97. Vice Admiral Day to F. T. Hallet, Office of the Minister of Science, 15 Apr 1962, CAB 124/2166, "Working Group on Oceanography."

98. Programme for International Oceanographic Research in the Northern Sections of the Atlantic and Pacific Ocean in 1964–66 (presented by the Soviet National Oceanographic Committee to the meeting of the Bureau of the Intergovernmental Oceanographic Commission held in Paris, 10–12 Apr 1962), 11 Apr 1962, UNESCOR, NS/IOC/INF-20.

99. Vice Admiral Archibald Day, "IOC Bureau and Consultative Committee, 10th to 12th April 1962," n.d., CAB 124/2166, "Working Group on Oceanography."

100. Milner B. Schaefer to Richard C. Vetter, 23 Jul 1962, ODWHOI, box 32, folder "Act. [ICO] 1961, 1962."

101. H. Arnold Karo to James H. Wakelin, Jr., 20 Aug 1962, OSTSF, box 180, folder "ICO Corr-1962 File #2."

102. "Intergovernmental Oceanographic Commission, Second Session, Paris, Sep 20–29, 1962, Position Paper, Dynamic Study of the Northern Oceans," 16 Aug 1962, UNESCOR, NS/OCEAN(62).

103. Royal Society, British National Committee on Oceanic Research, "Memorandum on USSR Proposals for International Oceanographic Research in the Northern Atlantic and Pacific Oceans 1964–66," NOR/17(62), CAB 124/2173, "Working Group on Oceanography: Papers."

104. H. Arnold Karo to James H. Wakelin, Jr., 20 Aug 1962, OSTSF, box 180, folder "ICO Corr-1962 File #2."

105. John Swallow to Warren Wooster, 27 Jul 1962, UNESCOR, box 551.46 A02 IOC "-66," pt. 3.

106. K. N. Fedorov to G. F. Humphrey, 8 Nov 1962, and G. F. Humphrey to K. N. Fedorov, 15 Nov 1962, UNESCOR, box 551.46 A02 IOC "-66," pt. 3.

107. W. M. Cameron to Vice Admiral V. A. Tchekourov, 23 Jul 1962, UNESCOR, box 551.46 A02 IOC 025 "-66," pt. 1.

108. Royal Society, "British National Committee for Oceanic Research," Report of Informal Meeting Called to Consider Ways of Improving Liaison with the Government Interdepartmental Working Group on Oceanography, NOR/2(63), n.d., CAB 124/2166, "Working Group on Oceanography."

109. W. M. Cameron to Warren Wooster, 1 Aug 1962, UNESCOR, box 551.46 A02 IOC 025 "-66," pt. 1.

110. W. M. Cameron to Vice Admiral V. A. Tchekourov, 23 Jul 1962, UNESCOR, box 551.46 A02 IOC 025 "-66," pt. 1.

6 / OCEANOGRAPHY, EAST AND WEST

1. IOC Secretariat, Item 3 of the Provisional Agenda, Second Session of Bureau meeting, Paris, 20–29 Sep 1962, 31 Jul 1962, CAB, 124/2173 "Working Group on Oceanography: Papers."

2. William S. von Arx, "A Science in Bondage," Feb 1965, ODWHOI, folder "Personnel: 'V' 1966."

3. E. Ll. Evans, Office of the Minister of Science, to Warren Wooster, 28 Feb 1963, and Warren Wooster to E. Ll. Evans, Office of the Minister of Science, 6 Mar 1963, CAB, 124/2162, "International Indian Ocean Expedition."

4. Royal Society, "British National Committee for Oceanic Research," report of informal meeting called to consider ways of improving liaison with the Government Interdepartmental Working Group on Oceanography, NOR/2 (63), n.d., CAB, 124/2168, "Working Group on Oceanography."

5. J. G. Liverman to R. N. Quirk, 8 Jan 1963, CAB, 124/2168, "Working Group on Oceanography."

6. G. E. R. Deacon to D. C. Martin, 16 Jan 1963, GERDP, G1/1, "British National Committee for Oceanic Research, Correspondence 1959–69."

7. G. E. R. Deacon to Admiral Archibald Day, 2 Jan 1963, GERDP, H2/3, "Intergovernmental Oceanographic Commission: Correspondence and Papers, 1963."

8. T. Garrett, "Visit by the UK Scientific Attaché, Moscow, to Professor V. G. Kort, Institute of Oceanology, Moscow, Notes on the Discussion," n.d., CAB, 124/2174, "Working Group on Oceanography: Papers."

9. G. E. R. Deacon to J. G. Liverman, 16 Apr 1963, CAB, 124/2168, "Working Group on Oceanography."

10. Ibid.

11. Ibid.

12. Ibid.

13. G. E. R. Deacon to Warren Wooster, 18 Apr 1963, GERDP, H2/3, "Intergovernmental Oceanographic Commission: Correspondence and Papers, 1963."

14. Ibid.

15. Vice Admiral Archibald Day, report, "IOC Bureau and Consultative Committee meeting—Moscow, 6–8th May 1963," n.d., CAB, 124/2168, "Working Group on Oceanography."

16. Japanese Oceanographic Data Center, Hydrographic Division, Maritime Safety Agency, Tokyo, Newsletter, Cooperative Study of the Kuroshio, no. 1, Jul 1965, CAB, 124/2177, "Working Group on Oceanography: Papers."

17. Y. Takenouti to K. Fedorov, 29 Jan 1965, "First Meeting of the International Coordinating Group for the CSK," UNESCOR, box 551.46(265.5) A53/02/06–1,2,3, Cooperative Study of the Kuroshio.

18. Intergovernmental Oceanographic Commission, Meeting of the International Coordination Group for the International Indian Ocean Expedition, Paris, 22–24 Jan 1964, CAB, 124/2175, "Working Group on Oceanography: Papers."

19. Roger Revelle to Edward C. Bullard, 25 Feb 1964, ECBP, folder C.18.

20. Edward C. Bullard to Roger Revelle, 29 Feb 1964, ECBP, folder C.18.

21. G. E. R. Deacon to D. C. Martin, 7 Oct 1964, GERDP, G1/1, "British National Committee for Oceanic Research, Correspondence, 1959–69."

Notes to Pages 187–191

22. G. E. R. Deacon to R. W. J. Keay, Royal Society, 11 Oct 1965, GERDP, G17/1, "Committee on International Scientific Co-operation/Committee on Overseas Scientific Relations, CISC Working Group on Oceanography: Correspondence and Papers, 1961–66."

23. "A Review of the Instrumented Buoys Project," NATO Sub-Committee on Oceanography, n.d., submitted by Working Group, 21 May 1964, CAB, 124/2175, "Working Group on Oceanography: Papers."

24. S. Byard, Ministry of Defence, to M. J. Tucker, National Institute of Oceanography, 14 Oct 1966, and M. J. Tucker to S. Byard, 21 Oct 1966, GERDP, D5/2, 1958–66."

25. A. J. Lee to D. C. Martin, 15 Nov 1968, GERDP, G1/2, "British National Committee for Oceanic Research, Correspondence 1966–69."

26. Edward C. Bullard to A. Potts, Ministry of Defence, 29 Oct 1968; Edward C. Bullard to D. C. Martin, 29 Oct 1968; and D. C. Martin to Edward C. Bullard, 8 Nov 1968, ECBP, folder E.135.

27. G. E. R. Deacon to Mrs. V. L. Hopper, Natural Environment Research Council, 23 Aug 1968; A. S. Laughton to G. E. R. Deacon, 21 Aug 1968; and A. Potts to Edward C. Bullard, 22 Nov 1968, ECBP, folder E.135.

28. Roll, *Intergovernmental Oceanographic Commission (IOC)*, 38.

29. Paper by the Admiralty for use of Vice Admiral Sir A. Day at meeting of the Intergovernmental Oceanographic Commission Bureau in Apr 1962, Instructions for the IOC representatives at the IHB meeting in May 1962, n.d., CAB, 124/2172, "Working Group on Oceanography: Papers."

30. H. H. Hess to Milner B. Schaefer, 13 Jul 1964, ODWHOI, box 39, folder "Act. [ICO] 1964."

31. Arthur E. Maxwell to Director, U.S. Coast and Geodetic Survey [H. Arnold Karo], 19 Aug 1964, ODWHOI, box 39, folder "Act. [ICO] 1964."

32. Committee on International Scientific Co-operation, Working Group on Oceanography, Minutes, Fourth Meeting, on 14 Nov 1962, 26 Nov 1962, CAB, 124/2170, "Working Group on Oceanography: Meetings." See also Roll, *Intergovernmental Oceanographic Commission (IOC)*, 46. Udintsev's chart is mentioned in many of the surveys conducted by the National Academy of Sciences after the Second Oceanographic Congress. See NASCPF 1966–72, folder "IR: International Congresses: Oceanography: Second: Moscow."

33. Vice Admiral Archibald Day, report, "IOC Bureau and Consultative Committee meeting—Moscow, 6–8th May 1963," n.d., CAB, 124/2168, "Working Group on Oceanography."

34. J. R. Rossiter, Director, Permanent Service for Mean Sea Level, IAPO, to Cmdr. D. P. D. Scott, Hydrographic Department, Admiralty, 3 Jan 1964, CAB, 124/2169, "Working Group on Oceanography."

35. These observations were made to the author on 10 May 2002 by Russian marine scientist Alexei Suzyumov, who was a student at the Institute of Oceanology during the 1960s.

36. Arthur E. Maxwell, "Minutes of 18 December 1964 Meeting, ICO Panel on International Programs," 5 Jan 1965, ODWHOI, folder "Act. [ICO] 1964."

37. Wertenbaker, *The Floor of the Sea*, 147.

38. Udintsev to Menard, n.d., HWMP, box 3, folder 3, "Correspondence, 1960."

39. Menard to Udintsev, 25 Feb 1960, HWMP, box 3, folder 2, "Correspondence, 1959."

40. Menard, *The Ocean of Truth*, 59–60.

41. Ibid., 111.

42. Ibid., 58–63.

43. Ibid., 90, 111.

44. "Interview with Sir Edward Crisp Bullard of Cambridge University," at U.S. Naval Institute, Annapolis, Maryland, 12 Dec 1969, by Glen B. Ruh, ECBP, folder "A.8."

45. See Loren R. Graham's introduction in Graham, *Science and the Soviet Social Order*. For a Marxist perspective, see Lecourt, *Proletarian Science?* 102.

46. Beloussov, "An Open Letter to J. Tuzo Wilson," 17.

47. Graham, *What Have We Learned about Science and Technology from the Russian Experience?* 93–97.

48. Hess, "History of the Ocean Basins," 599.

49. Salop and Scheinmann, "Tectonic History and Structures of Platforms and Shields."

50. Laudan, *Beyond Positivism and Relativism*, 240.

51. See Vine and Matthews, "Magnetic Anomalies over Oceanic Ridges."

52. For a discussion of the results of the *Eltanin* expedition, see Glen, *The Road to Jaramillo*.

53. "'Drifting Continents' Theory Attacked: No Evidence, Says Sir Harold," *Guardian*, 26 Oct 1963, 4, JTWP, B86–0066/019, "S. K. Runcorn."

54. J. T. Wilson to S. K. Runcorn, 5 Nov 1963, JTWP, B86–0066/019, "S. K. Runcorn."

55. These reflections on Beloussov's influence are drawn from a conversation on 10 May 2002 between the author and Alexei Suzyumov, who was a student at the Institute of Oceanology in the 1960s.

56. See Stewart, *Drifting Continents and Colliding Paradigms*, for a comprehensible account of Beloussov's theory.

57. Lynn Sykes, "Papers on Plate Tectonics," HWMP, box 70, folder 24, "[Ocean of Truth] Sykes, 1984."

58. Minutes of the Joint NASCO Ocean-wide Surveys Panel Meeting with the ICO Ocean Survey Advisory Panel, 15 May 1967, ODWHOI, box 54, folder "Act. NASCO-SCOR, 1967."

59. Ambassador [to Canada] of the Union of Soviet Socialist Republics to J. T. Wilson, 8 Mar 1961, JTWP, B86–0066/021, "USSR General."

Notes to Pages 197–204 297

60. J. T. Wilson to Nathaniel C. Gerson, 24 Mar 1959, JTWP, B86–0066/021, "US National Academy of Sciences."

61. J. T. Wilson, "Note on Possible Exchanges with South American Universities," n.d., JWTP, B93–0055/023, 4.

62. Harry Hess to J. Tuzo Wilson, 25 Mar 1968, HHHP, box 12, folder "J. Tuzo Wilson."

63. Harry Hess to Editor of *Science*, attached to letter from Hess to J. Tuzo Wilson, 25 Mar 1968, HHHP, box 12, folder "J. Tuzo Wilson."

64. Wilson, "A Revolution in Earth Science," 16.

65. Beloussov, "An Open Letter to J. Tuzo Wilson," 17–18.

66. Hitoshi Hattori to J. Tuzo Wilson, 17 Jun 1970, JTWP, B93–0050/006.

67. Bullard, "The Emergence of Plate Tectonics," 21.

68. Menard, *The Ocean of Truth*, 234.

69. See Kuhn, *The Structure of Scientific Revolutions*.

70. Menard, *The Ocean of Truth*, 273.

71. Wilson, "A Revolution in Earth Science," 12.

72. Beloussov, "An Open Letter to J. Tuzo Wilson," 19.

73. Wilson, "A Reply to V. V. Beloussov," 22.

74. Wilson, "A Revolution in Earth Science," 16.

75. Wood, *The Dark Side of the Earth*, 160.

76. See Wilson, "Preface."

77. Intergovernmental Oceanographic Commission, Seventh Session, Unesco, Paris, 26 Oct-6 Nov 1971, Cooperative Systematic Studies in the North Atlantic (Statement by the Delegation of the USSR), Paris, 29 Oct 1971, UNESCOR, SC/IOC/45.

78. A. J. Lee to D. C. Martin, 28 Jul 1967, GERDP, folder G1/2, "British National Committee for Oceanic Research, Correspondence 1966–69."

79. Captain (1st Rank) P. S. Mitrofanov, "Analysis of the Operating Costs of the Scientific Research Work on Oceanographic Survey Ships," *Morskoy Sbornik* (Naval Proceedings) 46 (116), no. 1 (Jan 1963), translated by G. Matzureff for Office of Naval Intelligence, ODWHOI, box 36, folder "Act. ICO 2 of 3."

80. "Minutes of the PIPICO Meeting, 4 November 1964," ODWHOI, box 39, folder "ICO [PIP] 1964, 1965, 1 of 2."

81. Edward C. Bullard to A. Potts, 14 Dec 1965, ECBP, folder "E.131."

82. A. Potts to Edward C. Bullard, 5 Jan 1966, ECBP, folder "E.132."

83. Vice Admiral Day, report, "Third Session—Intergovernmental Oceanographic Commission, Paris—10th to 19th June 1964," n.d., CAB, 124/2175, "Working Group on Oceanography: Papers."

84. Robert H. B. Wade, U.S. Permanent Representative to Unesco, to René Maheu, 25 May 1965, and Soo Young Lee, Republic of Korea Permanent Representative to Unesco, to Director-General, Unesco, 8 Jun 1966, UNESCOR, box 551.46 A06 II, pt. 7 [Second Oceanographic Congress]; and René Maheu to Robert H. B. Wade,

n.d., UNESCOR, box 551.46 A06 II, pt. 8 [Second Oceanographic Congress]. Regarding the 1963 incident, see Intergovernmental Oceanographic Commission, Bureau and Consultative Committee, Second Meeting, Moscow, 6–8 May 1963, Summary Report, UNESCOR, NS/IOC/B-7.

85. NASCO Notes: Activities of the NAS Committee on Oceanography and Oceanography News Items, Jul 1966 (no. 19), ODWHOI, folder "Act. NASCO Jul–Dec 66."

86. P. J. Hart [comments to NAS on file for Second International Oceanographic Congress, Moscow, 1966], n.d., NASCPF 1966–72, folder "IR: International Congresses: Oceanography: Second: Moscow."

87. Hugh Bradner to Richard C. Vetter, 8 Jul 1966, NASCPF 1966–72, folder "IR: International Congresses: Oceanography: Second: Moscow."

88. Robert E. Stevenson, "Second International Oceanographic Congress," n.d., NASCPF 1966–72, folder "IR: International Congresses: Oceanography: Second: Moscow," 14.

89. Ibid., 12–13.

90. Ibid., 13–14.

91. Hugh Bradner to Richard C. Vetter, 8 Jul 1966, NASCPF 1966–72.

92. Fred B. Phleger [comments to NAS on file for Second International Oceanographic Congress, Moscow, 1966], n.d., NASCPF 1966–72, folder "IR: International Congresses: Oceanography: Second: Moscow."

93. Robert E. Stevenson, "Second International Oceanographic Congress," n.d., NASCPF 1966–72, 10.

94. See Zenkevich, *Biology of the Seas of the USSR.*

95. Gilbert L. Voss to Richard C. Vetter, 30 Aug 1966, NASCPF 1966–72, folder "IR: International Congresses: Oceanography: Second: Moscow."

96. Robert E. Stevenson, "Second International Oceanographic Congress," n.d., NASCPF 1966–72, 6.

97. Robert L. Fisher to Richard C. Vetter, 6 Jul 1966, NASCPF 1966–72, folder "IR: International Congresses: Oceanography: Second: Moscow."

98. Robert E. Stevenson, "Second International Oceanographic Congress," n.d., NASCPF 1966–72, 1.

99. Ibid., 2.

100. Ibid., 18–20.

101. Margaret Robinson to Richard C. Vetter, 25 Jul 1966, NASCPF 1966–72, folder "IR: International Congresses: Oceanography: Second: Moscow."

102. J. Lamar Worzel [comments to NAS on file for Second International Oceanographic Congress, Moscow, 1966], n.d., NASCPF 1966–72, folder "IR: International Congresses: Oceanography: Second: Moscow."

103. Fred B. Phleger [comments to NAS on file for Second International Oceanographic Congress, Moscow 1966].

104. T. H. van Andel [comments to NAS on file for Second International Oceanographic Congress, Moscow 1966].

105. Robert E. Stevenson, "Second International Oceanographic Congress," n.d., NASCPF 1966–72, 3.
106. Ibid., 4.
107. Robert L. Fisher to Richard C. Vetter, 6 Jul 1966, NASCPF 1966–1972.
108. A. E. J. Engel [comments to NAS on file for Second International Oceanographic Congress, Moscow 1966], n.d., NASCPF 1966–72, folder "IR: International Congresses: Oceanography: Second: Moscow."
109. Robert E. Stevenson, "Second International Oceanographic Congress," n.d., NASCPF 1966–72, 4.
110. Robert L. Fisher to Richard C. Vetter, 6 Jul 1966, NASCPF 1966–72.
111. "Symposium on Continental Drift Emphasizing the History of the South Atlantic Area, List of Participants," JTWP, B93–0050/026, "Symposium on Continental Drift." V. V. Beloussov to L. Knopoff, 12 Feb 1968, NASCPF 1966–72, "IR: International Unions: Geological Sciences: General Assembly: Second: Prague: US Delegation Report 1968."
112. Lockwood, "Soviet Geology and Geologists," 15.
113. Thomas B. Nolan and William Thurston, "Report of the United States Delegation to the XXIII Meeting of the International Geological Congress and of the Quadrennial General Assembly of the International Union of Geological Sciences, both held at Prague in August 1968," n.d., NASCPF 1966–72, "IR: International Unions: Geological Sciences: General Assembly: Second: Prague: US Delegation Report 1968," 2.
114. William Thurston, notes on Meeting of United States Delegations, 19 Aug 1968, NASCPF 1966–72, "IR: International Unions: Geological Sciences: General Assembly: Second: Prague: US Delegation Report 1968."
115. Anonymous, "Late Late News."
116. Wood, *The Dark Side of the Earth*, 214–16. The papers (some of them only abstracts) were eventually published in *Tectonophysics*, vol. 7 (1969), as a special issue devoted to the congress's proceedings.
117. H. D. Hedberg, notes on Special Meeting of the Bureau of the Congress, 21 Aug 1968, NASCPF 1966–72, "IR: International Unions: Geological Sciences: General Assembly: Second: Prague: US Delegation Report 1968."
118. H. D. Hedberg, notes on Special Meeting of the Bureau of the Congress, 22 Aug 1968, NASCPF 1966–72.
119. Ibid.
120. Ibid.
121. H. D. Hedberg, notes on Closing Session of 23rd International Geological Congress, 23 Aug 1968, NASCPF 1966–72, "IR: International Unions: Geological Sciences: General Assembly: Second: Prague: US Delegation Report 1968."
122. Thomas B. Nolan and William Thurston to Frederick Seitz, 29 Oct 1968, NASCPF 1966–72, "IR: International Unions: Geological Sciences: General Assembly: Second: Prague: US Delegation Report 1968."

123. Hoover, "Debacle in Prague," 9.
124. See *Tectonophysics*, vol. 7 (1969).
125. Heezen, "The World Rift System," 275.
126. Lynn Sykes, "Papers on Plate Tectonics," HWMP, box 70, folder 24, "[Ocean of Truth] Sykes, 1984."
127. See Morgan, "Rises, Trenches, Great Faults, and Crustal Blocks"; and McKenzie and Parker, "The North Pacific."
128. Oliver, Sykes, and Isacks, "Seismology and the New Global Tectonics," 527.
129. Salop and Scheinmann, "Tectonic History and Structures of Platforms and Shields," 595.

7 / MARINE SCIENCE AND MARINE AFFAIRS

1. W. M. Chapman to George F. Humphrey, 29 Dec 1961, CAB, 124/2161, "International Indian Ocean Expedition" (hereafter cited as "IIOE").
2. Mr. Hiscocks to R. N. Quirk, Ministry of Science, 6 Jan 1960, CAB, 124/2159, "IIOE." The letter is posted in January, but Quirk claims to have received it just prior to Christmas.
3. J. G. Liverman to R. N. Quirk, 28 Nov 1961, "Oceanography—Interdepartmental Committees," CAB, 124/2166, "Working Group on Oceanography."
4. H. Marshall, Australia House, London, to Mr. Crawford, Ministry of Science, 21 Jan 1960, "S.C.O.R.P.I.O.," CAB, 124/2159, "IIOE."
5. Office of the Minister of Science, Extract from the Minutes of CISC (61) 17th, 20 Sep 1961, "Indian Ocean Expedition," CAB, 124/2159, "IIOE."
6. Commander F. W. Hunt, Hydrographic Department, Admiralty, to J. G. Liverman, Office of the Minister for Science, 25 Aug 1961, CAB, 124/2159, "IIOE."
7. W. M. Chapman to David Hall, East African Marine Research Organisation, n.d., CAB, 124/2161, "IIOE."
8. W. M. Chapman to George F. Humphrey, 29 Dec 1961, CAB, 124/2161, "IIOE."
9. Ibid.
10. George Deacon to F. T. Hallet, Office of the Minister of Science, 12 Feb 1962, CAB, 124/2161, "IIOE."
11. George Deacon to J. G. Liverman, 22 Feb 1962, CAB, 124/2161, "IIOE."
12. R. G. Williams, Secretary, National Institute of Oceanography, to J. G. Liverman, 26 Feb 1962, CAB, 124/2161, "IIOE."
13. R. N. [Quirk] to J. G. Liverman, 25 Mar 1962, CAB, 124/2161, "IIOE."
14. F. T. Hallet to Mrs. Z. M. Read, Ministry of Education, 14 May 1962, CAB, 124/2161, "IIOE."
15. "The Parliamentary Debates, Official Report," 5th ser., vol. 667, 2d vol. of Session 1962–63, House of Commons, 12 Nov 1962, CAB, 124/2162, "IIOE."
16. Extract from the House of Commons Report: vol. 679, no. 129, 19 Jun 1963, CAB, 124/2163, "IIOE."

17. A. K. Russell to J. S. Goldsmith, Treasury, 4 Sep 1963, CAB, 124/2163, "IIOE."
18. E. Ll. Evans to C. Wigfull, Treasury, 30 Dec 1963, CAB, 124/2163, "IIOE."
19. Paul Webster, "Solving the Mysteries of an Ocean," *Daily Telegraph*, 3 May 1963, CAB, 124/2163, "IIOE."
20. G. E. R. Deacon to C. Wigfull, 26 May 1964, CAB, 124/2178, "National Institute of Oceanography: Finance and Programme."
21. C. Wigfull, Department of Education and Science, to Miss Senior, 29 May 1964, CAB, 124/2178, "National Institute of Oceanography: Finance and Programme."
22. Edward C. Bullard to Roger Quirk, Department of Education and Science, 13 Aug 1964, and Roger Quirk to Admiral E. G. Irving, Hydrographer to the Navy, 27 Aug 1964, CAB, 124/2178, "National Institute of Oceanography: Finance and Programme."
23. G. E. R. Deacon to Håkon Mosby, 5 Aug 1964, GERDP, H8/6, "Correspondence and Papers, 1964–66."
24. James Faughn to Robert Snider, 14 Jul 1962, UNESCOR, box 551.46 (267) A57 "-66," pt. 3.
25. C. S. Ramage to George Humphrey, 22 Aug 1962, UNESCOR, box 551.46 (267) A57 "-66," pt. 3.
26. G. E. Hemmen, "Preliminary Summary, IIOE National Co-ordinators Meeting, Paris, 22–24 January 1964," n.d., UNESCOR, box 551.46 (267) A57/024 IIOE—International Coordination Group.
27. Johns Ryther to Warren Wooster, 27 Aug 1962, UNESCOR, box 551.46 (267) A57 "-66," pt. 3.
28. Ibid.
29. John Ryther to Warren Wooster, 22 Mar 1962, UNESCOR, box 551.46 (267) A57 "-66," pt. 1.
30. WWI-01 (not transcribed).
31. G. Hempel, Office of Oceanography, to Dr. Vagn Hansen, Denmarks Fiskeri, 4 May 1965, UNESCOR, box 551.46 (267) A031 10BC "-66," pt. 7, Indian Ocean Biological Centre.
32. R. S. Glover to David J. Tranter, CSIRO, Cronulla, Sydney, 11 Jan 1966, UNESCOR, box 551.46 (267) A031 10BC "-66," pt. 8, Indian Ocean Biological Centre.
33. Intergovernmental Oceanographic Commission, Third Meeting of the Bureau, Unesco, Paris, 28–31 Oct 1963, Secretariat Report to the Bureau, UNESCOR, NS/IOC/B-9.
34. Konstantin Fedorov to Edward Brinton, Scripps, 19 May 1965, UNESCOR, box 551.46 (267) A031 10BC "-66," pt. 7, Indian Ocean Biological Centre.
35. Vagn Hansen to G. Hempel, 2 Jun 1965, UNESCOR, box 551.46 (267) A031 10BC "-66," pt. 7, Indian Ocean Biological Centre.
36. Intergovernmental Oceanographic Commission, International Indian Ocean Expedition, Information Paper no. 22, app. 2, Report of the Curator to the Sixth Meeting of the Consultative Committee for the IOBC, UNESCOR, NS/IOC/INF-149.

37. Konstantin Fedorov to J. W. Brodie, New Zealand Oceanographic Institute, Wellington, 28 Sep 1965, UNESCOR, box 551.46 A02 IOC "-66," pt. 5.

38. Warren Wooster to Editor, *Science,* 14 Sep 1965, UNESCOR, box 551.46 A02 IOC "-66," pt. 5.

39. Intergovernmental Oceanographic Commission, Meeting of the International Coordination Group for the International Indian Ocean Expedition, Paris, 22–24 Jan 1964, CAB, 124/2175, "Working Group on Oceanography: Papers."

40. John Barbour, Associated Press, "Fish Killed Blamed on Oxygen Lack," *Washington Post,* 14 May 1964, UNESCOR, box 551.46 (267) A57 "-66," pt. 5 [IIOE].

41. Roll, *Intergovernmental Oceanographic Commission (IOC),* 22–24.

42. Krige, "NATO and the Strengthening of Western Science in the Post-Sputnik Era," 85–86.

43. Ibid., 98–100.

44. Columbus Iselin to George Deacon, 10 Apr 1959, GERDP, H8/3, "Correspondence and Minutes of Meetings, 1959–67."

45. George Deacon to M. J. Hanman, M Branch I, Admiralty, 15 Sep 1960, GERDP, H8/1, "NATO Subcommittee on Oceanographic Research, Correspondence and Papers, 1959–61."

46. ALUSNA HELSINKI to CNO, n.d., ODWHOI, box 21, folder "Act [CNO]."

47. Paul M. Fye to Admiral Arleigh A. Burke, 16 Nov 1960, ODWHOI, box 21, folder "Act [CNO]."

48. John A. Knauss to Paul M. Fye, 9 Nov 1960, ODWHOI, box 21, folder "Act [CNO]."

49. J. Tuzo Wilson to Chief of Naval Operations, U.S. Navy Department, 9 Nov 1960, ODWHOI, box 21, folder "Act [CNO]."

50. Arleigh Burke to Paul Fye, 28 Nov 1960, ODWHOI, box 21, folder "Act [CNO]."

51. Håkon Mosby to A. J. Lee, Fisheries Laboratory, Lowestoft, 9 Nov 1959, GERDP, H8/1, "NATO Subcommittee on Oceanographic Research, Correspondence and Papers, 1959–61."

52. A. J. Lee to George Deacon, 8 Jan 1960, GERDP, H8/1, "NATO Subcommittee on Oceanographic Research, Correspondence and Papers, 1959–61."

53. George Deacon to Håkon Mosby, 9 Jan 1961, GERDP, H8/1, "NATO Subcommittee on Oceanographic Research, Correspondence and Papers, 1959–61."

54. A. J. Lee to George Deacon, 28 Feb 1962, GERDP, H8/4, "Correspondence and Papers, 1962–63."

55. George Deacon to A. J. Lee, 1 Mar 1962, GERDP, H8/4, "Correspondence and Papers, 1962–63."

56. Ibid.

57. For the opposition of Jeffreys to continental drift, see Oreskes, *The Rejection of Continental Drift.* Henry Stommel to Jerome Wiesner, 8 Mar 1962, OSTSF, box 161, folder "Oceanography—Title Folder."

58. "Report of Meeting Held at Reykjavik on April 26–27, 1962 to Discuss a NATO Supported Irminger Sea Project," n.d., GERDP, H8/4, "Correspondence and Papers, 1962–63."

59. George Deacon to Hans Jørgen Helms, Scientific Affairs Division, NATO, 20 Apr 1964, GERDP, H8/6, "Correspondence and Papers, 1964–66."

60. Memorandum from Icelandic Delegation, Proposal for Oceanographic Research in the Irminger Sea, n.d., CAB, 124/2172, "Working Group on Oceanography: Papers."

61. Working Group on Oceanography, "Oceanography in NATO," note by the Office of the Minister of Science, WG (o)(62) 28, 27 Apr 1962, CAB, 124/2172, "Working Group on Oceanography: Papers."

62. Extract from NATO document AC/137–D/130, dated 14 Mar 1962. "Memorandum on Major Project in Oceanography," by Sub-committee on Oceanographic Research, Nov 1961, CAB, 124/2172, "Working Group on Oceanography: Papers."

63. "National Oceanographic Council," *Congressional Record, Proceedings and Debates of the 89th Congress, First Session,* 2 Feb 1965, ODWHOI, box 45, folder "Act. US Senate."

64. Ibid.

65. These efforts by Congress are explored in far greater detail in Wenk, *The Politics of the Ocean,* esp. chap. 2.

66. Hastings Keith to the Honorable Herbert C. Bonner, 18 Mar 1965, ODWHOI, box 43, folder "Act. House of Representatives 1965."

67. "Marine Exploration and Development," *Congressional Record, Proceedings and Debates of the 89th Congress, First Session,* 10 Feb 1965, ODWHOI, box 43, folder "Act. House of Representatives 1965."

68. Ibid.

69. Hastings Keith to the Honorable Herbert C. Bonner, 18 Mar 1965, ODWHOI.

70. Ibid.

71. "Marine Exploration and Development," *Congressional Record, Proceedings and Debates of the 89th Congress, First Session,* 10 Feb 1965, ODWHOI.

72. Paul M. Fye to the Honorable Hastings Keith, 2 Mar 1965, ODWHOI, box 43, folder "Act. House of Representatives 1965."

73. Hastings Keith to Paul M. Fye, 9 Mar 1965, ODWHOI, box 43, folder "Act. House of Representatives 1965."

74. Wenk, *Making Waves,* 101.

75. Wenk, *The Politics of the Ocean,* 68.

76. Ibid., 79.

77. Ibid., 111.

78. Ibid., 150.

79. Ibid., 119–28.

80. Wenk, *Making Waves,* 98–100.

81. Ibid., 104–26.

82. The report and the president are quoted in NASCO Notes: Activities of the NAS Committee on Oceanography and Oceanography News Items, Jul 1966 (no. 19), ODWHOI, folder "Act. NASCO Jul–Dec 66."

83. "Comments on the Recommendations in *Effective Uses of the Sea*," as prepared by Committee on Oceanography, Division of Earth Sciences, National Research Council, Dec 1966, ODWHOI, box 54, folder "Act. NASCO mtg, La Jolla, CA, Feb 15–16, 1967."

84. Wenk, *The Politics of the Ocean*, 111.

85. Ibid., 107.

86. Ibid., 214.

87. Wenk, *Making Waves*, 107.

88. Committee on Oceanography and U.S. National Committee to SCOR of the NAS/NRC Division of Earth Sciences, Minutes of the Fifty-sixth Meeting, La Jolla, Calif., 14–21 Jun 1968, Preliminary Minutes for Committee Use Only, ODWHOI, box 58, folder "Act. NASCO mtg., La Jolla, CA, June 14–15 '68."

89. John D. Isaacs to Richard C. Vetter, 13 Jan 1967, ODWHOI, box 54, folder "Act. NASCO mtg, La Jolla, CA, Feb 15–16, 1967."

90. Gordon A. Riley to Richard C. Vetter, 29 Dec 1966, ODWHOI, box 54, folder "Act. NASCO mtg, La Jolla, CA, Feb 15–16, 1967."

91. Committee on Oceanography and U.S. National Committee to SCOR of the NAS/NRC Division of Earth Sciences, Minutes of the Fifty-second Meeting, Washington, D.C., 26–28 Oct 1967, Preliminary Minutes for Committee Use Only, ODWHOI, box 54, folder "Act. NASCO mtg., NAS, Washington, DC, Oct 26–28, 1967."

92. Wenk, *The Politics of the Ocean*, 151.

93. Committee on Oceanography and U.S. National Committee to SCOR of the NAS/NRC Division of Earth Sciences, Minutes of the Fifty-sixth Meeting, 14–21 Jun 1968, ODWHOI.

94. Wenk, *Making Waves*, 112.

95. Wenk, *The Politics of the Ocean*, 240–41.

96. Ibid., 243. The interpretation that scientists saw *An Oceanic Quest* as an opportunity to put science into an otherwise top-down policy comes from Warren Wooster, who chaired the committee that prepared the report. Wooster mentioned this to this author in his comments on an early manuscript of this book.

97. Wenk, *The Politics of the Ocean*, 246–48.

98. Ibid., chap. 8.

99. Committee on Oceanography and U.S. National Committee to SCOR of the NAS/NRC Division of Earth Sciences, Minutes of the Fifty-third Meeting, Washington, D.C., 13–14 Dec 1967, Preliminary Minutes for Committee Use Only, ODWHOI, box 54, folder "Act. NASCO mtg., NAS, Wash., DC, Dec 13–14 '67."

100. Intergovernmental Oceanographic Commission, Fifth Session, Unesco, Paris, 19–28 Oct 1967, Address of Welcome at the Opening of the Fifth Session of

Notes to Pages 251–260

the Intergovernmental Oceanographic Commission by Professor A. N. Matveyev, Assistant Director-General for Science, Unesco, UNESCOR, IOC/V-INF.128.

101. Wenk, *The Politics of the Ocean*, 260–61.
102. Ibid., 222–24.
103. Ibid., 235.
104. Ibid., 232–34.
105. Wenk, *Making Waves*, 109.
106. Ibid., 118.
107. Wenk, *The Politics of the Ocean*, 232.
108. Pardo, "Who Will Control the Seabed?" 136.
109. Intergovernmental Oceanographic Commission, Comprehensive Outline of the Scope of the Long-Term and Expanded Programme of Oceanic Exploration and Research, as approved by the Sixth Session, Unesco, Paris, 2–13 Sept 1969, UNESCOR, SC/IOC/VI/-.
110. Wenk, *The Politics of the Ocean*, 235.
111. Intergovernmental Oceanographic Commission, Comprehensive Outline and the Scope of the Long-Term and Expanded Programme of Oceanic Exploration and Research, as approved by the Sixth Session, Unesco, Paris, 2–13 Sept 1969, UNESCOR, SC/IOC/VI/-
112. Roll, *Intergovernmental Oceanographic Commission (IOC)*, 25.
113. Intergovernmental Oceanographic Commission, Long-Term Expanded Programme of Oceanographic Research and International Decade of Ocean Exploration, Sixth Session, Unesco, Paris, 2–13 Sept 1969, UNESCOR, SC/IOC-VI/5(5).
114. Ibid.
115. Intergovernmental Oceanographic Commission, Eighth Session of the Assembly, Unesco, Paris, 5–17 Nov 1973, Compilation of Component Programmes of the International Decade of Ocean Exploration, UNESCOR, SC/IOC-VIII/11.
116. Ibid.
117. Intergovernmental Oceanographic Commission, Sixth Session, Unesco, Paris, Summary Report of the First Meeting of the Working Committee for an Integrated Global Ocean Station System (IGOSS), Unesco, Paris, 2–13 Apr 1968, UNESCOR, SC/IOC-VI/11.
118. Integrated Global Ocean Station System, IGOSS, programme information circular, Mar 1973, circular no. 6, "An Evaluation of the First Year of IGOSS," UNESCOR, IOC/IGOSS/6.
119. Intergovernmental Oceanographic Commission, Seventh Session, Unesco, Paris, 26 Oct–5 Nov 1971, Draft Summary Report, UNESCOR, SC/IOC-VII/46 Rev.

CONCLUSION

1. Intergovernmental Oceanographic Commission, Seventh Session, Unesco, Paris, 26 Oct–5 Nov 1971, Draft Summary Report, UNESCOR, SC/IOC-VII/46 Rev.

2. See Sapolsky, *Science and the Navy;* Weir, *Forged in War;* Weir, *An Ocean in Common;* Rainger, "Science at the Crossroads"; Forman, "Behind Quantum Electronics."

3. Revelle, "The Oceanographic and How It Grew," 22.

4. Polanyi, "The Republic of Science," 15.

BIBLIOGRAPHY

ARCHIVES AND SPECIAL COLLECTIONS

ADM   Admiralty Records, Public Record Office, Kew, England.
CAB   Cabinet Records, Public Record Office, Kew, England.
CLHP   Carl Leavitt Hubbs Papers, Archives of Scripps Institution of Oceanography, La Jolla, California.
ECBP   Edward Crisp Bullard Papers, Churchill College Archives, Cambridge, England.
FO   Foreign Office Records, Public Record Office, Kew, England.
GERDP   George E. R. Deacon Papers, Deacon Library, Southampton Oceanography Centre, Southampton, England.
HHHP   Harry Hammond Hess Papers, Firestone Library, Princeton University, Princeton, New Jersey.
HPRC   H. P. Robertson Collection, Archives of California Institute of Technology, Pasadena, California.
HWMP   H. W. Menard Papers, Archives of Scripps Institution of Oceanography, La Jolla, California.
IDABR   Records of the International Development Advisory Board, Office of the Administrator, Technical Cooperation Administration, Record Group 469, National Archives and Records Administration, College Park, Maryland.
IGYC   International Geophysical Year Collection, Archives of the National Academy of Sciences, Washington, District of Columbia.
IIAACF   Institute for Inter-American Affairs, Administrative Office, Country Files (Central Files), 1942–53, Record Group 469, National Archives and Records Administration, College Park, Maryland.
JTWP   John Tuzo Wilson Papers, University of Toronto Archives, Toronto, Canada.
MPC   Mohole Project Collection, Archives of the National Academy of Sciences, Washington, District of Columbia.
NASCPF   National Academy of Sciences–National Research Council Central Policy Files, Archives of the National Academy of Sciences, Washington, District of Columbia.

ODWHOI	Records of the Office of the Director, Woods Hole Oceanographic Institution, Archives of Woods Hole Oceanographic Center, Woods Hole, Massachusetts.
OSTSF	Office of Science and Technology, Subject Files, Record Group 359, National Archives and Records Administration, College Park, Maryland.
RRP	Roger Randall Dougan Revelle Papers, Archives of Scripps Institution of Oceanography, La Jolla, California.
RSDP	Robert Sinclair Dietz Papers, Archives of Scripps Institution of Oceanography, La Jolla, California.
SIOOD	Office of the Director, Scripps Institution of Oceanography, Records 1904–92, Archives of Scripps Institution of Oceanography, La Jolla, California.
SIOSF	Scripps Institution of Oceanography, Subject Files, Archives of Scripps Institution of Oceanography, La Jolla, California.
SPDR	Records of the Strategic Plans Division, United States Navy, OP-30S/OP-60S Subject and Serial Files, Series 16, Naval Historical Center, Washington, District of Columbia.
UNESCOR	Records of the United Nations Educational, Scientific and Cultural Organization, Unesco Archives, Paris, France.

ORAL HISTORY INTERVIEWS

ASI-76	Oral History Interview of Athelstan Spilhaus, conducted by Robert A. Calvert in 1976, Niels Bohr Library, American Institute of Physics, College Park, Maryland.
ASI-89	Oral History Interview of Athelstan Spilhaus, conducted by Ronald E. Doel in 1989, Niels Bohr Library, American Institute of Physics, College Park, Maryland.
JLWI-96	Oral History Interview of J. Lamar Worzel, conducted by Ronald E. Doel in 1996, Niels Bohr Library, American Institute of Physics, College Park, Maryland.
RRI-84	Oral History Interview of Roger Revelle, conducted by Sarah L. Sharp in 1984, Regional Oral History Office, Bancroft Library, University of California, Berkeley, California.
VBI-64	Oral History Interview of Vannevar Bush, conducted by Eric Hodgins in 1964, Massachusetts Institute of Technology, Archives and Special Collections, Cambridge, Massachusetts.
WHMI-97	Oral History Interview with Walter H. Munk, conducted by Ronald E. Doel in 1997, Niels Bohr Library, American Institute of Physics, College Park, Maryland.
WHMI-98	Oral History Interview with Walter H. Munk, conducted by

Bibliography

Ronald E. Doel in 1998, Niels Bohr Library, American Institute of Physics, College Park, Maryland.

WWI-01 Interview with Warren Wooster, conducted by Jacob Darwin Hamblin, 20 Feb 2001, recorded by telephone.

PUBLISHED WORKS

Aaserud, Finn. "Sputnik and the 'Princeton Three': The National Security Laboratory That Was Not to Be." *Historical Studies in the Physical and Biological Sciences* 25 (1995).

Acheson, Dean. *Present at the Creation: My Years in the State Department.* New York: Norton, 1969.

Adams, Douglas P. "A Primary Influence Ignored." *Bulletin of the Atomic Scientists* 3 (Aug 1947): 229.

Agassi, Joseph. *Science and Society: Studies in the Sociology of Science.* Dordrecht: D. Reidel, 1981.

Albritton, Claude C., Jr., ed. *Philosophy of Geohistory: 1785–1970.* Stroudsburg, PA: Dowden, Hutchinson & Ross, 1975.

Alcalde, Javier Gonzalo. *The Idea of Third World Development.* Lanham: University Press of America, 1987.

Anonymous. "Khrushchev, Stressing Peace, Plans No Military Aides on Trip." *New York Times,* 1 Sep 1959, 6.

Anonymous. "Late Late News." *Geotimes* 13 (Sep 1968): 21.

Anonymous. "Race to Inner Space: Will Russia Dig 'Stairsteps' through the Earth's Surface?" *World Oil,* May 1962, 15–17.

Anonymous. "Report on Science and Foreign Relations Released." *Department of State Bulletin* 22 (Jun 1950): 982–83.

Anonymous. "Science and Foreign Relations: Berkner Report to the U. S. Department of State." *Bulletin of the Atomic Scientists* 6 (Aug 1950): 293–98.

Anonymous. "Science and Public Policy: A Report by the President's Scientific Research Board." *Bulletin of the Atomic Scientists* 4 (Jan 1948): 23–29, 31.

Anonymous. "Soviet Considers 5 Earth Debates." *New York Times,* 10 Sep 1961, 13.

Anonymous. "US Extends Ban on Atomic Tests," *New York Times,* 27 Aug 1959, 1.

Baker, F. W. G. "The First International Polar Year, 1882–83." *Polar Record* 21 (Sep 1982): 1–11.

Barlow, Jeffrey G. *Revolt of the Admirals: The Fight for Naval Aviation, 1945–1950.* Washington, DC: Naval Historical Center, 1995.

Barron, Arthur S. "Why Do Scientists Read Science Fiction?" *Bulletin of the Atomic Scientists* 13 (Feb 1957): 62–65.

Bartholomew, James R. *The Formation of Science in Japan: Building a Research Tradition.* New Haven: Yale University Press, 1989.

Basalla, George. "The Spread of Western Science." *Science* 156 (5 May 1967): 611–22.

Bascom, Willard. *The Crest of the Wave: Adventures in Oceanography.* New York: Harper & Row, 1988.

———. *A Hole in the Bottom of the Sea: The Story of the Mohole Project.* Garden City, NY: Doubleday, 1961.

———. "John Dove Isaacs III." *Biographical Memoirs of the National Academy of Sciences* 57 (1987): 89–122.

Behrman, Daniel. *Assault on the Largest Unknown: The International Indian Ocean Expedition, 1959–65.* Paris: Unesco, 1981.

———. *The New World of the Oceans: Men and Oceanography.* Boston: Little, Brown, 1969.

———. *They Can't Afford to Wait.* Paris: Unesco, 1952.

Bell, David E. "The Quality of Aid." *Foreign Affairs* 44 (Jul 1966): 601–7.

Beloussov, V. V. "An Open Letter to J. Tuzo Wilson." *Geotimes* 13 (1968): 17–19.

Berkner, Lloyd V. "Is Secrecy Effective?" *Bulletin of the Atomic Scientists* 11 (Feb 1955): 62–63, 68.

———. "Power of Freedom." *Bulletin of the Atomic Scientists* 12 (May 1956): 174–76.

———. "Science and Military Power." *Bulletin of the Atomic Scientists* 9 (Dec 1953): 359–65.

Berliner, Joseph S. *Soviet Economic Aid: The New Aid and Trade Policy in Underdeveloped Countries.* New York: Praeger, 1958.

Bezrunov, P. L. "Research in the Indian Ocean by the Survey Vessel *Vitiaz* on Its Thirty-third Voyage." *Deep-Sea Research* 10 (1963): 59–66. Translated from *Okeanologiya* 1 (1961): 745–53.

Bigelow, Henry B. *Oceanography: Its Scope, Problems, and Economic Importance.* Cambridge, MA: Riverside Press, 1931.

Black, Lloyd D. *The Strategy of Foreign Aid.* Princeton: Van Nostrand, 1968.

Blanpied, William A. *Impacts of the Early Cold War on the Formulation of U.S. Science Policy.* Washington, DC: AAAS, 1995.

Bogorov, V. G. "Marine Scientific Organizations in the Indian Ocean." *Deep-Sea Research* 10 (1963): 507–9. Translated from *Okeanologiya* 1 (1961): 937–39.

Bowen, Harold G. *Ships, Machinery, and Mossbacks: The Autobiography of a Naval Engineer.* Princeton: Princeton University Press, 1954.

Brands, H. W. *The Specter of Neutralism: The United States and the Emergence of the Third World, 1947–1960.* New York: Columbia University Press, 1989.

Brink, Frank, Jr. "Detlev Wulf Bronk." *Biographical Memoirs of the National Academy of Sciences* 50 (1979): 3–87.

Bibliography

Brode, Wallace R. "National and International Science." *Department of State Bulletin* 42 (9 May 1960): 735–39.

———. "The Role of Science in Foreign Policy Planning." *Department of State Bulletin* 42 (22 Feb 1960): 271–76.

Brooks, H., C. Freeman, L. Gunn, L. Saint-Geours, and J. Spaey. *Government and Allocation of Resources to Science.* Paris: Organisation for Economic Co-operation and Development, 1966.

Brooks, Harvey. *The Government of Science.* Cambridge: MIT Press, 1968.

Brush, Stephan G. "Whole Earth History." *Historical Studies in the Physical and Biological Sciences* 17 (1987).

Bruun, Anton Fr. "An Oceanographic Organization for the Indo-Pacific Region." *Deep-Sea Research* 2 (1954): 84–86.

Brynes, Asher. *We Give to Conquer.* New York: Norton, 1966.

Bulkeley, Rip. *The Sputnik Crisis and Early United States Space Policy: A Critique of the Historiography of Space.* Bloomington: Indiana University Press, 1991.

Bullard, Edward Crisp. "The Emergence of Plate Tectonics: A Personal View." *Annual Review of Earth and Planetary Sciences* 3 (1975).

———. "William Maurice Ewing." *Biographical Memoirs of the National Academy of Sciences* 51 (1980): 119–93.

Bullis, Harold. *The Political Legacy of the International Geophysical Year.* Washington, DC: Government Printing Office, 1973.

Bush, Vannevar. "Dangers to Research, If Recognized, Can Be Avoided." *Bulletin of the Atomic Scientists* 3 (Aug 1947): 228.

———. *Science, the Endless Frontier.* Washington, DC: Government Printing Office, 1945.

Buzzati-Traverso, Adriano A. "Scientific Research: The Case for International Support." *Science* 148 (Jun 1965): 1440–44.

Caidin, Martin. *Hydrospace.* New York: E. P. Dutton & Co., 1964.

Calluther, Nick. "Development? It's History." *Diplomatic History* 24 (Fall 2000): 641–54.

Capshew, J. H., and K. A. Rader, "Big Science: Price to the Present." *Osiris* 7 (1992).

Castle, Eugene W. *Billions, Blunders, and Baloney: The Fantastic Story of How Uncle Sam Is Squandering Your Money Overseas.* New York: Devin-Adair, 1955.

———. *The Great Giveaway: The Realities of Foreign Aid.* Chicago: Henry Regnery Company, 1957.

Chambers, David Wade. "Does Distance Tyrannize Science?" In *International Science and National Scientific Identity,* ed. R. W. Home and S. G. Kohlstedt, 19–38. Dordrecht: Kluwer, 1991.

Chapman, Sydney. *IGY: Year of Discovery.* Ann Arbor: University of Michigan Press, 1959.

———. "Scientific Programme of the International Geophysical Year 1957–58." *Nature* 175, no. 4453 (5 Mar 1955): 402–6.
Charnock, H. "George Edward Raven Deacon." *Biographical Memoirs of the Royal Society* 31 (1985): 113–42.
Coleman-Cooke, John. *Discovery II in the Antarctic: The Story of British Research in the Southern Seas.* London: Odhams Press, 1963.
Congressional Research Service, Library of Congress. *U.S. Scientists Abroad: An Examination of Major Programs for Nongovernmental Scientific Exchange.* Washington, DC: Government Printing Office, 1974.
———. *United Nations Conference on Science and Technology for Development— A Background Paper.* Washington, DC: Government Printing Office, 1979.
Crawford, Elisabeth. *Nationalism and Internationalism in Science, 1880–1939: Four Studies of the Nobel Population.* Cambridge: Cambridge University Press, 1992.
Crawford, Elisabeth, Terry Shinn, and Sverker Sörlin, eds. *Denationalizing Science: The Contexts of International Scientific Practice.* Dordrecht: Kluwer, 1993.
Cueto, Marcos, ed. *Missionaries of Science: The Rockefeller Foundation and Latin America.* Bloomington: Indiana University Press, 1994.
———. "Visions of Science and Development: The Rockefeller Foundation's Latin American Surveys of the 1920s." In Cueto, *Missionaries of Science,* 1–22.
Cumings, Bruce. *The Origins of the Korean War.* Vol. 2, *The Roaring of the Cataract, 1947–1950.* Princeton: Princeton University Press, 1990.
Curti, Merle, and Kendall Birr. *Prelude to Point Four: American Technical Missions Overseas, 1838–1938.* Madison: University of Wisconsin Press, 1954.
Damms, Richard V. "James Killian, the Technological Capabilities Panel, and the Emergence of President Eisenhower's 'Scientific-Technological Elite.'" *Diplomatic History* 24 (Winter 2000): 57–78.
Daston, Lorraine. "The Moral Economy of Science." *Osiris* 10 (1995): 3–24.
Deacon, G. E. R. "Commonwealth Oceanographic Conference." *Nature* 174, no. 4443 (25 Dec 1954): 1175–76.
———. "Dedication of the Laboratory of Oceanography, Woods Hole." *Nature* 174, no. 4420 (17 Jul 1954): 107–8.
———. "Expansion of Marine Research in the United States." *Nature* 183, no. 4667 (11 Apr 1959): 1005–6.
———. "Hurricane Research in the United States." *Nature* 178, no. 4526 (28 Jul 1956): 181.
———. "The Indian Ocean Expedition." *Nature* 187, no. 4737 (13 Aug 1960): 561–62.
———. "International Control of Pelagic Fisheries." *Nature* 177, no. 4517 (26 May 1956): 966.

Bibliography

———. "International Co-operation in Marine Research." *Nature* 180, no. 4592 (2 Nov 1957): 894–95.

———. "The International Geophysical Year." *Nature* 188, no. 4750 (12 Nov 1960): 529–32.

———. "International Indian Ocean Expedition." *Nature* 201, no. 4919 (8 Feb 1964): 561–62.

———. "International Oceanographic Congress." *Nature* 184, no. 4699 (21 Nov 1959): 1605–6.

———. "Marine Science in the Pacific Area." *Nature* 177, no. 4504 (25 Feb 1956): 353–55.

———. "The National Institute of Oceanography." *Nature* 173, no. 4413 (29 May 1954): 1014–16.

———. "Ocean Waves and Swell." *Occasional Papers of the Challenger Society* 1 (Apr 1946): 1–13. Reprinted in M. Deacon, *Oceanography*, 196–97.

———, ed. *Seas, Maps, and Men: An Atlas-History of Man's Exploration of the Oceans.* New York: Doubleday, 1962.

Deacon, Margaret, ed. *Oceanography: Concepts and History.* Stroudsburg, PA: Dowden, Hutchinson & Ross, 1978.

———. *Scientists and the Sea, 1650–1900: A Study of Marine Science.* London: Academic Press, 1971.

Dees, Bowen. *The Allied Occupation and Japan's Economic Miracle: Building the Foundations of Japanese Science and Technology 1945–52.* Surrey: Japan Library, 1997.

Dennis, Michael Aaron. "Historiography of Science: An American Perspective." In Krige and Pestre, *Science in the Twentieth Century*, 1–30.

———. "'Our First Line of Defense': Two University Laboratories in the Postwar American State." *Isis* 85 (1994): 427–55.

Dietz, Robert S., Henry W. Menard, and Edwin L. Hamilton. "Echograms of the Mid-Pacific Expedition." *Deep-Sea Research* 1 (1954): 258–72.

Divine, Robert A. *Eisenhower and the Cold War.* New York: Oxford University Press, 1981.

———. *The Sputnik Challenge.* New York: Doubleday, 1993.

Doel, Ronald E. "The Earth Sciences and Geophysics." In Krige and Pestre, *Science in the Twentieth Century*, 391–416.

———. "Evaluating Soviet Lunar Science in Cold War America." *Osiris* 7 (1992): 238–64.

———. "Scientists as Policymakers, Advisors, and Intelligence Agents: Linking Contemporary Diplomatic History with the History of Contemporary Science." In Söderqvist, *The Historiography of Contemporary Science and Technology*, 215–44.

Doel, Ronald E., and Allan A. Needell. "Science, Scientists, and the CIA: Balancing International Ideals, National Needs, and Professional Opportunities." In Rhodri Jeffreys-Jones and Christopher Andrew, eds., *Eternal Vigilance? 50 Years of the CIA*. London: Frank Cass, 1997.

Doumani, George A. *Exploiting the Resources of the Seabed*. Washington, DC: Government Printing Office, 1971.

DuBridge, Lee. "Policy and the Scientists." *Foreign Affairs* 41 (Apr 1963): 571–87.

Dufek, George J. *Operation Deepfreeze*. New York: Harcourt, Brace, & World, 1957.

Dupree, A. Hunter. "The Great Instauration of 1940: The Organization of Scientific Research for War." In *The Twentieth Century Sciences: Studies in the Biography of Ideas*, edited by Gerald Holton, 443–67. New York: Norton, 1972.

Eberstadt, Nicholas. *Foreign Aid and American Purpose*. Washington, DC: American Enterprise Institute for Public Policy Research, 1988.

Einstein, Albert. "The Military Mentality." *Bulletin of the Atomic Scientists* 3 (Aug 1947): 223–24.

Eisenhower, Dwight D. *Waging Peace, 1956–1961*. New York: Doubleday, 1965.

Elkin, Adolphus Peter. *Pacific Science Association: Its History and Role in International Cooperation*. Honolulu: Bishop Museum Press, 1961.

Elliot, William Y. "Facts and Values." *Bulletin of the Atomic Scientists* 3 (Aug 1947): 227.

Elzinga, Aant, Jan Nolin, Rob Pranger, and Sune Sunesson, eds. *Moral and Political Issues of Science in Society*. Lund, Sweden: Lund University Press, 1990.

Emery, William J. "The *Meteor* Expedition, an Ocean Survey." In Sears and Merriman, *Oceanography*, 690–702.

England, J. Merton. *A Patron for Pure Science: The National Science Foundation's Formative Years, 1945–57*. Washington, DC: National Science Foundation, 1982.

Evan, William M. *Knowledge and Power in a Global Society*. London: Sage, 1981.

Ewing, Gifford C., ed. *Oceanography from Space*. Woods Hole, MA: Woods Hole Oceanographic Institution, 1965.

Fisher, Robert L., and Edward D. Goldberg. "Henry William Menard." *Biographical Memoirs of the National Academy of Sciences* 64 (1994): 267–76.

Fleming, John A. "The International Scientific Unions." *Proceedings of the American Philosophical Society* 91, no. 1 (Feb 1947): 121–25.

Forman, Paul. "Behind Quantum Electronics: National Security as Basis for Physical Research in the United States, 1940–1960." *Historical Studies in the Physical and Biological Sciences* 18 (1987): 149–229.

Forman, Paul, and José M. Sánchez-Ron, eds. *National Military Establishments and the Advancement of Science and Technology*. Dordrecht: Kluwer, 1996.

Franck, James. "The Social Task of the Scientist." *Bulletin of the Atomic Scientists* 3 (Jan 1947): 70.

Frankel, Henry. "The Development, Reception, and Acceptance of the Vine-Matthews-Morley Hypothesis." *Historical Studies in the Physical and Biological Sciences* 13 (1982): 1–39.

Friedman, Robert Marc. *Appropriating the Weather: Vilhelm Bjerknes and the Construction of Modern Meteorology.* Ithaca, NY: Cornell University Press, 1989.

Gaddis, John Lewis. *Strategies of Containment: A Critical Appraisal of Postwar American National Security Policy.* New York: Oxford University Press, 1982.

Galey, Margaret E. "The Intergovernmental Oceanographic Commission: Its Capacity to Implement an International Decade of Ocean Exploration." *Occasional Papers Series, Law of the Sea Institute* 20 (Dec 1973).

Galison, Peter, and Bruce Hevly, eds. *Big Science: The Growth of Large-Scale Research.* Stanford: Stanford University Press, 1992.

Geelhoed, E. Bruce. *Charles E. Wilson and Controversy at the Pentagon, 1953 to 1957.* Detroit: Wayne State University Press, 1979.

Geiger, Roger L. *Research and Relevant Knowledge: American Research Universities since World War II.* New York: Oxford University Press, 1993.

———. "What Happened after Sputnik? Shaping University Research in the United States." *Minerva* 35 (1997): 349–67.

Gilpin, Robert, and Christopher Wright, eds. *Scientists and National Policy-Making.* New York: Columbia University Press, 1964.

Glen, William. *The Road to Jaramillo: Critical Years of the Revolution in Earth Science.* Stanford: Stanford University Press, 1982.

Glennon, John P., ed. *Foreign Relations of the United States, 1958–1960.* Vol. 11: *United Nations and General International Matters.* Washington, DC: Department of State, 1992.

Goldstein, Jack S. *A Different Sort of Time: The Life of Jerrold R. Zacharias, Scientist, Engineer, Educator.* Cambridge: MIT Press, 1992.

Good, Gregory A., ed. *The Earth, the Heavens and the Carnegie Institution of Washington.* Washington, DC: American Geophysical Union, 1994.

Gould, Laurence M. "The History of the Scientific Committee on Antarctic Research (SCAR)." In *Research in the Antarctic,* edited by Louis O. Quam, 47–55. Washington, DC: AAAS, 1971.

Graham, Loren R. *Science and Philosophy in the Soviet Union.* New York: Knopf, 1972.

———. *What Have We Learned about Science and Technology from the Russian Experience?* Stanford: Stanford University Press, 1998.

Graham, Loren R., ed. *Science and the Soviet Social Order.* Cambridge: Harvard University Press, 1990.

Greenaway, Frank. *Science International: A History of the International Council of Scientific Unions.* Cambridge: Cambridge University Press, 1996.

Greenberg, D. S. "International Programs: Frankel Resigns from State." *Science* 158 (Dec 1967): 1436.

———. "Mohole: The Project That Went Awry." *Science* 143 (Jan 1964).

———. *The Politics of Pure Science.* New York: New American Library, 1967.

———. "Pollack to Head State Science Office." *Science* 157 (Jul 1967): 292.

———. "Science and Foreign Affairs: New Effort Under Way to Enlarge Role of Scientists in Policy Planning." *Science* 138 (Oct 1962): 122–24.

———. "Science Attachés: U.S. Aides Meet to Report on International Scene." *Science* 161 (Sep 1968): 1114–16.

Grobstein, Clifford, John Conly, Irving Feister, Laurence Heilprin, Robert D. Siehler, Francis J. Weiss, and Laurence A. Wood. "National Science Foundation: The Steelman Report Misses the Point." *Bulletin of the Atomic Scientists* 4 (Jan 1948): 30–31.

Grodzins, Morton, and Eugene Rabinowitch, eds. *The Atomic Age: Scientists in National and World Affairs.* New York: Simon & Schuster, 1965.

Hackman, Willem D. "Sonar Research and Naval Warfare 1914–1954: A Case Study of a Twentieth Century Establishment Science." *Historical Studies in the Physical and Biological Sciences* 16 (1986).

Hales, Anton L. "Lloyd Viel Berkner." *Biographical Memoirs of the National Academy of Sciences* 61 (1992): 3–25.

Hallam, A. *A Revolution in the Earth Sciences: From Continental Drift to Plate Tectonics.* Oxford: Clarendon, 1973.

Harris, Seymour E. "An Economist Views the Problem." *Bulletin of the Atomic Scientists* 3 (Aug 1947): 226–27.

Haskins, Caryl P. "Technology, Science and American Foreign Policy." *Foreign Affairs* 40 (Jan 1962): 224–43.

Heezen, B. C. "The World Rift System: An Introduction to the Symposium." *Tectonophysics* 8 (1969): 269–80.

Heezen, Bruce C., and John E. Nafe, "Vema Trench: Western Indian Ocean." *Deep-Sea Research* 11 (1964): 79–84.

Herken, Gregg. "In the Service of the State: Science and the Cold War." *Diplomatic History* 24 (Winter 2000): 107–15.

Hess, H. H. "History of the Ocean Basins." In A. E. J. Engel, Harold L. James, and B. F. Leonard, eds., *Petrologic Studies: A Volume to Honor A. F. Buddington,* 599–620. New York: Geological Society of America, 1962.

Hevly, Bruce. "The Tools of Science: Radio, Rockets, and the Science of Naval Warfare." In Forman and Sánchez-Ron, *National Military Establishments and the Advancement of Science and Technology,* 215–32.

Hewlett, Richard G., and Francis Duncan. *Atomic Shield, 1947–1952: A History of the United States Atomic Energy Commission.* University Park: Pennsylvania State University Press, 1969.

Hewlett, Richard G., and Jack M. Holl. *Atoms for Peace and War, 1953–1961: Eisenhower and the Atomic Energy Commission.* Berkeley: University of California Press, 1989.

Holton, Gerald, and William A. Blanpied, eds. *Science and Its Public: The Changing Relationship.* Dordrecht: D. Reidel, 1976.

Home, R. W., and S. G. Kohlstedt, eds. *International Science and National Scientific Identity.* Dordrecht: Kluwer, 1991.

Hoover, Linn. "Debacle in Prague." *Geotimes* 13 (Oct 1968): 9.

Hunsaker, Jerome C. "International Scientific Congresses." *Proceedings of the American Philosophical Society* 91, no. 1 (Feb 1947): 126–28.

Huxley, Aldous. "A Positive Program of Research for Peace." *Bulletin of the Atomic Scientists* 3 (Aug 1947): 225.

Intergovernmental Oceanographic Commission. *Intergovernmental Oceanographic Commission (Five Years of Work).* Paris: Unesco, 1966.

———. *Legal Problems Associated with Ocean Data Acquisition Systems (ODAS): A Study of Existing National and International Legislation.* Paris: Unesco, 1969.

Isaacson, Walter, and Evan Thomas. *The Wise Men: Six Friends and the World They Made: Acheson, Bohlen, Harriman, Kennan, Lovett, McCloy.* New York: Simon & Schuster, 1986.

Ivanov, M. M. "The Work of the Non-magnetic Vessel *Zarya* in the Pacific Ocean." *Deep-Sea Research* 10 (1963): 645–47. Translated from *Okeanologiya* 1 (1961): 920–22.

James, Harold L. "Harry Hammond Hess." *Biographical Memoirs of the National Academy of Sciences* 43 (1973): 109–28.

Jeffreys, Harold. *The Earth: Its Origin, History, and Physical Constitution.* Cambridge: Cambridge University Press, 1924.

Jonas, Gerald. *The Circuit Riders: Rockefeller Money and the Rise of Modern Science.* New York: Norton, 1989.

Jones, Greta. *Science, Politics, and the Cold War.* New York: Routledge, 1988.

Kaplan, Norman, ed. *Science and Society.* New York: Rand McNally, 1965.

Kaufman, Burton I. *Trade and Aid: Eisenhower's Foreign Economic Policy, 1953–1961.* Baltimore: Johns Hopkins University Press, 1982.

Kay, Lily E. "Rethinking Institutions: Philanthropy as an Historiographic Problem of Knowledge and Power." *Minerva* 35 (1997): 283–93.

Kempe, D. R. C., and H. A. Buckley. "Fifty Years of Oceanography in the Department of Mineralogy, British Museum (Natural History)." *Bulletin of the British Museum (Natural History)* 15, no. 2 (26 Nov 1987).

Kevles, Daniel J. "Cold War and Hot Physics: Science, Security, and the American State, 1945–56." *Historical Studies in the Physical and Biological Sciences* 20 (1990): 239–64.

———. "K1S2: Korea, Science, and the State." In Galison and Hevly, *Big Science*, 312–33.

———. *The Physicists: The History of a Scientific Community in America.* New York: Knopf, 1977.

———. "Scientists, the Military, and the Control of Postwar Defense Research: The Case of the Research Board for National Security, 1944–1946." *Technology and Culture* 16 (1975): 20–47.

Killian, James R. *Sputnik, Scientists and Eisenhower: A Memoir.* Cambridge: MIT Press, 1977.

Kistiakowsky, George B. "Science and Foreign Affairs." *Department of State Bulletin* 42 (22 Feb 1960): 276–83.

———. *A Scientist in the White House.* Cambridge: Harvard University Press, 1976.

Kleinman, Daniel. *Politics on the Endless Frontier: Postwar Research Policy in the United States.* Durham: Duke University Press, 1995.

Kline, Ronald. "Construing 'Technology' and 'Applied Science': Public Rhetoric of Scientists and Engineers in the United States, 1880–1945." *Isis* 86 (1995): 194–221.

Knopoff, Leon. "Beno Gutenberg." *Biographical Memoirs of the National Academy of Sciences* 76 (1999): 115–47.

Koczy, F. F. "A Survey on Deep-Sea Features Taken during the Swedish Deep-Sea Expedition." *Deep-Sea Research* 1 (1954): 176–84.

Kramers, H. A. "The Scientists' Role in International Relations." *Bulletin of the Atomic Scientists* 2 (Dec 1946): 12–13.

Krige, John. "NATO and the Strengthening of Western Science in the Post-Sputnik Era." *Minerva* 38 (2000): 81–108.

Krige, John, and Dominique Pestre, eds. *Science in the Twentieth Century.* Amsterdam: Harwood Academic Publishers, 1997.

Kuhn, Thomas S. *The Structure of Scientific Revolutions.* Chicago: University of Chicago Press, 1996.

Kuno, Hisashi, Robert L. Fisher, and Noriyuki Nasu. "Rock Fragments and Pebbles Dredged near Jimmu Seamount, Northwestern Pacific." *Deep-Sea Research* 8 (1956): 126–33.

LaFeber, Walter. "Technology and U.S. Foreign Relations." *Diplomatic History* 24 (Winter 2000): 1–19.

Lakoff, Sanford A., ed. *Knowledge and Power: Essays on Science and Government.* New York: Free Press, 1966.

Laudan, Larry. *Beyond Positivism and Relativism: Theory, Method, and Evidence.* Boulder, CO: Westview Press, 1996.

Lecourt, Dominique. *Proletarian Science? The Case of Lysenko.* Norfolk: New Left Books, 1977.

Lederer, William J., and Eugene Burdick. *The Ugly American.* New York: Norton, 1959.

Leffler, Melvyn P. *A Preponderance of Power: National Security, the Truman Administration, and the Cold War.* Stanford: Stanford University Press, 1992.

Legislative Reference Service, Library of Congress. *The Evolution of International Technology.* Washington, DC: Government Printing Office, 1970.

———. *Toward a New Diplomacy in a Scientific Age.* Washington, DC: Government Printing Office, 1970.

LeGrand, H. E. *Drifting Continents and Shifting Theories: The Modern Revolution in Geology and Scientific Change.* Cambridge: Cambridge University Press, 1988.

Lenz, Walter, and Margaret Deacon, eds. *Ocean Sciences: Their History and Relation to Man.* Hamburg: Bundesamt für Seeschiffahrt und Hydrographie, 1990.

Leslie, Stuart W. *The Cold War and American Science: The Military-Industrial-Academic Complex at MIT and Stanford.* New York: Columbia University Press, 1993.

———. "Playing the Education Game to Win: the Military and Interdisciplinary Research at Stanford." *Historical Studies in the Physical and Biological Sciences* 18 (1987).

Library of Congress Legislative Reference Service. *Abridged Chronology of Events Related to Federal Legislation for Oceanography, 1956–66.* Washington, DC: Government Printing Office, 1966.

Lill, Gordon G. "Office of Naval Research Laboratory of Oceanography and Hydraulics Laboratory, Woods Hole, Massachusetts." *Nature* 173, no. 4413 (29 May 1954): 1017–19.

Limburg, Peter R. *Oceanographic Institutions: Science Studies the Sea.* New York: Elsevier/Nelson Books, 1979.

Lisitzin, A. P., and A. V. Zhivago, "Marine Geological Work of the Soviet Antarctic Expedition, 1955–1957." *Deep-Sea Research* 6 (1960): 77–87.

Little, Arthur D. "Natural Resources in Their Relation to Military Supplies." In *Annual Report of the Board of Regents of the Smithsonian Institution, 1919.* Washington, DC: Government Printing Office, 1921, 211–38.

Lockwood, John P. "Soviet Geology and Geologists." *Geotimes* 13 (May–Jun 1968): 15.

Logue, John J., ed. *The Fate of the Oceans.* Villanova, PA: Villanova University Press, 1972.

London, Ivan D. "Toward a Realistic Appraisal of Soviet Science." *Bulletin of the Atomic Scientists* 13 (May 1957): 169–73, 176.

Lowen, Rebecca S. *Creating the Cold War University: The Transformation of Stanford.* Berkeley: University of California Press, 1997.

Macleod, Roy. "'Kriegsgeologen and Practical Men': Military Geology and Modern Memory, 1914–18." *British Journal for the History of Science* 28 (1995): 427–50.

———, ed. *Science and the Pacific War: Science and Survival in the Pacific, 1939–1945.* Dordrecht: Kluwer, 2000.

Macleod, Roy, and Philip F. Rehbock. "Developing a Sense of the Pacific: The 1923 Pan-Pacific Science Congress in Australia." *Pacific Science* 54, no. 3 (2000): 209–26.

———, eds. *Nature in Its Greatest Extent: Western Science in the Pacific.* Honolulu: University of Hawaii Press, 1988.

Malone, Thomas F., Edward D. Goldberg, and Walter H. Munk. "Roger Randall Dougan Revelle." *Biographical Memoirs of the National Academy of Sciences* 75 (1998): 289–309.

Manzione, Joseph. "'Amusing and Amazing and Practical and Military': The Legacy of Scientific Internationalism in American Foreign Policy, 1945–1963." *Diplomatic History* 24 (Winter 2000): 21–55.

McBride, William M. "The 'Greatest Patron of Science'? The Navy-Academia Alliance and U.S. Naval Research, 1896–1923." *Journal of Military History* 56 (Jan 1992): 7–33.

McDougall, Walter. "The Cold War Excursion of Science." *Diplomatic History* 24 (Winter 2000): 119–27.

———. *The Heavens and the Earth: A Political History of the Space Age.* New York: Basic Books, 1985.

McEvoy, Arthur F., and Harry N. Scheiber. "Scientists, Entrepreneurs, and the Policy Process: A Study of the Post-1945 California Sardine Depletion." *Journal of Economic History* 44 (1984): 393–406.

McKenzie, D. P., and R. L. Parker. "The North Pacific: An Example of Tectonics of a Sphere." *Nature* 216 (1967): 1276–80.

McMahon, Robert J. *The Cold War on the Periphery: The United States, India, and Pakistan.* New York: Columbia University Press, 1994.

Menard, H. W. "Consolidated Slabs on the Floor of the Eastern Pacific." *Deep-Sea Research* 7 (1960): 35–41.

———. *The Ocean of Truth: A Personal History of Global Tectonics.* Princeton: Princeton University Press, 1986.

———. *Science: Growth and Change.* Cambridge: Harvard University Press, 1971.

Mendelsohn, Everett, and Merritt Roe Smith. *Science, Technology and the Military.* Dordrecht: Kluwer, 1988.

Bibliography

Merriam, Charles. "Physics and Politics." *Bulletin of the Atomic Scientists* 1 (May 1946): 9–11.

Merton, Robert K. "The Seven Propositions of Professor Ridenour." *Bulletin of the Atomic Scientists* 3 (Aug 1947): 225.

Mikhal'tsev, I. Ye. "First Academy of Sciences U.S.S.R. Atlantic Expedition in the *Sergei Vavilov* and *Petr Lebedev*." *Deep-Sea Research* 10 (1963): 771–74. Translated from *Okeanologiya* 1 (1961): 1089–93.

Miles, Edward. "COSPAR and SCOR: The Political Influence of the Committee on Space Research and the Scientific Committee on Oceanic Research." In Evan, *Knowledge and Power in a Global Society*, 133–52.

Millikan, Robert A. "The Interchange of Men of Science." *Proceedings of the American Philosophical Society* 91, no. 1 (Feb 1947): 129–32.

Mills, Eric L. *Biological Oceanography: An Early History, 1870–1960*. Ithaca, NY: Cornell University Press, 1989.

———. "From Marine Ecology to Biological Oceanography." *Helgoländer Meeresuntersuchungen* 49 (1995): 29–44.

———. "The Historian of Science and Oceanography after Twenty Years." *Earth Sciences History* 12 (1993): 5–18.

———. "The Oceanography of the Pacific: George F. McEwen, H. U. Sverdrup and the Origin of Physical Oceanography on the West Coast of North America." *Annals of Science* 48 (1991): 241–66.

———. "Oceanography, Physical: Disciplinary History." In *Sciences of the Earth: An Encyclopedia of Events, People, and Phenomena*, edited by G. A. Good, 630–36. New York: Garland Publishing, 1998.

———. "'Physisches Meereskunde': From Geography to Physical Oceanography in the Institut für Meereskunde, Berlin, 1900–1935." *Historisch-Meereskundliches Jahrbuch* 4 (1997): 45–70.

———. "Socializing Solenoids: The Acceptance of Dynamic Oceanography in Germany around the Time of the 'Meteor' Expedition." *Historisch-Meereskundliches Jahrbuch* 5 (1998): 11–26.

———. "Useful in Many Capacities: An Early Career in American Physical Oceanography." *Historical Studies in the Physical and Biological Sciences* 20 (1990): 265–311.

Morgan, Jason. "Rises, Trenches, Great Faults, and Crustal Blocks." *Journal of Geophysical Research* 73 (1968): 1959–82.

Morrison, Philip. "Science Should Be Kept Free." *Bulletin of the Atomic Scientists* 3 (Aug 1947): 224.

Mukerji, Chandra. *A Fragile Power: Scientists and the State*. Princeton: Princeton University Press, 1990.

Munk, W. H. "Harald Ulrik Sverdrup (1888–1957)." *Deep-Sea Research* 4 (1957): 289–90.
Nakayama, Shigeru. *Science, Technology, and Society in Postwar Japan*. London: Kegan Paul, 1991.
Nakayama, Shigeru, David L. Swain, and Yagi Eri, eds. *Science and Society in Modern Japan*. Cambridge: MIT Press, 1974.
National Academy of Sciences–National Research Council. *Economic Benefits from Oceanographic Research*. Washington, DC: National Academy of Sciences–National Research Council, 1964.
———. *Oceanography 1960 to 1970: A Report by the Committee on Oceanography*. Washington, DC: National Academy of Sciences–National Research Council, 1959.
———. *Oceanography 1966: Achievements and Opportunities*. Washington, DC: National Academy of Sciences–National Research Council, 1967.
Needell, Allan A. "Preparing for the Space Age: University-based Research, 1946–1957." *Historical Studies in the Physical and Biological Sciences* 18 (1987).
———. *Science, Cold War, and the American State: Lloyd V. Berkner and the Balance of Professional Ideals*. Amsterdam: Harwood, 2000.
———. "'Truth Is Our Weapon': Project TROY, Political Warfare, and Government-Academic Relations in the National Security State." *Diplomatic History* 17 (1993): 399–420.
Nelkin, Dorothy. *The University and Military Research: Moral Politics at M.I.T.* Ithaca, NY: Cornell University Press, 1972.
Neushul, Peter. "Science, Technology, and the Arsenal of Democracy." Ph.D. diss., University of California, Santa Barbara, 1993.
Nicolet, M. "Historical Aspects of the IGY." *Eos* 64 (1983): 369–70.
Nierenberg, William A. "Harald Ulrik Sverdrup." *Biographical Memoirs of the National Academy of Sciences* 69 (1996): 339–74.
Noyes, Albert, Jr. "Do We Need a Foreign Policy in Science?" *Bulletin of the Atomic Scientists* 13 (Sep 1957): 234–37.
———. "The United Nations Educational, Scientific and Cultural Organization." *Bulletin of the Atomic Scientists* 2 (Aug 1946): 16–17.
———. "The United Nations Educational, Scientific and Cultural Organization." *Proceedings of the American Philosophical Society* 91, no. 1 (Feb 1947): 133–36.
Olchi-Oglu, N. I. "A Description of Some Foreign Research Vessels and Their Equipment." *Deep-Sea Research* 10 (1963): 87–91. Translated from *Okeanologiya* 1 (1961): 763–69.
Oldroyd, David R. *Thinking about the Earth: A History of Ideas in Geology*. Cambridge: Harvard University Press, 1996.
Oliver, J., L. Sykes, and B. Isacks. "Seismology and the New Global Tectonics." *Tectonophysics* 7 (1969): 527–41.

Bibliography

Oreskes, Naomi. "*Laissez-tomber:* Military Patronage and Women's Work in Mid-20th Century Oceanography." *Historical Studies in the Physical and Biological Sciences* 30 (2000): 373–92.

———. *The Rejection of Continental Drift: Theory and Method in American Earth Science.* New York: Oxford, 1999.

———. "Weighing the Earth from a Submarine: The Gravity Measuring Cruise of the U.S.S. S-21." In Good, *The Earth, the Heavens, and the Carnegie Institution of Washington.*

Oreskes, Naomi, and Ronald Rainger. "Science and Security before the Atomic Bomb: The Loyalty Case of Harald U. Sverdrup." *Studies in the History and Philosophy of Modern Physics* 31 (2000): 309–69.

Ovey, C. D. "Forward." *Deep-Sea Research* 1 (1953): 1–2.

———. "The Work of the Joint Commission on Oceanography 1951–54, of the International Council of Scientific Unions." *Deep-Sea Research* 2 (1954): 159.

Palladino, Paolo, and Michael Worboys. "Science and Imperialism." *Isis* 84 (1993): 91–102.

Pardo, Arvid. "Who Will Control the Seabed?" *Foreign Affairs* 47 (Oct 1968): 123–37.

Pells, Richard H. *The Liberal Mind in a Conservative Age: American Intellectuals in the 1940s and 1950s.* New York: Harper & Row, 1985.

Penick, James, Carroll Pursell, Morgan Sherwood, and Donald Swain, eds. *The Politics of American Science, 1939 to the Present.* Cambridge: MIT Press, 1972.

Pennington, Howard. *The New Ocean Explorers: Into the Sea in the Space Age.* Boston: Little, Brown, 1972.

Perkins, James A. "Foreign Aid and the Brain Drain." *Foreign Affairs* 44 (Jul 1966): 608–19.

Pernet, Ann, ed. *International Science.* Guernsey, British Isles: Francis Hodgson, 1976.

Pettersson, Hans. "The Swedish Deep-Sea Expedition, 1947–48." *Deep-Sea Research* 1 (1953): 17–24.

Polanyi, Michael. "The Republic of Science: Its Political and Economic Theory." *Minerva* 1 (1962): 54–73. Reprinted in *Minerva* 38 (2000): 1–32.

Potter, E. B. *Admiral Arleigh Burke.* New York: Random House, 1990.

Price, Derek J. de Solla. *Little Science, Big Science.* New York: Columbia University Press, 1963.

Price, Don K. "Money and Influence: The Links of Science to Public Policy." *Daedalus* 103 (1974).

———. *The Scientific Estate.* Cambridge: Harvard University Press, 1965.

Price, Matt. "Roots of Dissent: The Chicago Met Lab and the Origins of the Franck Report." *Isis* 86 (1995): 222–44.

Pursell, Carroll W., Jr., ed. *The Military-Industrial Complex.* New York: Harper & Row, 1972.

Pyenson, Lewis. "Assimilation and Innovation in Indonesian Science." *Osiris* 13 (1998): 34–47.

———. *Civilizing Mission: Exact Sciences and French Overseas Expansion, 1830–1940.* Baltimore: Johns Hopkins University Press, 1993.

———. *Cultural Imperialism and Exact Sciences: German Expansion Overseas, 1900–1930.* New York: Peter Lang, 1985.

———. "Cultural Imperialism and Exact Sciences Revisited." *Isis* 84 (1993): 103–8.

———. *Empire of Reason: Exact Sciences in Indonesia, 1840–1940.* New York: E. J. Brill, 1989.

Rabinowitch, Eugene. "Editorial: International Cooperation of Scientists." *Bulletin of the Atomic Scientists* 2 (Sep 1946): 1.

Rabkin, Yakov M. *Science between the Superpowers.* New York: Priority Press, 1988.

Rainger, Ronald. "Constructing a Landscape for Postwar Science: Roger Revelle, the Scripps Institution and the University of California, San Diego." *Minerva* 39 (2001): 327–52.

———. "Patronage and Science: Roger Revelle, the US Navy, and Oceanography at the Scripps Institution." *Earth Sciences History* 19 (2000): 58–89.

———. "Science at the Crossroads: The Navy, Bikini Atoll, and American Oceanography in the 1940s." *Historical Studies in the Physical and Biological Sciences* 30 (2000): 349–71.

Raitt, Helen, and Beatrice Moulton. *Scripps Institution of Oceanography: First Fifty Years.* Los Angeles: Ward Ritchie Press, 1967.

Rao, T. S. S., and Ray C. Griffiths. *Understanding the Indian Ocean: Perspectives on Oceanography.* Paris: Unesco, 1998.

Rehbock, Philip F. "Organizing Pacific Science: Local and International Origins of the Pacific Science Association." In Macleod and Rehbock, *Nature in Its Greatest Extent*, 195–221.

Reingold, Nathan. "Choosing the Future: The U.S. Research Community, 1944–1946." *Historical Studies in the Physical and Biological Sciences* 25 (1995): 301–28.

———. "Vannevar Bush's New Deal for Research, or The Triumph of the Old Order." *Historical Studies in the Physical and Biological Sciences* 17 (1987): 299–344.

Revelle, Roger. "Harrison Brown." *Biographical Memoirs of the National Academy of Sciences* 65 (1994): 41–55.

———. "The Oceanographic and How It Grew." In Sears and Merriman, *Oceanography*, 10–24.

Revelle, Roger, and Arthur E. Maxwell. "Heat Flow through the Floor of the Eastern North Pacific Ocean." *Nature* 170, no. 4318 (Aug 1952): 199–200.

Reynolds, Clark G. *History and the Sea: Essays on Maritime Strategies.* Columbia: University of South Carolina Press, 1989.

Richards, Adrian F. "Intergovernmental Oceanographic Commission Meeting, Paris, 10–19 June 1964." Office of Naval Research report ONRL-C-15-64, Sep 23, 1964.

Richardson, Philip L. "The Benjamin Franklin and Timothy Folger Charts of the Gulf Stream." In Sears and Merriman, *Oceanography,* 703–17.

Ridenour, Louis N. "Military Support of American Science, a Danger?" *Bulletin of the Atomic Scientists* 3 (Aug 1947): 221–23.

———. "Secrecy in Science." *Bulletin of the Atomic Scientists* 1 (Mar 1946): 3, 8.

Roland, Alex. "Science and War." *Osiris* 1 (1985): 247–72.

Roll, Hans Ulrich. *Intergovernmental Oceanographic Commission (IOC): History, Functions, Achievements.* Paris: Unesco, 1979.

Rosenberg, David Alan. "American Naval Strategy in the Era of the Third World War: An Inquiry into the Structure and Process of General War at Sea, 1945–90." In *Naval Power in the Twentieth Century,* edited by N. A. M. Rodger, 242–54. Annapolis: Naval Institute Press, 1996.

Rostow, W. W. *Eisenhower, Kennedy, and Foreign Aid.* Austin: University of Texas Press, 1985.

Rothenberg, Marc. "'In Behalf of the Science of the Country': The Smithsonian and the U.S. Navy in the North Pacific in the 1850s." *Pacific Science* 52 (1998): 301–7.

Rozwadowski, Helen. *The Sea Knows No Boundaries: A Century of Marine Science under ICES.* Seattle and London: University of Washington Press and International Council for the Exploration of the Sea, 2002.

Rusk, Dean. "Building an International Community of Science and Scholarship." *Department of State Bulletin* 44 (May 1961): 624–28.

Salop, L. I., and Yu. M. Scheinmann. "Tectonic History and Structures of Platforms and Shields." *Tectonophysics* 7 (1969): 565–97.

Sapolsky, Harvey. *Science and the Navy: A History of the Office of Naval Research.* Princeton: Princeton University Press, 1990.

Satofuka, Fumihiko, and Abdur Rahman. "Post War Science and Technology Policy of Japan: A View from a Developing Country." *Historia Scientiarum* 33 (1987).

Schlee, Susan. *The Edge of an Unfamiliar World: A History of Oceanography.* New York: E. P. Dutton, 1973.

———. *On Almost Any Wind: The Saga of the Oceanographic Research Vessel Atlantis.* Ithaca, NY: Cornell University Press, 1978.

Schneer, Cecil J., ed. *Two Hundred Years of Geology in America.* Hanover: University of New Hampshire Press, 1979.

Schrecker, Ellen W. *No Ivory Tower: McCarthyism and the Universities.* New York: Oxford University Press, 1986.

Schwarzbach, Martin, ed. *Alfred Wegener: The Father of Continental Drift.* Madison: Science Tech, Inc., 1986.

Sears, Mary, and Daniel Merriman, eds. *Oceanography: The Past.* New York: Springer-Verlag, 1980.

Seidel, Robert W. "A Home for Big Science: The Atomic Energy Commission's Laboratory System." *Historical Studies in the Physical and Biological Sciences* 16 (1986).

Shils, Edward A. *The Torment of Secrecy: The Background and Consequences of American Security Policies.* Glencoe, IL: Free Press, 1956.

Shor, Elizabeth Noble. *Scripps Institution of Oceanography: Probing the Oceans, 1936 to 1976.* San Diego: Tofua Press, 1978.

Simpson, J. A., Jr. "A Scientist's Visit to England and France." *Bulletin of the Atomic Scientists* 1 (Apr 1946): 16–17.

Siple, Paul. *90 South: The Story of the American South Pole Conquest.* New York: G. P. Putnam's Sons, 1959.

Skolnikoff, Eugene B. *The Elusive Transformation: Science, Technology, and the Evolution of International Politics.* Princeton: Princeton University Press, 1993.

———. *The International Imperatives of Technology: Technological Development and the International Political System.* Berkeley: Institute of International Studies, 1972.

———. *Science, Technology, and American Foreign Policy.* Cambridge: MIT Press, 1967.

———. "Scientific Advice in the State Department." *Science* 154 (Nov 1966): 980–85.

Smith, Alice Kimball. *A Peril and a Hope: The Scientists Movement in America, 1945–1947.* Chicago: University of Chicago Press, 1965.

Smith, Crosbie, and Jon Agar, eds. *Making Space for Science: Territorial Themes in the Shaping of Knowledge.* New York: St. Martin's Press, 1998.

Smith, Merritt Roe, ed. *Military Enterprise and Technological Change: Perspectives on the American Experience.* Cambridge: MIT Press, 1985.

Snead, David L. *The Gaither Committee, Eisenhower, and the Cold War.* Columbus: Ohio State University Press, 1999.

Söderqvist, Thomas, ed. *The Historiography of Contemporary Science and Technology.* Amsterdam: Harwood Academic Publishers, 1997.

Solo, Robert A. *Organizing Science for Technology Transfer in Economic Development.* East Lansing: Michigan State University Press, 1975.

Solov'yev, B. S. "Summary of Work by the Second Azcherniro Indian Ocean Expedition." *Deep-Sea Research* 11 (1964): 478–79.

Stevenson, Adlai E. "Science, Diplomacy, and Peace." *Department of State Bulletin* 45 (Sep 1961): 402–7.

Stewart, John A. *Drifting Continents and Colliding Paradigms: Perspectives on the Geoscience Revolution.* Bloomington: Indiana University Press, 1990.

Stockman, Robert H. *The Intergovernmental Oceanographic Commission: An Uncertain Future.* Seattle: Division of Marine Resources, University of Washington, 1974.
Strickland, Donald A. *Scientists in Politics: The Atomic Scientists Movement, 1945–46.* Purdue: Purdue Research Foundation, 1968.
Sullivan, Walter. *Assault on the Unknown: The International Geophysical Year.* New York: McGraw-Hill, 1961.
———. *Continents in Motion: The New Earth Debate.* New York: McGraw-Hill, 1974.
Sverdrup, Harald Ulrik. "New International Aspects of Oceanography." *Proceedings of the American Philosophical Society* 91, no. 1 (Feb 1947): 75–78.
Sverdrup, Harald Ulrik, Martin W. Johnson, and Richard H. Fleming. *The Oceans: Their Physics, Chemistry, and General Biology.* New York: Prentice Hall, 1942.
Taft, William H., III. "The United States Scientific Attaché Program." *Department of State Bulletin* 52 (Jan 1965): 113–15.
Tickner, Fred. *Technical Cooperation.* New York: Praeger, 1966.
Till, Geoffrey, ed. *Maritime Strategy and the Nuclear Age.* New York: St. Martin's Press, 1984.
Troebst, Cord-Christian. *Conquest of the Sea.* Translated from German by Brian C. Price and Elsbeth Price. New York: Harper & Row, 1962.
Turkevich, John. "Soviet Science Appraised." *Foreign Affairs* 44 (Apr 1966): 489–500.
United States Commission on Marine Science, Engineering and Resources. *Our Nation and the Sea: A Plan for National Action.* Washington, DC: Government Printing Office, 1969.
United States Congress. House of Representatives. Committee on Foreign Affairs. *Foreign Assistance Act of 1962, Hearings,* 87th Cong. Washington, DC: Government Printing Office, 1962.
———. Committee on Foreign Affairs. *Foreign Assistance Act of 1963: Hearings,* 88th Cong. Washington, DC: Government Printing Office, 1963.
———. Committee on Foreign Affairs. *Foreign Assistance Act of 1964: Hearings,* 88th Cong. Washington, DC: Government Printing Office, 1964.
———. Special Subcommittee on Oceanography of the Committee on Merchant Marine and Fisheries. *National Marine Sciences Program: Hearings,* 90th Cong., 1st sess. Washington, DC: Government Printing Office, 1968.
———. Special Subcommittee on Oceanography of the Committee on Merchant Marine and Fisheries. *Oceanography: Hearings,* 86th Cong., 2d sess. Washington, DC: Government Printing Office, 1960.
———. Special Subcommittee on Oceanography of the Committee on Merchant Marine and Fisheries. *Oceanography in the United States: Hearings,* 86th Cong., 1st sess. Washington, DC: Government Printing Office, 1959.
———. Special Subcommittee on Oceanography of the Committee on Merchant

Marine and Fisheries. *Oceanography Legislation: Hearings*, 90th Cong. Washington, DC: Government Printing Office, 1968.

———. Subcommittee of the Committee on Interstate and Foreign Commerce. *United States Postinternational Geophysical Year Scientific Program: Hearings*, 85th Cong., 2d sess., 26 Mar 1958.

———. Subcommittee on Earth Sciences of the Committee on Science and Astronautics. Hearings. *Education in the Field of Oceanography: Hearings on H.R. 6298*, 86th Cong., 1st sess., 25 Aug 1959. Washington, DC: Government Printing Office, 1959.

———. Subcommittee on Fisheries and Wildlife Conservation and the Subcommittee on Oceanography of the Committee on Merchant Marine and Fisheries. *Ocean Dumping of Waste Materials: Hearings*, 92d Cong., 1st sess. Washington, DC: Government Printing Office, 1971.

———. Subcommittee on Oceanography of the Committee on Merchant Marine and Fisheries. *Mohole Project: Hearings*, 88th Cong., 1st sess. Washington, DC: Government Printing Office, 1963.

———. Subcommittee on Oceanography of the Committee on Merchant Marine and Fisheries. *National Oceanographic Program—1965: Hearings*, 88th Cong., 2d sess. Washington, DC: Government Printing Office, 1964.

———. Subcommittee on Oceanography of the Committee on Merchant Marine and Fisheries. *National Oceanographic Program Legislation: Hearings*, 89th Cong., 1st sess. Washington, DC: Government Printing Office, 1965.

———. Subcommittee on Oceanography of the Committee on Merchant Marine and Fisheries. *Oceanography—Ships of Opportunity: Hearings*, 89th Cong., 1st sess. Washington, DC: Government Printing Office, 1965.

———. Subcommittee on Oceanography of the Committee on Merchant Marine and Fisheries. *Oceanography 1961—Phase 1: Hearings*, 87th Cong., 1st sess. Washington, DC: Government Printing Office, 1961.

———. Subcommittee on Oceanography of the Committee on Merchant Marine and Fisheries. *Oceanography 1961—Phase 2: Hearings*, 87th Cong., 1st sess. Washington, DC: Government Printing Office, 1961.

———. Subcommittee on Oceanography of the Committee on Merchant Marine and Fisheries. *Oceanography 1961—Phase 3: Hearings*, 87th Cong., 1st sess. Washington, DC: Government Printing Office, 1961.

———. Subcommittee on Oceanography of the Committee on Merchant Marine and Fisheries. *Study of the Effectiveness of the Committee on Oceanography of the Federal Council for Science and Technology: Hearings*, 87th Cong., 2d sess., 28 Feb and 1–2 March 1962. Washington, DC: Government Printing Office, 1962.

United States Congress. Senate. Committee on Interstate and Foreign Commerce. *Marine Science: Hearings,* 87th Cong., 1st sess., 15 Mar–2 May 1961.

———. Committee on Interstate and Foreign Commerce. *Marine Science: Hearings on S. 901 and S. 1189,* 87th Cong., 1st sess., 15–17 March and 2 May 1961. Washington, DC: Government Printing Office.

Van Allen, J. A. "Genesis of the International Geophysical Year." *Eos* 64 (1983).

Van Keuren, David K. "Science, Progressivism, and Military Preparedness: The Case of the Naval Research Laboratory, 1915–1923." *Technology and Culture* 33 (Oct 1992): 710–36.

Vessuri, Hebe M. C. "Foreign Scientists, the Rockefeller Foundation and the Origins of Agricultural Science in Venezuela." *Minerva* 32 (1994): 267–96.

Vine, F. J., and D. H. Matthews. "Magnetic Anomalies over Oceanic Ridges." *Nature* 199 (1963): 947–49.

Waggoner, Walter H. "State Department Urges U. S. Office of World Science." *New York Times,* 5 Jun 1950, 1.

Wallerstein, Mitchel B., ed. *Scientific and Technological Cooperation among Industrialized Countries: The Role of the United States.* Washington, DC: National Academy Press, 1984.

Wang, Jessica. *American Science in an Age of Anxiety: Scientists, Anticommunism, and the Cold War.* Chapel Hill: University of North Carolina Press, 1999.

———. "Liberals, the Progressive Left, and the Political Economy of Postwar American Science: The National Science Foundation Debate Revisited." *Historical Studies in the Physical and Biological Sciences* 26 (1995): 139–66.

———. "Science, Security, and the Cold War: The Case of E. U. Condon." *Isis* 83 (1992): 238–69.

Wang, Zuoyue. "American Science and the Cold War: The Rise of the US President's Science Advisory Committee." Ph.D. diss., University of California, Santa Barbara, 1994.

Waterman, Alan T., and Robert D. Conrad. "Office of Naval Research Discusses Ridenour's Views." *Bulletin of the Atomic Scientists* 3 (Aug 1947): 230.

Weinberg, Alvin. *Reflections on Big Science.* Cambridge: MIT Press, 1967.

Weir, Gary E. *Forged in War: The Naval-Industrial Complex and American Submarine Construction, 1940–1961.* Washington, DC: Naval Historical Center, 1993.

———. *An Ocean in Common: American Naval Officers, Scientists, and the Ocean Environment.* College Station: Texas A&M Press, 2001.

Wenk, Edward, Jr. *Making Waves: Engineering, Politics, and the Social Management of Technology.* Urbana: University of Illinois Press, 1995.

———. *The Politics of the Ocean.* Seattle: University of Washington Press, 1972.

Wertenbaker, William. *The Floor of the Sea: Maurice Ewing and the Search to Understand the Earth.* Boston: Little, Brown, 1974.

Wiener, Norbert. "The Armed Services Are Not Fit Almoners for Research." *Bulletin of the Atomic Scientists* 3 (Aug 1947): 228.

Wiesner, Jerome B. *Where Science and Politics Meet.* New York: McGraw-Hill, 1965.

Wilson, J. Tuzo, ed. *Continents Adrift: Readings from Scientific American.* San Francisco: W. H. Freeman, 1972.

———. *IGY: The Year of the New Moons.* New York: Knopf, 1961.

———. "Preface." In Wilson, *Continents Adrift.*

———. "A Reply to V.V. Beloussov." *Geotimes* 13 (1968): 21–22.

———. "A Revolution in Earth Science." *Geotimes* 13 (1968): 13–16.

Wiseman, John D. H. "International Collaboration in Deep-Sea Research." *Deep-Sea Research* 1 (1953): 3–10.

Wiseman, John D. H., and William R. Riedel, "Tertiary Sediments from the Floor of the Indian Ocean." *Deep-Sea Research* 7 (1961): 215–17.

Wohlstetter, Albert. "Scientists, Seers and Strategy." *Foreign Affairs* 41 (Apr 1963): 466–78.

Wolff, Torben. "The Creation and First Years of SCOR (Scientific Committee on Oceanic Research)." In Lenz and Deacon, *Ocean Sciences,* 337–43.

———. "The 50 Years' Anniversary of the Galathea Deep-Sea Expedition 1950–1952." *History of Oceanography* 12 (Sep 2000): 5–9.

Wolfle, Dael. *Renewing a Scientific Society: The American Association for the Advancement of Science from World War II to 1970.* Washington, DC: American Association for the Advancement of Science, 1989.

Wood, Robert Muir. *The Dark Side of the Earth: The Battle for the Earth Sciences, 1800–1980.* London: Allen & Unwin, 1985.

Woodward, Llewellyn. "Science and the Relations between States." *Bulletin of the Atomic Scientists* 12 (Apr 1956): 119–24.

Wooster, Warren S., ed. *Freedom of Oceanic Research.* New York: Crane, Russak & Company, 1973.

Wright, Quincy. "On the Application of Intelligence to World Affairs." *Bulletin of the Atomic Scientists* 4 (Apr 1948): 249–52.

Wüst, Georg. "Proposed International Indian Ocean Expedition, 1962–1963." *Deep-Sea Research* 6 (1959): 245–49.

Yang, Jing Yi, and David Oldroyd. "The Introduction and Development of Continental Drift Theory and Plate Tectonics in China: A Case Study in the Transference of Scientific Ideas from West to East." *Annals of Science* 46 (1989).

York, Herbert F. *Making Weapons, Talking Peace: A Physicist's Odyssey from Hiroshima to Geneva.* New York: Basic Books, 1987.

Yoshida, Kozo, ed. *Studies in Oceanography: A Collection of Papers Dedicated to Koji Hidaka.* Seattle: University of Washington Press, 1965.

Young, Donald. "Subsidies Acceptable with Proper Safeguards." *Bulletin of the Atomic Scientists* 3 (Aug 1947): 229.

Zachary, G. Pascal. *Endless Frontier: Vannevar Bush, Engineer of the American Century.* Cambridge: MIT Press, 1999.

Zaklinskii, G. B. "Deep Circulation of Water in the Indian Ocean." *Deep-Sea Research* 11 (1964): 286–92.

Zenkevich, L. A. *Biology of the Seas of the USSR.* New York: Interscience Publishers, 1963.

———. "Importance of Studies of the Ocean Depths." *Deep-Sea Research* 4 (1956): 67–70.

INDEX

Acheson, Dean, 20
Acoustics, xii, xx, xxvi, 9, 34, 42, 58, 78, 149
Afghanistan, 19, 86
Africa, 164–65. *See also* East African Marine Fisheries Research Organization; East African Rift; *specific countries*
Agnew, Spiro T., 243
Air Force Cambridge Research Center, 67
Akers, Frank, 39
Albatross, 79
American Association for the Advancement of Science (AAAS), 117–18
American Miscellaneous Society (AMSOC), 159–63, 165
American Revolution, xix
American Scholar, The, 34
Antarctica, xxvii, 7, 201; and the IGY, 56-57, 61–66, 70, 76, 80, 82, 88–89, 95, 97, 261; and public interest, 133–34, 265
Antarctic Convergence, 76
Antarctic Treaty, 61–62, 97
Anti-submarine warfare (ASW). *See* Undersea warfare
Anton Bruun, 132
Anton Dohrn, 69
Arctic Ice Reconnaissance Project, 78
Arctic Institute (Leningrad), 70, 88
Arctic Sealer, 47

Argentina, 64, 82
Arms race, 62
Asahi Press, 26
Asia, 5–6, 16, 18, 84, 105–7, 110, 122, 134, 185, 217, 224. *See also specific countries*
Atlantis, 69, 77, 79
Atomic bomb. *See* Nuclear weapons
Atomic energy, 51
Atomic Energy Commission. *See* United States government
Austria: expeditions, xix

Baird. See *Spencer F. Baird*
Banse, Karl, 131
Bardin, Ivan P., 92
Barron, Arthur S., 128
Bartlett, Bob, 238–39
Basalla, George, 4
Basaltification. *See* Oceanization
Bascom, Willard, 159–63, 168
Bathymetry. *See* Charts
Bathythermograph (BT). *See* Instruments
Bayonnaise rocks, 26–27
Bellingshausen, Fabian Gottlieb, 64–65, 89
Bell Telephone Laboratories, 11, 40
Beloussov, Vladimir, 63, 81, 87, 160, 164–66, 203, 211, 215; opposition to continental drift, 194–200. *See also* Oceanization

Bennett, Rawson, 71
Berkner, Lloyd: and Berkner report, 21; and classification, 53, 56; and the IGY, 61–63, 66, 73, 80, 85–86, 90, 94; and SCOR, 111, 114–15
Bezrukov, Panteleimon L., 209
Bigelow, Henry B., xxiv, 8
Bikini Scientific Resurvey, 11
Biological oceanography, xxiv, xxvii, 9, 22, 112, 131–32
Bjerknes, Vilhelm, 8
Bogdanov, Aleksei, 206–7, 210
Böhnecke, Günther, 67, 69–70, 80, 99, 111, 120, 174
Bolivia, 73
Bonner, Herbert C., 153
Booth, E. T., 230
Bowen, Harold, 36, 38
Bradner, Hugh, 205–6
Brattström, Hans, 104
Brazil, 85, 91, 163
Britain. *See* United Kingdom
British Museum, 52
Brode, Wallace, 129–30
Bronk, Detlev, 143, 156, 163
Brooks, Overton, 148
Brown, Harrison, 119, 143, 154, 247
Brown Bear, 77
Browne, B. C., 10–11
Bruun, Anton, 66, 68, 99, 226, 231; death of, 170; and the IIOE, 132–33; and Unesco, 102, 105–7, 110; and SCOR, 111, 114
Bullard, Edward, 117, 186, 223; and postwar marine geophysics, 8–12; and the IGY, 67; and Soviet activities, 86–87, 188; and continental drift, 198–99
Bulletin of the Atomic Scientists, 34–35, 96, 127
Bulletin of the Geological Society of America, 27
Buoys. *See* Unmanned stations
Burke, Arleigh, 38, 45–47, 54, 89, 97, 142, 231–32

Bush, Vannevar, xxvi, 35–36, 42, 45, 50–52, 55
Byrd, Richard E., 57, 65

Cabot, Operation, 44
Caesar, Project, 39, 47
Calhoun, John, 245
California Institute of Technology, 44
Cambridge University, 9–11, 48, 87, 186
Cameron, W. M., 171, 175
Capurro, Luis, 111
Carnegie, 83
Canada, 5–6, 28, 75
Carruthers, J. N., 108
Central Intelligence Agency. *See* United States government
Centre National de la Recherche Scientifique (CNRS), 114
CERN (European Center for Nuclear Research), 109
Ceylon, 121, 123–25
Chain, 77, 231
Challenger, xix
Challenger II, 11, 79, 87
Chapman, Sydney, 61, 63, 66
Chapman, Wilbert, 220, 259
Charles II, 223
Charts, xix–xx, 42, 52, 56, 71, 73, 98, 149, 158, 188–90, 192–93, 205
Chemical oceanography, xxiv
Chile, 64, 75, 82
China: and the IGY, 74, 90, 93, 197; and international meetings, 204, 259
Clark, John E., 93–94
Classification. *See* Classification of noise; Secrecy
Classification of noise, 39–40
Cold War, xvii–xviii, xx, 18, 62, 65, 85, 101, 118, 124, 155, 167, 262, 265; and scientific communities, xxi–xxii, 187–216 passim, 263. *See also* Foreign policy
Coleman, John, 247
Colombo Plan, 18–19
Columbia, 17

Index

Comité Speciale de l'Année Géophysique Internationale (CSAGI), 61, 63, 65–66
Commission on Marine Science, Engineering, and Resources (COMSER), 249
Committee on Space Research (COSPAR), 97, 134
Communism: containing or defeating, xxvii, 14–15, 19–20, 84; influence on science, 96, 157, 177, 209–10
Condon, E. U., 51
Conferences, 6; and travel restrictions, 51, 119, 204. *See also specific meetings*
Congress. *See* United States Congress
Continental drift, 193–202 passim, 210–11, 214; plate tectonics, xxiii, 194, 196, 200, 214–15, 260; ridge and rift systems 77–78, 214; seafloor spreading, 194–98, 203; transform faults, 196–97. *See also* Oceanization
Continental shelf and slope, 10, 80, 238–39, 246, 250
Coolidge, Harold J., 22, 24
Cooperative Investigations of the Caribbean and Adjacent Regions (CICAR), 229
Cooperative Investigations of the Mediterranean (CIM), 229
Cooperative Investigations of the Northern Part of the Eastern Central Atlantic (CINECA), 229
Cooperative Study of the Kuroshio and Adjacent Regions (CSK), 136, 180–81, 184–85, 187, 254
Cornell, S. D., 96
Coulomb, Jean, 63, 114
Crawford, 77
Creager, Joseph, 206
Cromwell, Townsend, 77
Cromwell Current, 77, 232
Crossroads, Operation, 11
Czechoslovakia: Soviet invasion of, 212–14. *See also* International Geological Congress

Daily Telegraph, 222
Dana, 7, 47, 105
Dartmouth College, 56
Darwin, Charles, 199
Data exchange, xviii, xx, 4, 27, 33, 44–45, 48–49, 70, 78, 83, 92–94, 137, 156, 178, 190–91, 215–16, 254, 256, 261
Day, Archibald, 171–72, 181, 184–85, 190, 259
Deacon, George, xxviii, 43, 47, 99, 101–4; and the IOC, 128, 181–82, 187; and creation of SCOR, 110–11, 114–16; and the IGY, 66, 68–70, 72, 80, 82–83, 88–89; and the IIOE, 120–21, 133–34, 183, 186, 221–24; and NATO, 230–36; on Soviet oceanography, 182–84, 216, 229–30
Deacon, Margaret, 223
Defense Department. *See* United States Department of Defense
Deep-Sea Research, 27, 79, 101, 137, 157
Defense Research and Intelligence Board, 38
Denfield, Louis, 37
Denmark, 7, 10, 47, 79
Developing countries, xviii, xxi–xxii, xxvi; agenda for oceanographers, 100, 103–5, 109, 122, 217, 238, 245, 262–63; difficulties with, 134, 181, 228–30; scientific communities, 30, 120, 126, 136; and seabed minerals, 250–53; technical assistance, 13–15, 20–21, 84; and United Nations goals, 129; and U.S.–Soviet competition, 124. *See also specific countries or regions*
Dialectical materialism, 193–94, 199–200
Dietrich, Günther, 65
Dietz, Robert S., 3, 7, 23, 25, 47, 65, 67, 71, 87–88, 199
Discovery (built 1901), 7
Discovery (built 1962), 222–23
Discovery II, 7, 77
Doldrums expedition, 77

Dolphin expedition, 77
Doodson, Arthur T., 113–14
Downwind expedition, 78, 94, 192
Dredging, 12
Droessler, Earl, 53
Dufek, George, 65, 76
Dynamical oceanography. *See* Physical oceanography

East African Marine Fisheries Research Organization (EAMFRO), 220-21
East African Rift, 164–65, 167
Eastern Europe, xxv. *See also specific countries*
Echo sounder. *See* Instruments
Ecuador, 75
Effective Uses of the Sea, 243–46
Egypt, 84–86
Einstein, Albert, 35
Eisenhower, Dwight, 56–57, 59–61, 64, 84, 91, 128, 140, 145, 154
Eisenhower administration, 56–57, 84, 145–46
Eltanin, 195
Emery, K. O., 205
Engel, A. E. J., 210
Environmental monitoring, xxiii, 255–57, 260
EQUAPAC, 30
Equatorial Countercurrent, 77
Ernest Holt, 48
Espionage: fear of, 62, 91, 207
Ethiopia, 165
Ewing, W. Maurice, 117, 152–53; and acoustics, 8–9, 11–12, 34, 39; and Atlantic expeditions, 40, 52; and the IGY, 70, 72, 77; and the IIOE, 132; and ocean rift system, 78
Exchange of scientists, 13, 70, 79, 83, 113, 138, 202
Expeditions, 10–11, 21, 76, 78; scientific value of, xxi, xxii–xxiii; prestige value of, xix, 7. *See also specific expeditions*
Eyriès, Marc, 102, 231

Faeroe-Shetland Channel, 232
FAMOUS (French-American Mid-ocean Undersea Study), 256
Federal Council for Science and Technology, 145, 154
Federal Republic of Germany. *See* Germany
Fedorov, Konstantin, 174, 205, 227–28
Field, Richard, 8–9
Finland, 68, 75
Fisher, Robert L., 127, 207, 210
Fisheries, xix, xxvii, 43, 70, 257, 259, 263; American, 12, 147; fear of dominance, 104; and ICES, 67; and the IIOE, 121, 131, 144, 219–29 passim; Japan, 23, 28, 30; Latin America, 14, 16–19; and NATO, 232–35; and radioactive waste, 62; and Southeast Asia, 106; and Soviet activities, 190
Fishery Society (Japan), 26
Fitzgerald, D. A., 86
Fleming, Richard, xxiv, 9, 91, 96
Folger, Timothy, xix
Food and Agriculture Organization (FAO), 103
Foreign policy, xviii, xxii, xxvi, 3, 14–22 passim, 30, 32, 60, 72, 86, 127, 129, 131, 134, 240, 243–46, 251, 260
Forman, Paul, 260
Fracture zones, 78, 192, 196
France: expeditions, xix, 30, 230, 256; and GEBCO, 190; and the IGY, 63, 65, 75, 82; and technical assistance, 19
Franklin, Benjamin, xix
Fraser, R., 111
Freedom of the seas, 250, 257, 260
Fye, Paul, xxvii, 148–52, 232–33, 239–41, 245–47

G. O. Sars, 48
Galathea, 10, 47, 79
Galtsoff, P., 111
General Bathymetric Chart of the Oceans (GEBCO), 188–90

Index

Geochemical Ocean Sections Study (GEOSECS), 256
Geological Society (Japan), 26
Geology, xxiv, 8–9, 14, 55, 197; bottom topography, 10; and military planning, 7. *See also* Continental drift
Geophysics, xx, xxiii–xxiv, xxvi, 3, 7–10, 12, 23, 25, 30, 55, 192, 197, 209, 211, 214; gravitational compensation, 7, 9. *See also* Continental drift
Geopoetry, 193–96. *See also* Continental drift; Hess, Harry
Geotimes, 198–99, 214
Germany, 67, 204, 251, 260; Admiralty, 7; data exchange, 48; expeditions, xix, 7, 230; and the IGY, 65, 68–69
Ghana, 165
Global Baseline Data Project, 256
Gorsline, D. S., 206
Gould, Laurence, 90, 95–96
Great Depression, 6
Greenberg, Daniel, 128, 159
Gregory, Herbert E., 5
Groves, Leslie, 50
Guatemala, 73
Gulf Stream, xix, 44, 68

Hamilton, Edwin, 117
Hansen, Vagn, 226–27
Harbor mining, 34
Harris, Oren, 95
Hartwell, Project, 37–39, 46, 50, 54
Hattori, Hitoshi, 198
Hayward, John T., 153
Hechler, Ken, 147
Hedberg, Hollis, 212–13
Heezen, Bruce, 77–78, 214
Hess, Harry, 8, 13, 189; and classified data, 52–53, 55, 57; and the IGY, 76, 161; and Mohole, 159–63; and continental drift, 194–98
Hevly, Bruce, 50
Hidaka, Koji, 29, 70, 82, 102
Hidalgo, 77

Hill, Maurice, 9–10, 111, 115–16
Hirohito, Emperor, 25–26
Hollister, John B., 84–85
Hong Kong, 105–6
Hoover, Herbert, Jr., 85–86
Hoover, Lynn, 214
Hubbs, Carl L., 5–6, 22–25, 124
Hudson Laboratories, 40, 142
Hughes, Hector, 222
Humphrey, George, 134, 174, 220
Humphrey, Hubert, 146, 240–41, 243
Huxley, Aldous, 34
Huxley, T. H., 199
Hydrographic Office (Japan) 26–27
Hydrographic Office (United States). *See* United States Navy

Ice Pick, Project, 46–49, 142
India, 15, 85–86, 107, 121–22, 158; and the IIOE, 125–27, 217, 219, 224; National Institute of Oceanography, 122, 227; sorting center at Cochin, 225–27
Indian Ocean Bubble, 134, 183
Indonesia, 6, 85–86, 102, 105–7, 119–22, 124
Institute of Oceanology (Moscow), 70, 88, 182, 192, 206–7
Institutions, oceanographic: access, 4, 52, 192; American, xxiv; Soviet, 70. *See also* Invisible college
Instruments: accelerometer, 11; bathythermograph (BT), 40, 42, 44, 46, 71, 76; for coring, 10, 12, 80; crystal chronometer, 11; echo sounder, 52, 71, 80, 98; expendable bathythermograph (XBT), 257; explosives, 9–10; geophones and hydrophones, 10–11; to measure heat flow, 11–12; microbarovariograph, 75; pendulum gravimeter, 9; Soviet, 71, 208; seismograph, 9, 80; submersible crafts, 256; tide gauge, 74, 223; wave recorder, 187. *See also* Unmanned stations

Integrated Global Ocean Station System (IGOSS), 256–57
Interagency Committee on Oceanography (ICO), 152–59, 168–73, 241–42; and Panel on International Programs (PIPICO), 170, 189, 191, 202
Intercontinental ballistic missiles (ICBMS). *See* Rocketry
Intergovernmental Oceanographic Commission (IOC), xxi–xxii, 100, 116, 128–29, 134–39, 144, 190, 200–202, 258–59, 261–64; American and Soviet plans, 169–76, 229; dissatisfaction with, 177–87, 217–18, 250; and the IDOE, 245, 254–55
International Advisory Committee on Marine Science (IACOMS). *See* United Nations Educational, Scientific and Cultural Organization
International Association of Physical Oceanography (IAPO), 101–4, 112–15, 190–91
International Commission of Northwest Atlantic Fisheries (ICNAF), 67–68, 189, 200, 233–35, 254
International Conference on the Peaceful Uses of Atomic Energy (Geneva), 70
International Congress on Oceanography: First, 100, 116–20, 156, 202; Second, 190, 202–10
International cooperation: disillusionment, xxi, 18, 93–94, 97–99, 162–63; easing tensions, xvii, xx–xxi, xxv, xxvii, 4, 7, 90, 97, 100, 177, 260, 262, 264; economic and social development, xxi–xxii, 100, 110, 129; leadership, xxi, xxvi, 97, 140–41, 164–76, 197; military support, xx, xxii, 32; motivations, xxvi–xxviii, 4, 23, 59–60, 151, 216, 238, 260, 265
International Cooperation Administration. *See* United States government
International Cooperative Investigations of the Tropical Atlantic (ICITA), 136, 171–76, 178, 180–82, 184, 187, 189, 200
International Coordination Group for the Southern Oceans (SOC), 229
International Council of Scientific Unions (ICSU), 14, 69, 73, 103, 114–16, 123, 126, 135, 164
International Council for the Exploration of the Sea (ICES), xix, 59, 67–68, 80, 103, 200, 221, 233–35, 254
International Decade of Ocean Exploration (IDOE), xxii, 201, 244–58 passim, 264
International Development Advisory Board (IDAB), 19–20
International Geological Congress, Twenty-third, 211–15
International Geophysical Cooperation (IGC), 93, 190–91
International Geophysical Year (IGY), xviii, xx–xxi, xxvii, 4, 29–30, 32, 53, 56–58, 99–100, 102–3, 116–17, 119–21, 123, 125, 139, 141, 147, 155, 157–58, 177–78, 189–91, 261–62, 264; diplomatic legacy, 61; origins and planning, 59–73 passim; political context, 61–65, 94-98, 156; scientific efforts, 65–79; Soviet plans, 79–98 passim
International Hydrographic Bureau, xx, 189
International Ice Patrol, xx
International Indian Ocean Expedition (IIOE), xviii, xxi, 84, 100, 120–27, 131–40, 171, 178–79, 182, 185, 191, 207, 217–29 passim, 262
International relations. *See* Foreign policy
International Union of Biological Sciences (IUBS), 66, 101, 104
International Union of Geodesy and Geophysics (IUGG), 63, 66, 101, 104, 126, 161, 164–65, 231–32
International Upper Mantle Committee, 164–65, 197, 210–12

Index 339

International vessel, 107–10, 137
Invisible college, 52, 193
Isaacs, John, 246
Isaacson, Walter, 20
Isacks, Bryan, 215
Iselin, Columbus, 8–9, 33, 44, 68, 111, 120, 231, 259
Islam, S. R., 123
Island Observatories Project, 68, 74–76
Italy, 230

Japan, 3, 18–19, 23–31 passim, 59, 69, 75, 78, 115, 147; and the CSK, 181, 185; and the IIOE, 121–22, 125–26, 228; militarism, 6; oceanographic community, 5–6
Jeffreys, Harold, 195
Jennings, F. D., 142
Jezebel, Project, 39, 47
Jimmu seamount, 27
Johnson, Felix, 54
Johnson, Louis, 37
Johnson, Lyndon Baines, xvii–xviii, xxii, xxviii, 163, 167, 238, 242, 244–55
Johnson, Martin W., xxiv, 9
Johnson administration, 243, 255
Joint Commission on Oceanography, 101
Jones, Harold Spencer, 114–15
Jordan, 19
Joseph, Arnold, 247

Kaplan, Joseph, 85, 89, 94
Karo, H. Arnold, 153, 170–73
Keith, Hastings, 239–40, 244
Kennedy, John F., 130, 140, 154–55, 238, 240
Kennedy, Robert F., 239
Kennedy administration, 156
Kenya, 165, 220-21
Khrushchev, Nikita, 84, 118
Kista Dan, 47
Kistiakowsky, George, 61, 129–30, 154, 159

Knauss, John, 232
Knopoff, Leon, 164, 205, 211
Korean War, 36, 38
Kort, Vladimir, 118, 168, 196, 207, criticized, 181–85; and the IGY, 70, 80–82, 88–89; and the IIOE, 126; and the IOC, 128–29, 135, 175; and SCOR, 99, 112
Kuhn, Thomas, xxiii, 199–200
Kullenberg, Börje, 10, 112
Kullenberg apparatus, 10, 12

Laclavère, Georges, 66, 82
Lacombe, Henri, 65
LaFond, Eugene, 134
Lamont Geological Observatory, 40, 52, 74–75, 77–78, 142, 152, 195, 215
Latin America, 14–21 passim, 26, 73, 85–86, 105, 136, 197. *See also specific countries*
Lee, A. J., 201, 232–35
Lehigh University, 8
Lena, 89, 95
Life, 159
Lill, Gordon, 49, 71, 75, 142, 148, 152, 161
Limited Test Ban Treaty, 61
Little America, 65
London, Ivan D., 96
Long-Term and Expanded Programme of Oceanic Exploration and Research (LEPOR), 253–55. *See also* International Decade of Ocean Exploration
Low Frequency Analysis and Recording (LOFAR), 39–40, 42, 54
Lyman, John, 128
Lysenko, Trofim, 194

Macdonald, Torbert H., 95
Madagascar, 165
Magnuson, Warren, 148, 153, 155–57, 237–38, 240
Maheu, René, 204
Mailer, Norman, 96
Malta, 251. *See also* Pardo, Arvid

Manhattan Project, 50–51
Manihine, 219–23
Manila Oceanographic Institute, 22
Maps and mapping. *See* Charts
Marine affairs, xxii, 236–49
Marine biology, 25, 106
Marine geology. *See* Geology
Marine geophysics. *See* Geophysics
Marine Resources and Engineering Development Act, 240
Marine science. *See* Oceanography
Marine Sciences Council, xxii–xxiii, 145, 154, 240–49, 251–55
Marshall, N., 111
Marshall Islands, 11
Marshall Plan, 15, 84
Martin, D. C., 181, 186, 188
Massachusetts Institute of Technology (MIT), 37, 42, 130
Massive retaliation, 146
Matthews, Drummond, 195
Matveyev, A. N., 250
Maxwell, Arthur, 12, 142, 170, 191
McCain, J. S., 40
Menard, H. W., 41, 52, 78, 94, 138, 192, 196, 199, 203
Merz, Alfred, xix
Meteor, xix, 7
Meteorology: data, 43, 48, 76; services and organizations, 124, 135, 224; weather and climate change, 43, 58, 74, 144–45, 149, 236, 255–57. *See also* World Meteorological Organization
Mexico, 75
Mid-Atlantic ridge. *See* Mid-ocean ridge
Mid-ocean ridge, 53, 77–78, 164, 194–95, 239, 256
MIDPAC expedition, 11–12, 25, 41, 58
Mikhail Lomonosov, 118, 201, 204
Military: dominance in research, 34–35, 141–52 passim; inter-service rivalry, xx, xxvi, 36–37, 57. *See also* Patronage
Military-industrial complex, xvii

Miller, George P., 153–54
Millikan, Robert, 44
Mills, Eric L., xviii
Mirny, 65, 89
Miyake, Y., 111
Mohole, Project, xxi, 141, 159–67, 262–63
Mohorovičić discontinuity, 159–60, 162
Monin, Andrei S., 196, 207
Monroe Doctrine, 97
Monsoon expedition, 138
Montevideo Symposium on Continental Drift, 210
Morocco, 107
Mosby, Håkon, 70, 99, 104, 109–11, 223, 231, 233–34
Moscow University, 206
Munk, Walter, 43, 156, 159, 168, 205
Murmansk Hydrometeorological Service, 84
Muromtsev, A. M., 84
Murphy, Garth, 248
Mussolini, Benito, 18
Mutual Security Program, 85

Naga expedition, 110, 178
Nagasaki Marine Observatory, 26
NASCO. *See* National Academy of Sciences (United States)
Nasu, Noriyuki, 30
National Academy of Engineering (United States), 247
National Academy of Sciences (United States), 22, 43–44, 55, 72–73, 92, 138, 140, 143, 162–64, 166, 169, 197, 246–49; Committee on Oceanography (NASCO), 72, 99, 119, 132, 140–48, 152, 154, 157, 161, 170, 201, 204, 208, 237, 240–49, 262; Division of Earth Science, 72; National Research Council, 22–23, 115, 247
National Aeronautics and Space Administration (NASA), 94, 153
National Council on Marine Resources

and Engineering Development. *See* Marine Sciences Council
National Institute of Oceanography (United Kingdom), 47, 66–67, 77, 83, 88, 151, 181, 186–87, 223
National Oceanic and Atmospheric Administration (NOAA), 243, 249
National Research Council (United States). *See* National Academy of Sciences (United States)
National Research Council of Japan, 5
National Science Foundation, xx, 57, 59–60, 64, 69, 72, 76, 85, 90–94, 129, 132, 144, 155, 163, 166, 169, 249, 254
National security, 33, 37-38, 42, 51, 53, 56–58, 85–86, 141–52 passim, 157, 237, 244
National Security Act, 86
NATO. *See* North Atlantic Treaty Organization
Nature, 27, 199, 227
Navy. *See* United States Navy
Needell, Allan, 61
Netherlands, 6, 75
New York Times, 160
Nicolet, Marcel, 63
Nielsen, E., 111
Nigeria, 165
Nixon, Richard, 243
Nobska, Project, 141, 231
Nolan, Thomas, 212, 214
Normandy, Allied landings at, 43
NORPAC, 21, 28–32, 59, 178, 185
NORPAX, 255–56
North Atlantic Treaty Organization (NATO): NATO Science Committee, xxi, 184, 218, 229–36
Northern Holiday expedition, 192
North Korea, 260
Norway: Defense Research Establishment, 44; expeditions, xix; and IDOE, 251; influence of, 8; Norwegian Polar Institute, 47, 67, 104; University of Bergen, 104; whaling, 97
Noyes, Albert, 127

Nuclear-powered submarines, 150
Nuclear weapons: hydrogen bomb, 51; testing, 11, 18, 35, 65, 70, 118, 144, 236
Nudibranchs, 26

Ob, 80, 88-89, 95
Ocean circulation, 67, 74, 77, 83–84, 112, 116, 194, 234
Ocean data acquisition systems. *See* Unmanned stations
Ocean drilling. *See* Mohole
Oceanic Quest, An, 247–48
Oceanization, 196, 198
Oceanographer, 244
Oceanographic Society (Japan), 26
Oceanography: competition with other sciences, 109, 135, 143, 162, 237, 240, 264; defining, xxiii–xxiv, xxviii, 8, 73, 132–33, 242, 263; economic and social development, 12, 16, 112, 121, 126, 130–36, 179, 236, 242–51, 264; leadership, 123, 140–43, 147–52, 232, 237, 244, 264; military and strategic importance, xvii–xix, 144–52, 177, 237; tradition of internationalism, xvii–xix, 3–4, 167. *See also specific disciplines, bodies, and projects*
Oceans, The, xxiv, 9
Ocean Science News, 158
Ocean Sciences and National Security, 148
Odishaw, Hugh, 91
Office of Naval Research, 9, 11, 24–25, 35, 47, 57, 71–72, 142, 148–49, 169, 223
Oil industry, 9
Okeanologiya, 157
Oliver, Jack, 215
Oppenheimer, J. Robert, 51
Oreskes, Naomi, 8
Our Nation and the Sea, 249
Ovey, C. D., 101
Owen, 222–23

Pacific Oceanographic Group (POG), 28
Pacific Science Association (PSA), 5, 7

Pacific Science Congresses, 5–6, 22, 24, 44
Pacific Tuna Conference, Fifth, 28
Pakistan, 121–23
Panikkar, N. K., 122, 225–27
Pardo, Arvid, 251–54
Patronage, xx, xxii–xxiv, xxvi–xxviii, 3–4, 8, 21–23, 32, 34, 52, 59, 127–35 passim, 147–52, 218, 260–62, 265
People's Republic of China. *See* China
Peru, 16–17, 21, 104
Petrie, William L., 160, 163
Pettersen, Hans, 10–11
Philippines, 105–7
Phleger, Fred, 206, 209
Physical oceanography, xx, xxiv, xxvi, 3, 8–9, 12, 22–23, 25, 29–30, 67, 77, 101–2, 114, 200, 217, 263; and the IIOE, 131–32; and NATO, 184, 187, 193, 230, 234–36. *See also* International Association of Physical Oceanography; Ocean circulation
Piston corer, 10, 12. *See also* Instruments
Plate tectonics. *See* Continental drift
Point Four programs, 15–21, 26, 122
Polanyi, Michael, 4–5, 263–64
Polar fronts, 67
Polaris, xxvii, 141–42, 150–51, 261
Polar Year, First and Second, 61
Pollution, xxiii. *See also* Environmental monitoring; Radioactive waste
Potomac oceanographers, 169–71, 173, 189, 233
Potts, A., 188
President's Science Advisory Committee (PSAC), xxiii, 94, 96, 128, 130, 145, 159, 243–46, 248
Proudman, Joseph, 101, 114
Pyenson, Lewis, 14

Quarles, Donald, 91

Radioactive waste, 62–63, 70, 72, 77, 109, 112, 144, 150, 253; in the Black Sea, 62–63

Radioactivity, 112, 195
Raitt, Russell, 88
Rakestraw, Norris, 111
Ramanatham, K. R., 126
Rao, T. S. S., 121
Ray, Dixy Lee, 132
Real-time data, 256–57
Reid, Joseph, 28–29
Republic of science, 4–5, 263–64
Revelle, Roger, 12, 117–19, 155-57, 242; early international bodies, 101–2; and classified data, 53, 57; creation of SCOR, 108–11, 115–16, 135, 186, 220; directorship of Scripps, 12, 111; and the IGY, 68–72, 76, 90; and the IIOE, 120, 122, 132–33, 138–39; and the IOC, 128–29, 139, 170; and Japan, 23, 27–29, 178; and Navy patronage, 9, 11, 32, 34–35, 41, 58–59, 261
Revolt of the Admirals, 37–38, 46
Ridenour, Louis, 34, 36, 51–52
Riley, Gordon A., 246
Robinson, Margaret, 208
Rockall Bank, 11
Rockefeller, Nelson A., 20, 89–90
Rockefeller Foundation, xxiv, 13
Rocketry, 39, 61, 86, 92, 95; ICBMS, 62, 91, 98
Roll, Hans, 254
Rollefson, R., 166
Romania, 259
Rossby, Carl-Gustaf, 99, 152
Royal Society (United Kingdom), 67, 180–82, 219, 229
Runcorn, Keith, 195
Rusk, Dean, 130
Russia: expeditions, xix. *See also* Soviet Union
Ryther, John, 225–26
Ryzhikov, Konstantin, 171

S-21 expedition, 7
Sakurai, Joji, 5
Satellites. *See* Rocketry
Schaefer, Milner, 172, 259

Index

Science, 27, 128, 159, 198, 228
Science: American strengths and weaknesses, xviii, xx, 60, 94–95, 140–76 passim; basic and applied, xxvi–xxvii, 9, 28, 32, 35–36, 42, 49, 51, 54-60, 73, 93, 100, 124, 126, 202, 230, 240–42, 249; and colonialism, 4, 13–14, 18–19; and Cold War competition, xxi, xxvi, 92, 99, 140–76 passim, 244; and development, xxi–xxii, xxvii, 13–17, 21, 23, 127–31, 133, 218–29, 265; and international relations, 5, 14–21 passim, 44, 139; and social responsibility, xxv, xxviii, 5–7, 14, 128; its universality, xxii, 4, 193–202, 263
Science, the Endless Frontier, 42
Science Council of Japan, 18, 26
Science fiction, 128
Science liaison offices, 21, 67, 86–87
Science policy, xxvi, 127–31, 149, 251
Scientific American, 200
Scientific Committee on Antarctic Research (SCAR), 97, 134
Scientific Committee on Oceanic Research (SCOR), xxi, 73, 97, 100, 102, 110–21 passim, 126, 128–31, 134–36, 138, 144, 174–75, 186, 220, 246, 261–62
Scripps Institution of Oceanography (SIO), xxiv, 3, 8–9, 11–12, 21–23, 25–27, 29–30, 34–35, 40–41, 43, 46, 49, 52, 58, 142, 193, 215; and the IGY, 68–69, 74–75, 77, 90; and the IIOE, 134
Seafloor spreading. *See* Continental drift
Sea-launched nuclear missiles, xvii, xx, 37-38, 141, 145, 261. See also *Polaris*
Sears, Mary, 101, 117
Secrecy, xxvii, 4, 11, 60, 138, 192, 241; classification of data, xx, xxii, xxvi, 33, 42, 50–59, 123, 191, 261
Security classification. *See* Secrecy
Seitz, Frederick, 246

Severyanka, 169
Shapley, Alan H., 92
Shellback expedition, 21
Sherman, Forrest, 37
Shils, Edward, 55
Sierra Leone, 165
Silas Bent, 204
Singapore, 105–6, 121, 124
Smith, Edward H., 8, 44; and the IGY, 65–66, 68–72, 80, 82
Smyth, Henry, 51
Snider, Robert G., 122–27, 133–34, 185, 220, 225
Somov, Mikhail, 70
Sonar, 40. *See also* Acoustics
Sound classification. *See* Classification of noise
Sound Fixing and Ranging (SOFAR), 39
Sound Surveillance System (SOSUS), 39–40, 46
Sound transmission. *See* Acoustics
Southeast Asian Treaty Organization (SEATO), 19
South Korea, 204
South Sea Bubble, 133–34
Southern Rhodesia, 165
Soviet Fleet, 65
Soviet Union, xviii, xx, xxiii, 54, 177–216 passim; rejection of continental drift, xxii–xxiii, 193–202 passim, 215, 263; and the IDOE, 251–53; and the IGY, 63–64, 70–71, 75, 78–98 passim; and the IIOE, 124–27, 158; and the IOC, 129; scientific strengths and weaknesses, xx–xxii, xxvii, 67, 71, 92, 129, 140, 143, 146–47, 156–67 passim, 202, 262; style of oceanography, 178–87, 200–202, 229–30, 255; suspicions of, 187–93. *See also* Antarctica
SOVMOHOLE, Project, 159–60, 165
Space science and exploration, xxvi, 62, 91, 96, 133–34, 139, 240, 262, 265
Space Treaty, 61
Spencer F. Baird, 25–26, 30

Spilhaus, Athelstan, 42, 68, 143, 155, 158–59
Sputnik, xx–xxii, xxvii, 60, 62, 65, 89, 91–99, 127-28, 261; post-*Sputnik*, 139–41, 143, 146, 148, 157, 238, 262
Stalin, Joseph: death of, 60, 63
Stassen, Harold, 84
State Department. *See* United States government
Steinbeck, John, 159
Stevenson, Adlai, 129–30
Stevenson, Robert E., 204, 207–10
Stommel, Henry, xxvii, 68, 77, 84, 109, 120, 134, 151–52, 174, 194, 235
Stratton, Julius A., 249
Structure of Scientific Revolutions, 199. *See also* Kuhn, Thomas
Submarine cables, 67
Submarine-launched nuclear missiles. *See* Sea-launched nuclear missiles
Submarines. *See* Nuclear-powered submarines; Undersea Warfare
Suda, Kanji, 26–27, 74, 82
Sullivan, John, 37
Surveys. *See* Expeditions
Sverdrup, Harald, xxiv, 8–9, 12, 43–44, 47, 49, 59, 67, 104, 109
Swallow, John, 67, 174
Sweden, 10, 12, 79
Swedish Deep Sea Expedition, 10, 12
Swell forecasting, 13, 42
Sykes, Lynn, 196, 214
Synoptic studies, 28, 31, 59, 66, 73, 178
Syria, 86
Sysoev, N. N., 82–83

Taiwan. *See* China
Tanganyika, 220
Tarawa, amphibious landing at, 43
Task Force 43: and the IGY, 65, 76
Tata Foundation, 126
Tchekourov, V. A., 175, 190, 203
Technical assistance, 14-21 passim, 84–86
Technical Panel on Oceanography. *See*

United States National Committee (IGY)
Technology, xx–xxi, xxvi, 33, 36, 38, 41–42, 49–50, 54–60, 91, 93, 150, 230, 239–41, 246, 256; and Cold War competition, xxi, 50, 128, 139–63 passim, 187, 260. *See also* Instruments *and other specific technologies*
Tectonophysics, 215
TENOC report, 141–42, 148–50, 152
Terada, Kazuhiko, 134
Texas A&M University, 74, 77
Thailand, 15, 105–6, 121–22, 224
Tharp, Marie, 78
Thermocline, 40–42
Thomas, Evan, 20
Titanic, xx
Torment of Secrecy, The, 55
TRANSPAC expedition, 21–32, 45, 59, 192
Treaty of Versailles, xix, 7
Trenches, 79, 87, 214–15
Truman, Harry, 14–15, 20, 37, 132
Tugarinov, Alexei, 210
Tully, John P., 28
Tuve, Merle, 166

Udintsev, Gleb, 138, 190, 192, 205
Uganda, 220
Umitaka Maru, 69
Undersea warfare, xvii, xx, 9, 25, 34, 38–41, 55, 142, 145–46, 152, 230
Unesco. *See* United Nations Educational, Scientific and Cultural Organization
Union of Soviet Socialist Republics. *See* Soviet Union
United Kingdom, 59, 75, 147, 173, 179, 251; Admiralty, 42, 189, 219, 222, 231; data, 48; expeditions, xix, 7, 10, 79; Foreign Office, 19, 63–64; and the IIOE, 180, 219–24; Ministry of Defence, 86–88, 187–88; Ministry of Science, 171, 218–22, 236; National Committee for the IGY, 66, 68;

Parliament, 222; techniques, 12; wartime oceanography, 43. *See also* Deacon, George; National Institute of Oceanography; Royal Society

United Nations, 20, 117, 127, 129–30, 136, 250–52, 255; Committee on Peaceful Uses of the Seabed and the Ocean Floor Beyond the Limits of National Jurisdiction, 254; Convention on the Law of the Sea, 238, 250; Economic and Social Council (ecosoc), 250; Special Fund, 125, 220

United Nations Educational, Scientific and Cultural Organization (Unesco), xxii, xxvii–xxviii, 14–15, 20, 22, 68–69, 87, 99–139 passim, 165, 203–4, 225–27, 244, 250, 263; International Advisory Committee on Marine Science (iacoms), 102–10, 116–17, 120, 135; Marine Sciences Programme, 103

United States (supercarrier), 37

United States Congress, xxi, xxvii, 36, 64, 85, 90–91, 140, 145, 148, 152–59, 236–40; House Committee on Interstate and Foreign Commerce, 95; House Committee on Merchant Marine and Fisheries, 146–47, 239; House Committee on Science and Astronautics, 148; House Committee on Un-American Activities (huac), 51; Senate Committee on Interstate and Foreign Commerce, 148, 157; Senate Committee on Foreign Relations, 130

United States government: Air Force, 37; Army, 37, 91; Atomic Energy Commission, 69, 72, 132, 154–55; Bureau of Commercial Fisheries, 154, 204, 237; Central Intelligence Agency, 92, 97, 160; Coast and Geodetic Survey, 74–75, 148, 153, 237, 244; Coast Guard, xx, 44, 75, 148, 237, 249; Department of Commerce, 70, 155; Department of Defense, 65, 84, 129, 155; Department of Health, Education and Welfare, 70, 155; Department of Interior, 70, 155; Department of State, 16, 19–20, 51, 61, 64, 70, 84, 91, 97–98, 119, 128–29, 154, 166, 169, 212, 230, 251; Fish and Wildlife Service, 72; Foreign Operations Administration, 84; Geological Survey, 148, 237; International Cooperation Administration (ica), 84–86, 125; Joint Chiefs of Staff, 97; National Institutes of Health, 129. *See also* National Science Foundation; United States Congress; United States Navy

United States National Committee for the igy (usnc-igy), 68, 71; Technical Panel on Oceanography, 71–73, 78, 85, 89

United States Navy: attitude toward international cooperation, 20, 27–28, 42–49, 261; Bureau of Ships, 9, 75; Hydrographic Office, 43, 46, 52–53; influence on research, xxvii, 12, 25, 28, 33, 58; inter-service rivalry, xx, xxvi, 36–37, 57; logistical support, xx, 32, 64, 71, 75–76, 78, 89; naval intelligence, 39, 54; naval operations and planning, xvii, 31, 33, 39, 45–49, 54; Naval Electronics Laboratory, 41, 52, 78, 199; Naval Research Laboratory, 36, 38; patronage of oceanography, xx, xxiv, xxvi–xxviii, 3, 7–12, 20–21, 23, 32-58 passim, 72, 141–43, 153, 178, 240, 260. *See also* Office of Naval Research; Secrecy

University of Washington, 74, 77, 132

Unmanned stations, xxiii, 10, 67, 187–88, 201, 232, 257

Upper Mantle Committee. *See* International Upper Mantle Committee

Upper Mantle Project, xxi, 159, 164–67, 196–97, 199, 263

Urey, Harold, 101
Uruguay, 17

V-1 and V-2 rockets, 39
Van Allen, James, 61
Van Andel, T. H., 209
Van Camp Foundation, 220
Vaughan, Thomas Wayland, 5, 8
Velikovsky, Immanuel, 76
Vema, 40, 77
Venezuela, 8, 13–14
Vening Meinesz, Felix Andries, 7–9, 11
Vetter, Richard C., 147, 154, 161, 246
Vietnam, 18–19, 22, 105–8, 110, 259–60
Vietnam War, 163, 242, 244–45
Vine, Allyn C., 151
Vine, Fred, 195
Vityaz, 69, 71, 79, 88, 91, 96, 126–27, 158, 201
Von Arx, William, 149, 179
Voss, Gilbert, 131, 207

Wade, Charles B., 16–17
Wakelin, James, Jr., 172–73, 233
Washington Post, 93
Waterman, Alan T., 57, 69, 90, 166
Weather. *See* Meteorology
Wegener, Alfred, 198
Wenk, Edward, Jr., 145, 148, 241–43, 245, 248, 251–53
Wexler, Harry, 95–96
Whales: sound, 40; whaling, 97
Wiener, Norbert, 35
Wiesner, Jerome, 140, 152, 154–55, 166, 235
Williamsburg, 132
Willmore, P. L., 10

Wilson, Charles E., 36
Wilson, E. Bright, Jr., 151
Wilson, J. Tuzo, 92, 164, 195–200, 232
Wiseman, John D. H., 52, 101
Wolfle, Dael, 117–18
Woods Hole Oceanographic Institution (WHOI), xxiv, xxvii, 8–9, 11–12, 32, 34, 40, 44, 46–47, 55, 58, 65, 135, 179; and the IGY, 67–69, 74, 77, 80; and the IIOE, 134; and TENOC recommendations, 142, 148–52
Wooster, Warren, 26, 136, 170, 179, 184, 226–27, 250
Working group on oceanography (IGY), 66, 72, 82, 89, 113
World data centers, 74, 137, 190, 206, 224, 256
World Meteorological Organization, 135, 144, 256
World War I, xix–xx, 7, 67
World War II, xviii–xx, xxiv–xxv, 3–4, 7, 9, 13–14, 22, 32–33, 35, 42, 48, 51, 60, 67, 71, 106, 259–60
Worzel, J. Lamar, 209
Wyrtki, Klaus, 102, 105

York, Herbert, 129
Yoshida, Masao, 107
Yugoslavia, 86

Zaklinskii, G. B., 84
Zanzibar, 221
Zarya, 83, 91, 157, 169
Zenkevich, Lev, 79, 81–82, 102, 111–12, 126, 133, 207, 259
Zenkevich, Nikita, 94
ZoBell, Claude, 22–24
Zuckerman, Solly, 218

CPSIA information can be obtained
at www.ICGtesting.com
Printed in the USA
JSHW051414010922
30040JS00001B/19